D0841862

Regression Models for Categorical

Dependent Variables Using Stata

Revised Edition

Regression Models for Categorical Dependent Variables Using Stata

Revised Edition

J. SCOTT LONG
Department of Sociology
Indiana University
Bloomington, Indiana

JEREMY FREESE
Department of Sociology
University of Wisconsin-Madison
Madison, Wisconsin

A Stata Press Publication
STATA CORPORATION
College Station, Texas

Stata Press, 4905 Lakeway Drive, College Station, Texas 77845

To our parents

Contents

Preface

Our goal in writing this book was to make it routine to carry out the complex calculations necessary to fully interpret regression models for categorical outcomes. Interpreting these models is complex because the models are nonlinear. Most software packages that fit these models do not provide options that make it simple to compute the quantities that are useful for interpretation. In this book, we briefly describe the statistical issues involved in interpretation, and then we show how Stata can be used to make these computations. As you read this book, we strongly encourage you to be at your computer so that you can experiment with the commands as you read. To facilitate this, we include two appendices. Appendix A summarizes each of the commands that we have written for interpreting regression models. Appendix B provides information on the datasets that we use as examples.

Many of the commands that we discuss are not part of official Stata, but instead they are commands (in the form of ado-files) that we have written. To follow the examples in this book, you will have to install these commands. Details on how to do this are given in Chapter 2. While the book assumes that you are using Stata 8 or later, most commands will work in Stata 7 or Stata 6, although some of the output will appear differently. For details, see

http://www.indiana.edu/~jslsoc/spost.htm

The screen shots that we present are from Stata 8 for Windows. If you are using a different operating system, your screen might appear differently. See the StataCorp publication *Getting Started with Stata* for your operating system for further details. All of the examples, however, should work on all computing platforms that support Stata.

We use several conventions throughout the manuscript. Stata commands, variable names, filenames, and output are all presented in a typewriter-style font; for example, `logit lfp age wc hc k5`. Italics are used to indicate that something should be substituted for the word in italics. For example, `logit` *variablelist* indicates that the command `logit` is to be followed by a specific list of variables. When output from Stata is shown, the command is preceded by a period (which is the Stata prompt). For example,

```
. logit lfp age wc hc k5, nolog
Logit estimates                              Number of obs   =      753
   (output omitted )
```

If you want to reproduce the output, you do *not* type the period before the command. And, as just illustrated, when we have deleted part of the output, we indicate this with (*output omitted*). Keystrokes are set in this font. For example, alt-f means that you are to hold down the alt key and press f. The headings for sections that discuss advanced topics are tagged with an asterick (*). These sections can be skipped without any loss of continuity with the rest of the book.

As we wrote this book and developed the accompanying software, many people provided their suggestions and commented on early drafts. In particular, we would like to thank Simon Cheng, Ruth Gassman, Claudia Geist, Lowell Hargens, and Patricia Mc-Manus. David Drukker at StataCorp provided valuable advice throughout the process. Lisa Gilmore and John Williams, both at StataCorp, typeset and proofread the book.

Finally, while we will do our best to provide technical support for the materials in this book, our time is limited. If you have a problem, please read the conclusion of Chapter 8 and check our web page before contacting us. Thanks.

Part I

General Information

Our book is about using Stata for fitting and interpreting regression models with categorical outcomes. The book is divided into two parts. Part I contains general information that applies to all the regression models that are considered in detail in Part II.

- **Chapter 1** is a brief orienting discussion that also includes *critical information* about installing a collection of Stata commands that we have written to facilitate the interpretation of regression models. Without these commands, you will not be able to do many of the things we suggest in the later chapters.

- **Chapter 2** includes both an introduction to Stata for those who have not used the program and more advanced suggestions for using Stata effectively for data analysis.

- **Chapter 3** considers issues of estimation, testing, assessing fit, and interpretation that are common to all of the models considered in later chapters. We discuss both the statistical issues involved and the Stata commands that carry out these operations.

Chapters 4 through 7 of Part II are organized by the type of outcome being modeled. Chapter 8 deals primarily with complications on the right-hand side of the model, such as including nominal variables and allowing interactions. The material in the book is supplemented on our web site at *http://www.indiana.edu/~jslsoc/spost.htm*, which includes data files, examples, and a list of Frequently Asked Questions (FAQs). While the book assumes that you are running Stata 8, most of the information also applies to Stata 7 and Stata 6; our web site includes special instructions for users of these releases.

1 Introduction

1.1 What is this book about?

Our book shows you efficient and effective ways to use regression models for categorical and count outcomes. It is a book about data analysis and is not a formal treatment of statistical models. To be effective in analyzing data, you want to spend your time thinking about substantive issues and not laboring to get your software to generate the results of interest. Accordingly, good data analysis requires good software and good technique.

While we believe that these points apply to all data analysis, they are particularly important for the regression models that we examine. The reason is that these models are *nonlinear* and consequently the simple interpretations that are possible in linear models are no longer appropriate. In nonlinear models, the effect of each variable on the outcome depends on the level of *all* variables in the model. As a consequence of this nonlinearity, which we discuss in more detail in Chapter 3, no single method of interpretation can fully describe the relationships among the independent variables and the outcome. Rather, a series of *post-estimation* explorations are needed to uncover the most important aspects of these relationships. In general, if you limit your interpretations to the standard output, that output constrains and can even distort how you understand your results.

In the linear regression model, most of the work of interpretation is complete once the estimates are obtained. You simply read off the coefficients, which can be interpreted as: for a unit increase in x_k, y is expected to increase by β_k units, holding all other variables constant. In nonlinear models, such as logit or negative binomial regression, a substantial amount of additional computation is necessary after the estimates are obtained. With few exceptions, the software that fits regression models does not provide much help with these analyses. Consequently, the computations are tedious, time-consuming, and error-prone. All in all, it is not fun work. In this book, we show how post-estimation analysis can be accomplished easily using Stata and the set of new commands that we have written. These commands make sophisticated, post-estimation analysis routine and even enjoyable. With the tedium removed, the data analyst can focus on the substantive issues.

1.2   Which models are considered?

Regression models analyze the relationship between an explanatory variable and an outcome variable while controlling for the effects of other variables. The linear regression model (LRM) is probably the most commonly used statistical method in the social sciences. As we have already mentioned, a key advantage of the LRM is the ease of interpreting results. Unfortunately, this model applies only to cases in which the dependent variable is continuous.[1] Using the LRM when it is not appropriate produces coefficients that are biased and inconsistent, and there is nothing advantageous about the simple interpretation of results that are incorrect.

Fortunately, a wide variety of appropriate models exists for categorical outcomes, and these models are the focus of our book. We cover cross-sectional models for four kinds of dependent variables. *Binary* outcomes (dichotomous or dummy variables) have two values, such as whether a citizen voted in the last election or not, whether a patient was cured after receiving some medical treatment or not, or whether a respondent attended college or not. *Ordinal* or *ordered* outcomes have more than two categories, and these categories are assumed to be ordered. For example, a survey might ask if you would be "very likely", "somewhat likely", or "not at all likely" to take a new subway to work, or if you agree with the president on "all issues", "most issues", "some issues", or "almost no issues". *Nominal* outcomes also have more than two categories but are not ordered. Examples include the mode of transportation a person takes to work (e.g., bus, car, train) or an individual's employment status (e.g., employed, unemployed, out of the labor force). Finally, *count* variables count the number of times something has happened, such as the number of articles written by a student after receiving the Ph.D., or the number of patents a biotechnology company has obtained. The specific cross-sectional models that we consider, along with the corresponding Stata commands, are

Binary outcomes: binary logit (`logit`) and binary probit (`probit`).
Ordinal outcomes: ordered logit (`ologit`) and ordered probit (`oprobit`).
Nominal outcomes: multinomial logit (`mlogit`) and conditional logit (`clogit`).
Count outcomes: Poisson regression (`poisson`), negative binomial regression (`nbreg`), zero-inflated Poisson regression (`zip`), and zero-inflated negative binomial regression (`zinb`).

While this book covers models for a variety of different types of outcomes, they are all models for cross-sectional data. We do not consider models for survival or event history data, even though Stata has a powerful set of commands for dealing with these data (see the entry for `st` in the *Survival Analysis Reference Manual*). Likewise, we do not consider any models for panel data, even though Stata contains several commands for fitting these models (see the entry for `xt` in the *Cross-Sectional Time-Series Reference Manual*).

---

[1] The use of the LRM with binary dependent variables leads to the linear probability model (LPM). We do not consider the LPM further, given the advantages of models such as logit and probit. See Long (1997, 35–40) for details.

1.3   Who is this book for?

We expect that readers of this book will vary considerably in both their knowledge of statistics and their knowledge of Stata. With this in mind, we have tried to structure the book in a way that best accommodates the diversity of our audience. Minimally, however, we assume that readers have a solid familiarity with OLS regression for continuous dependent variables and that they are comfortable using the basic features of the operating system of their computer. While we have provided sufficient information about each model so that you can read each chapter without prior exposure to the models discussed, we strongly recommend that you do *not* use this book as your sole source of information on the models (Section 1.6 recommends additional readings). Our book will be most useful if you have already studied the models considered or are studying these models in conjunction with reading our book.

We assume that you have access to a computer that is running Stata 8 or later and that you have access to the Internet to download commands, datasets, and sample programs that we have written (see Section 1.5 for details on obtaining these). For information about obtaining Stata, see the StataCorp web site at *http://www.stata.com*. While most of the commands in later chapters also work in Stata 7 and Stata 6, there are some differences. For details, check our web site at *http://www.indiana.edu/~jslsoc/spost.htm*.

1.4   How is the book organized?

Chapters 2 and 3 introduce materials that are necessary for working with the models we present in the later chapters:

Chapter 2: Introduction to Stata reviews the basic features of Stata that are necessary to get new or inexperienced users up and running with the program. This introduction is by no means comprehensive, so we include information on how to get additional help. New users should work through the brief tutorial that we provide in Section 2.17. Those who are already skilled with Stata can skip this chapter, although even these readers might benefit from quickly reading it.

Chapter 3: Estimation, Testing, Fit, and Interpretation provides a review of using Stata for regression models. It includes details on how to fit models, test hypotheses, compute measures of model fit, and interpret results. We focus on those issues that apply to all the models considered in Part II. We also provide detailed descriptions of the add-on commands that we have written to make these tasks easier. Even if you are an advanced user, we recommend that you look over this chapter before jumping ahead to the chapters on specific models.

Chapters 4 through 7 each cover models for a different type of outcome:

Chapter 4: Binary Outcomes begins with an overview of how the binary logit and probit models are derived and how they can be fitted. After the model has been fitted, we show how Stata can be used to test hypotheses, compute residuals and influence statistics, and calculate scalar measures of model fit. Then, we describe post-estimation commands that assist in interpretation using predicted probabilities, discrete and marginal change in the predicted probabilities, and, for the logit model, odds ratios. Because binary models provide a foundation on which some models for other kinds of outcomes are derived, and because Chapter 4 provides more detailed explanations of common tasks than later chapters do, we recommend reading this chapter even if you are mainly interested in another type of outcome.

Chapter 5: Ordinal Outcomes introduces the ordered logit and ordered probit models. We show how these models are fitted and how to test hypotheses about coefficients. We also consider two tests of the parallel regression assumption. In interpreting results, we discuss similar methods as those described in Chapter 4, as well as interpretation in terms of a latent dependent variable.

Chapter 6: Nominal Outcomes focuses on the multinomial logit model. We show how to test a variety of hypotheses that involve multiple coefficients and discuss two tests of the assumption of the independence of irrelevant alternatives. While the methods of interpretation are again similar to those presented in Chapter 4, interpretation is often complicated due to the large number of parameters in the model. To deal with this complexity, we present two graphical methods of representing results. We conclude the chapter by introducing the conditional logit model, which allows characteristics of both the alternatives and the individual to vary.

Chapter 7: Count Outcomes begins with the Poisson and negative binomial regression models, including a test to determine which model is appropriate for your data. We also show how to incorporate differences in exposure time into the estimation. Next, we consider interpretation in terms of changes in the predicted rate and changes in the predicted probability of observing a given count. The last half of the chapter considers fitting and interpretating zero-inflated count models, which are designed to account for the large number of zero counts found in many count outcomes.

Chapter 8 returns to issues that affect all models.

Chapter 8: Additional Topics deals with several topics, but the primary concern is with complications among independent variables. We consider the use of ordinal and nominal independent variables, nonlinearities among the independent variables, and interactions. The proper interpretation of the effects of these types of variables requires special adjustments to the commands considered in earlier chapters. We then comment briefly on how to modify our commands to work with other estimation commands. Finally, we discuss several features in Stata that we think make data analysis easier and more enjoyable.

1.5   What software do you need?

To get the most out of this book, you should read it while you are at a computer where you can experiment with the commands as they are introduced. We assume that you are using Stata 8 or later. If you are running Stata 7 or Stata 6, most of the commands work, but some things must be done differently, and the output will look slightly different. For details, see *http://www.indiana.edu/~jslsoc/spost.htm*. If you are using Stata 5 or earlier, the commands that we have written will not work.

Advice to New Stata Users If you have never used Stata, you might find the instructions in this section to be confusing. It might be easier if you only skim the material now and return to it after you have read the introductory sections of Chapter 2.

1.5.1   Updating Stata 8

Before you work through our examples in later chapters, we strongly recommend that you have the latest version of `wstata.exe` and the official Stata ado-files. *You should do this even if you have just installed Stata* because the CD that you received might not have the latest changes to the program. If you are connected to the Internet and are in Stata, you can update Stata by selecting Official Updates from the Help menu. Stata responds with the following screen:

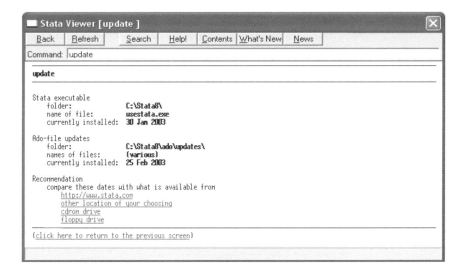

This screen tells you the current dates of your files. You can update your files to the latest versions by clicking on `http://www.stata.com`. We suggest that you do this every few months. Or, if you encounter something that you think is a bug in Stata or in our commands, update your copy of Stata to see if the problem has been resolved.

1.5.2   Installing SPost

From our point of view, one of the best things about Stata is how easy it is to add your own commands. This means that if Stata does not have a command you need or some command does not work the way you like, you can program the command yourself, and it will work as if it were part of official Stata. Indeed, we have created a suite of programs, referred to collectively as SPost (for Stata Post-estimation commands), for the post-estimation interpretation of regression models. **These commands must be installed before you can try the examples in later chapters.**

What is an ado-file? Programs that add commands to Stata are contained in files that end in the extension `.ado` (hence the name, ado-files). For example, the file `prvalue.ado` is the program for the command `prvalue`. Hundreds of ado-files are included with the official Stata package, but experienced users can write their own ado-files to add new commands. However, for Stata to use a command implemented as an ado-file, *the ado-file must be located in one of the directories where Stata looks for ado-files*. If you type the command `sysdir`, Stata lists the directories that Stata searches for ado-files in the order that it searches them. However, if you follow our instructions below, you should not have to worry about managing these directories.

Installing SPost using net search

Installation should be simple, although you must be connected to the Internet. Type `net search spost`. The `net search` command accesses an online database that StataCorp uses to keep track of user-written additions to Stata. Typing `net search spost` brings up the names and descriptions of several packages (a package is a collection of related files) in the Results window. One of these packages is labeled `spostado from http://www.indiana.edu/~jslsoc/stata`. The label is in blue, which means that it is a link that you can click on.[2] After you click on the link, a window opens in the Viewer (this is a new window that will appear on your screen) that provides information about our commands and another link saying "click here to install". If you click on this link, Stata attempts to install the package. After a delay during which files are downloaded, Stata responds with one of the following messages:

---

[2]If you click on a link and immediately get a beep with an error message saying that Stata is busy, the problem is probably that Stata is waiting for you to press a key. Most often, this occurs when you are scrolling output that does not fit on one screen.

`installation complete` means that the package has been successfully installed and that you can now use the commands. Just above the "installation complete" message, Stata tells you the directory where the files were installed.

`all files already exist and are up-to-date` means that your system already has the latest version of the package. You do not need to do anything further.

`the following files exist and are different` indicates that your system already has files with the same names as those in the package being installed, and that these files differ from those in the package. The names of those files are listed and you are given several options. Assuming that the files listed are earlier versions of our programs, you should select the option "Force installation replacing already-installed files". This might sound ominous, but it is not. Since the files on our web site are the latest versions, you want to replace your current files with these new files. After you accept this option, Stata updates your files to newer versions.

`cannot write in directory` *directory-name* means that you do not have write privileges to the directory where Stata wants to install the files. Usually, this occurs only when you are using Stata on a network. In this case, we recommend that you contact your network administrator and ask if our commands can be installed using the instructions given above. If you cannot wait for a network administrator to install the commands or to give you the needed write access, you can install the programs to any directory where you have write permission, including a zip disk or your directory on a network. For example, suppose that you want to install SPost to your directory called `d:\username` (which can be any directory where you have write access). You should use the following commands:

```
. cd d:\username
d:\username
. mkdir ado
. sysdir set PERSONAL "d:\username\ado"
. net set ado PERSONAL
. net search spost
(contacting http://www.stata.com)
```

Then, follow the installation instructions that we provided earlier for installing SPost. If you get the error "could not create directory" after typing `mkdir ado`, then you probably do not have write privileges to the directory.

If you install ado-files to your own directory, each time you begin a new session you must tell Stata where these files are located. You do this by typing `sysdir set PERSONAL` *directory*, where *directory* is the location of the ado-files you have installed. For example,

```
. sysdir set PERSONAL d:\username
```

Installing SPost using net install

Alternatively, you can install the commands entirely from the Command window. (If
you have already installed SPost, you do not need to read this section.) While you are
online, enter

> `. net from http://www.indiana.edu/~jslsoc/stata/`

The available packages will be listed. To install `spostado`, type

> `. net install spostado`

`net get` can be used to download supplementary files (e.g., datasets, sample do-files)
from our web site. For example, to download the package `spostst8`, type

> `. net get spostst8`

These files are placed in the current working directory (see Chapter 2 for a full discussion
of the working directory).

1.5.3   What if commands do not work?

This section assumes that you have installed SPost but that some of the commands do
not work. Here are some things to consider:

1. If you get the error message `unrecognized command`, there are several possibilities.

   a. If you discover that commands that used to work do not work anymore,
      you could be working on a different computer or on a different station in a
      computer lab. Because user-written ado-files work seamlessly in Stata, you
      might not realize that these programs must be installed on each machine that
      you use.

   b. If you sent a do-file that contains SPost commands to another person, and
      they cannot get the commands to work, let them know that they need to
      install SPost.

   c. If you get the error message `unrecognized command:` *strangename* after typ-
      ing one of our commands, where *strangename* is not the name of the command
      that you typed, it means that Stata cannot find an ancillary ado-file that the
      command needs. We recommend that you install the SPost files again.

2. If you are getting an error message that you do not understand, click on the blue
   return code beneath the error message for more information about the error.

3. Make sure that Stata is properly installed and up to date. Typing `verinst` will
   verify that Stata has been properly installed. Typing `update query` will tell you
   if the version you are running is up to date and what you need to type to update
   it. If you are running Stata over a network, your network administrator may need
   to do this for you.

4. Often, what appears to be a problem with one of our commands is actually a mistake you have made (we know because we make them, too). For example, make sure that you are not using = when you should be using ==.

5. Because our commands work after you have fitted a model, make sure that there were no problems with the last model fitted. If Stata was not successful in fitting your model, our commands will not have the information needed to operate properly.

6. Irregular value labels can cause Stata programs to fail. We recommend using labels that have fewer than 8 characters and contain no spaces or special characters other than underscores (_). If your variables (especially your dependent variable) do not meet this standard, try changing your value labels with the `label` command (details are given in Section 2.15).

7. Unusual values of the outcome categories can also cause problems. For ordinal or nominal outcomes, some of our commands require that all of the outcome values be integers between 0 and 99. For these types of outcomes, we recommend using consecutive integers starting with 1.

In addition to this list, we recommend that you check our Frequently Asked Questions (FAQ) page at *http://www.indiana.edu/~jslsoc/spost.htm*. This page contains the latest information on problems that users have encountered.

1.5.4   Uninstalling SPost

Stata keeps track of the packages that it has installed, which makes it easy for you to uninstall them in the future. If you want to uninstall our commands, simply type: `ado uninstall spostado`.

1.5.5   Additional files available on the web site

In addition to the SPost commands, we have provided other packages that you might find useful. For example, the package called `spostst8` contains the do-files and datasets needed to reproduce the examples from this book. The package `spostrm4` contains the do-files and datasets to reproduce the results from Long (1997). To obtain these packages, type `net search spost` and follow the instructions you will be given. **Important**: if a package does not contain ado-files, Stata will download the files to the current working directory. Consequently, you need to change your working directory to wherever you want the files to go *before* you select "click here to get". More information about working directories and changing your working directory is provided in Section 2.5.

1.6   Where can I learn more about the models?

There are many valuable sources for learning more about the regression models that are covered in this book. Not surprisingly, we recommend

> Long, J. Scott. 1997. *Regression Models for Categorical and Limited Dependent Variables.* Thousand Oaks, CA: Sage Publications.

This book provides further details about all the models discussed in the current book. In addition, we recommend the following:

> Cameron, A. C. and P. K. Trivedi. 1998. *Regression Analysis of Count Data.* Cambridge: Cambridge University Press. This is the definitive reference for count models.

> Greene, W. H. 2003. *Econometric Analysis.* 5th ed. Upper Saddle River, NJ: Prentice–Hall. While this book focuses on models for continuous outcomes, several later chapters deal with models for categorical outcomes.

> Hosmer, D. W., Jr., and S. Lemeshow. 2000. *Applied Logistic Regression.* 2d ed. New York: John Wiley & Sons. This book, written primarily for biostatisticians and medical researchers, provides a great deal of useful information about logit models for binary, ordinal, and nominal outcomes. In many cases, the authors discuss how their recommendations can be executed using Stata.

> Powers, D. A. and Y. Xie. 2000. *Statistical Methods for Categorical Data Analysis.* San Diego: Academic Press. This book considers all the models discussed in our book, with the exception of count models, and also includes loglinear models and models for event history analysis.

2 Introduction to Stata

This book is about fitting and interpreting regression models using Stata, and to earn our pay we must get to these tasks quickly. With that in mind, this chapter is a relatively concise introduction to Stata 8 for those with little or no familiarity with the package. Experienced Stata users can skip this chapter, although a quick reading might be useful. We focus on teaching the reader what is necessary to work through the examples later in the book and to develop good working techniques for using Stata for data analysis. By no means are the discussions exhaustive; in many cases, we show you either our favorite approach or the approach that we think is simplest. One of the great things about Stata is that there are usually several ways to accomplish the same thing. If you find a better way than what we have shown you, use it!

You cannot learn how to use Stata simply by reading. Accordingly, we strongly encourage you to try the commands as we introduce them. We have also included a tutorial in Section 2.17 that covers many of the basics of using Stata. Indeed, you might want to try the tutorial first and then read our detailed discussions of the commands.

While people who are new to Stata should find this chapter sufficient for understanding the rest of the book, if you want further instruction, look at the resources listed in Section 2.3. We also assume that you know how to load Stata on the computer you are using and that you are familiar with your computer's operating system. By this, we mean that you should be comfortable copying and renaming files, working with subdirectories, closing and resizing windows, selecting options with menus and dialog boxes, and so on.

(*Continued on next page*)

2.1 The Stata interface

Figure 2.1: Opening screen in Stata for Windows.

When you launch Stata, you will see a screen in which several smaller windows are located within the larger Stata window, as shown in Figure 2.1. This screen shot is for Windows using the default windowing preferences. If the defaults have been changed or you are running Stata under Unix or the MacOS, your screen will look slightly different.[1] Figure 2.2 shows what Stata looks like after several commands have been entered and data have been loaded into memory. In both figures, four windows are shown.

---

[1] Our screen shots and descriptions are based on Stata for Windows. Please refer to the books *Getting Started with Stata for Macintosh* or *Getting Started with Stata for Unix* for examples of the screens for those operating systems.

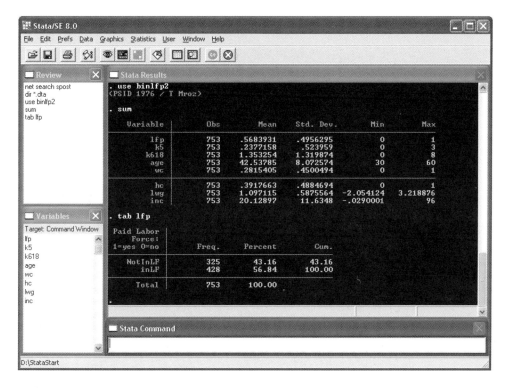

Figure 2.2: Example of Stata windows after several commands have been entered and data have been loaded.

The Command window is where you type commands that are executed when you press Enter. As you type commands, you can edit them at any time *before* pressing Enter. Pressing PageUp brings the most recently used command into the Command window; pressing PageUp again retrieves the command before that; and so on. Once a command has been retrieved to the Command window, you can edit it and press Enter to run the modified command.

The Results window contains output from the commands entered in the Command window. The Results window also echoes the command that generated the output, where the commands are preceded by a "." as shown in Figure 2.2. The scroll bar on the right lets you scroll back through output that is no longer on the screen. Only the most recent output is available this way; earlier lines are lost unless you have saved them to a log file (discussed below).

The Review window lists the commands that have been entered from the Command window. If you click on a command in this window, it is pasted into the Command window, where you can edit it before execution of the command. If you double-click on a command in the Review window, it is pasted into the Command window and immediately executed.

The Variables window lists the names of variables that are in memory, including those loaded from disk files and those created with Stata commands. If you click on a name, it is pasted into the Command window.

The Command and Results windows illustrate the important point that Stata has its origins in a command-based system. This means that you tell Stata what to do by typing commands that consist of a single line of text and then pressing Enter.[2] Beginning with Stata 8, there is a complete graphical user interface (GUI) for accessing all nonprogramming commands. At the risk of seeming old-fashioned, however, we still prefer the command-based interface. While it can take longer to learn, once you learn it, you should find it much faster to use. If you currently prefer using pull-down menus, stick with us, and you will likely change your mind.

Because Stata 8 and later have a complete GUI, you can do almost anything in Stata by pointing and clicking. Some of the most important tasks can be performed by clicking on icons on the toolbar at the top of the screen. While we on occasion mention the use of these icons, for the most part we stick with text commands. Indeed, even if you do click on an icon or issue a command from a dialog box, Stata shows you how this could be done with a text command. For example, if you click on the browse button ![icon], Stata opens a spreadsheet for examining your data. Meanwhile, ". `browse`" is written to the Results window. This means that instead of clicking the icon, you could have typed `browse`. Overall, not only is the range of things you can do with menus limited, but almost everything you can do with the mouse can also be done with commands, often more efficiently. It is for this reason, and also because it makes things much easier to automate later, that we describe things mainly in terms of commands. Even so, we encourage you to explore the tasks available through menus and the toolbar and to use them when preferred.

Changing the scrollback buffer size

How far back you can scroll in the Results window is controlled by the command

    set scrollbufsize #

where $10,000 \le \# \le 500,000$. By default, the buffer size is 32,000 bytes. When you change the size of the scroll buffer using `set scrollbufsize`, the change will take effect the next time you launch Stata.

Changing the display of variable names in the Variables window

The Variables window displays both the names of variables in memory and their variable labels. By default, 32 columns are reserved for the name of the variable. The maximum number of characters to display for variable names is controlled by the command

---

[2]For now, we only consider entering one command at a time, but in Section 2.9 we show you how to run a series of commands at once using "do-files".

```
set varlabelpos #
```

where $8 \leq \# \leq 32$. By default, the size is 32. In Figure 2.2, none of the variable labels are shown since the 32 columns take up the entire width of the window. If you use short variable names, it is useful to `set varlabelpos` to a smaller number so that you can see the variable labels.

Tip: Changing Defaults We prefer a larger scroll buffer and less space for variable names. We could enter the command `set varlabelpos 14` at the start of each Stata session, but it is easier to add the commands to `profile.do`, a file that is automatically run each time Stata begins. We show you how to do this in Chapter 8.

2.2 Abbreviations

Commands and variable names can often be abbreviated. For variable names, the rule is easy: *any variable name can be abbreviated to the shortest string that uniquely identifies it.* For example, if there are no other variables in memory that begin with a, the variable `age` can be abbreviated as `a` or `ag`. If you have the variables `income` and `income2` in your data, neither of these variable names can be abbreviated.

There is no general rule for abbreviating commands, but as one would expect, typically the most common and general command names can be abbreviated. For example, four of the most often used commands are `summarize`, `tabulate`, `generate`, and `regress`, and these can be abbreviated as `su`, `ta`, `g`, and `reg`, respectively. From now on, when we introduce a Stata command that can be abbreviated, we underline the shortest abbreviation (e.g., `generate`). But, while very short abbreviations are easy to type, when you are getting started the short abbreviations can be confusing. Accordingly, when we use abbreviations, we stick with at least three-letter abbreviations.

2.3 How to get help

2.3.1 Online help

If you find our description of a command incomplete, or if we use a command that is not explained, you can use Stata's online help to get further information. The `help`, `search`, and `net search` commands, described below, can be typed in the Command window with results displayed in the Results window. Or, you can open the Viewer by clicking on ⬜. At the top of the Viewer, there is a line labeled Command, where you can type commands such as `help`. The Viewer is particularly useful for reading help files that are long. Here is further information on commands for getting help:

help lists a shortened version of the documentation in the manual for any command. You can even type help help for help on using help. The output from help often makes reference to other commands, which are shown in blue. *Anything in the Results window that is in blue type is a link that you can click on.* In this case, clicking on a command name in blue type is the same as typing help for that command.

search is handy when you do not know the specific name of the command that you need information about. search *word* [*word* ...] searches Stata's online index and lists the entries that it finds. For example, search gen lists information on generate but also on many related commands. Or, if you want to run a truncated regression model but cannot remember the name of the command, you could try search truncated to get information on a variety of possible commands. These commands are listed in blue, so you can click on the name, and details appear in the Viewer. If you keep your version of Stata updated on the Internet (see Section 1.5 for details), search with the all option also provides current information from the Stata web site FAQ (Frequently Asked Questions) and articles in the *Stata Journal* (often abbreviated as SJ). findit ... is equivalent to search *word* [*word* ...], all.

net search is a command that searches a database at *http://www.stata.com* for information about commands written by users (accordingly, you have to be online for this command to work). This is the command to use if you want information about something that is not part of official Stata. For example, when you installed the SPost commands, you used net search spost to find the links for installation. To get a further idea of how net search works, try net search truncated and compare the results with those from search truncated.

Tip: Help with error messages Error messages in Stata can be terse and sometimes confusing. While the error message is printed in red, errors also have a *return code* (e.g., r(199)) listed in blue. Clicking on the return code provides a more detailed description of the error.

2.3.2   Manuals

The Stata manuals are extensive, and it is worth taking an hour to browse them to get an idea of the many features in Stata. In general, we find that learning how to read the manuals (and use the help system) is more efficient than asking someone else, and it allows you to save your questions for the really hard stuff. For those new to Stata, we recommend the *Getting Started* manual (which is specific to your platform) and the first part of the *User's Guide*. As you become more acquainted with Stata, the *Reference* manuals will become increasingly valuable for detailed information about commands, including a discussion of the statistical theory related to the commands and references for further reading.

2.3.3   Other resources

The *User's Guide* also discusses additional sources of information about Stata. Most importantly, the Stata web site (*http://www.stata.com*) contains many useful resources, including links to tutorials and an extensive FAQ section that discusses both introductory and advanced topics. You can also get information on the NetCourses offered by Stata, which are four- to seven-week courses offered over the Internet. Another excellent set of online resources is provided by UCLA's Academic Technology Services at *http://www.ats.ucla.edu/stat/stata/*.

There is also a Statalist listserver that is independent of StataCorp, although many programmers and statisticians from StataCorp participate. This list is a wonderful resource for information on Stata and statistics. You can submit questions and usually receive answers very quickly. Monitoring the listserver is also a quick way to pick up insights from Stata veterans. For details on joining the list, go to *http://www.stata.com*, follow the link to *User Support*, and click on the link to *Statalist*.

2.4   The working directory

The *working directory* is the default directory for any file operations such as using data, saving data, or logging output. If you type `cd` in the Command window, Stata displays the name of the current working directory. To load a data file stored in the working directory, you just type `use` *filename* (e.g., `use binlfp2`). If a file is not in the working directory, you must specify the full path (e.g., `use d:\spostdata\examples\binlfp2`).

At the beginning of each Stata session, we like to change our working directory to the directory where we plan to work, since this is easier than repeatedly entering the path name for the directory. For example, typing `cd d:\spostdata` changes the working directory to `d:\spostdata`. If the directory name includes spaces, you must put the path in quotation marks (e.g., `cd "d:\my work\"`).

You can list the files in your working directory by typing `dir` or `ls`, which are two names for the same command. With this command, you can use the * wildcard. For example, `dir *.dta` lists all files with the extension `.dta`.

2.5   Stata file types

Stata uses and creates many types of files, which are distinguished by extensions at the end of the filename. Some of the extensions used by Stata are

.ado	Programs that add commands to Stata, such as the SPost commands.
.class	Files that define classes in the Stata class system.
.dlg	Programs that define the appearance and functionality of dialog boxes.
.do	Batch files that execute a set of Stata commands.
.dta	Data files in Stata's format.
.gph	Graphs saved in Stata's proprietary format.
.hlp	The text displayed when you use the help command. For example, fitstat.hlp has help for fitstat.
.log	Output saved as plain text by the log using command.
.smcl	Output saved in the SMCL format by the log using command.
.wmf	Graphs saved as Windows Metafiles.

The most important of these for a new user are the .smcl, .log, .dta, and .do files, which we will now discuss.

2.6   Saving output to log files

Stata does not automatically save the output from your commands. To save your output to print or examine later, you must open a *log file*. Once a log file is opened, both the commands and the output they generate are saved. Because the commands are recorded, you can tell exactly how the results were obtained. The syntax for the log command is

log using *filename* [, append replace [smcl | text]]

By default, the log file is saved to your working directory. You can save it to a different directory by typing the full path (e.g., log using d:\project\mylog, replace).

Options

append means that if the file exists, new output should be added to the end of the existing file.

replace indicates that you want to replace the log file if it already exists. For example, log using mylog creates the file mylog.smcl. If this file already exists, Stata generates an error message. So, you could use log using mylog, replace, and the existing file would be overwritten by the new output.

smcl is the default option that requests that the log be written using the Stata Markup and Control Language (SMCL) with the file suffix .smcl. SMCL files contain special codes that add solid horizontal and vertical lines, bold and italic typefaces, and hyperlinks to the Result window. The disadvantage of SMCL is that the special features can only be viewed within Stata. If you open a SMCL file in a text editor, your results will appear amidst a jumble of special codes.

`text` specifies that the log should be saved as plain text (ASCII), which is the preferred format for loading the log into a text editor for printing. Instead of adding the `text` option, such as `log using mywork, text`, you can specify plain text by including the `.log` extension (for example, `log using mywork.log`).

Tip: Plain text logs by default We prefer plain text for output rather than SMCL. Typing `set logtype text` at the beginning of a Stata session makes plain text the default for log files for the current session. Typing `set logtype text, permanently` makes plain text the default for future sessions.

2.6.1   Closing a log file

To close a log file, type

```
. log close
```

Also, when you exit Stata, the log file closes automatically. Because you can only have one log file open at a time, any open log file must be closed before you can open a new one.

2.6.2   Viewing a log file

Regardless of whether a log file is open or closed, a log file can be viewed by selecting File→Log→View from the menu, and the log file will be displayed in the Viewer. When in the Viewer, you can print the log by selecting File→Print Viewer.... You can also view the log file by clicking on , which opens the log in the Viewer. If the Viewer window "gets lost" behind other windows, you can click on to bring the Viewer to the front.

2.6.3   Converting from SMCL to plain text or PostScript

If you want to convert a log file in SMCL format to plain text, you can use the `translate` command. For example,

```
. translate mylog.smcl mylog.log, replace
(file mylog.log written in .log format)
```

tells Stata to convert the SMCL file `mylog.smcl` to a plain text file called `mylog.log`. Or, you can convert a SMCL file to a PostScript file, which is useful if you are using TeX or LaTeX or if you want to convert your output into Adobe's Portable Document Format. For example,

```
. translate mylog.smcl mylog.ps, replace
(file mylog.ps written in .ps format)
```

Converting can also be done via the menus by selecting File→Log→Translate.

2.7   Using and saving datasets

2.7.1   Data in Stata format

Stata uses its own data format with the extension .dta. The use command loads such data into memory. Pretend that we are working with the file nomocc2.dta in directory d:\spostdata. We can load the data by typing

```
. use d:\spostdata\nomocc2, clear
```

where the .dta extension is assumed by Stata. The clear option erases all data currently in memory and proceeds with loading the new data. Stata does not give an error if you include clear when there are no data in memory. If d:\spostdata was our working directory, we could use the simpler command

```
. use nomocc2, clear
```

If you have changed the data by deleting cases, merging in another file, or creating new variables, you can save the file with the save command. For example,

```
. save d:\spostdata\nomocc3, replace
```

where again we did not need to include the .dta extension. Also notice that we saved the file with a different name so that we can use the original data later. The replace option indicates that if the file nomocc3.dta already exists, Stata should overwrite it. If the file does not already exist, replace is ignored. If d:\spostdata was our working directory, we could save the file with

```
. save nomocc3, replace
```

save stores the data in a format that can be read only by Stata 8 or later. If you use the command saveold instead of save, the dataset is written so that it can be read by Stata 7, but if your data contain multiple missing-value codes, a feature that only became available in Stata 8, all the missing-value codes will be mapped to the smallest missing value (.).

Tip: compress **before saving** Before saving a file, run compress, which checks each variable to determine if it can be saved in a more compact form. For instance, binary variables fit into the byte type, which takes up only one-fourth of the space of the float type. If you run compress, it might make your data file much more compact, and at worst it will do no harm.

2.7.2   Data in other formats

To load data from another statistical package, such as SAS or SPSS, you need to convert it into Stata's format. The easiest way to do this is with a conversion program such as Stat/Transfer (*http://www.stattransfer.com*). We recommend obtaining one of these programs if you are using more than one statistical package or if you often share data with others who use different packages.

Alternatively but less conveniently, most statistical packages allow you to save and load data in ASCII format. You can load ASCII data with the `infile` or `infix` commands and export it with the `outfile` command. The *Reference* manual entry for `infile` contains an extensive discussion that is particularly helpful for reading in ASCII data, or you can type `help infile`.

2.7.3   Entering data by hand

Data can also be entered by hand using a spreadsheet-style editor. While we do not recommend using the editor to change existing data (because it is too easy to make a mistake), we find that it is very useful for entering small datasets. To enter the editor, click on or type `edit` on the command line. The *Getting Started* manual has a tutorial for the editor, but most people who have used a spreadsheet before will be immediately comfortable with the editor.

As you use the editor, *every* change that you make to the data is reported in the Results window and is captured by the log file, if it is open. For example, if you change `age` for the fifth observation to 32, Stata reports `replace age = 32 in 5`. This tells you that instead of using the editor, you could have changed the data with a `replace` command. When you close the editor, you are asked if you really want to keep the changes or revert to the unaltered data.

2.8   Size limitations on datasets*

If you get the error message `r(900): no room to add more observations` when trying to load a dataset or the message `r(901): no room to add more variables` when trying to add a new variable, you may need to allocate more memory. Typing `memory` shows how much memory Stata has allocated and how much it is using. You can increase the amount of memory by typing `set memory #k` (for KB) or `#m` (for MB). For example, `set memory 32000k` or `set memory 32m` sets the memory to 32MB.[3] Note that if you have variables in memory, you must type `clear` before you can set the memory.

If you get the error `r(1000): system limit exceeded--see manual` when you try to load a dataset or add a variable, your dataset might have too many variables or the

---

[3]Stata can use virtual memory if you need to allocate memory beyond that which is physically available on a system, but we find that virtual memory makes Stata unbearably slow.

width of the dataset might be too large. Intercooled Stata is limited to a maximum of 2,047 variables, and the dataset can be up to 24,564 units wide (a binary variable has width 1, a double-precision variable width 8, and a string variable has width equal its length). Stata/SE allows 32,767 variables, and the dataset can be up to 393,192 units wide. String variables can be up to 80 characters in Intercooled Stata and 244 characters in Stata/SE. File transfer programs such as Stat/Transfer can drop specified variables and optimize variable storage. You can use these programs to create multiple datasets that each only contain the variables necessary for specific analyses.

2.9   do-files

You can execute commands in Stata by typing one command at a time into the Command window and pressing Enter, as we have been doing. This interactive mode is useful when you are learning Stata, exploring your data, or experimenting with alternative specifications of your regression model. Alternatively, you can create a text file that contains a series of commands and then tell Stata to execute all the commands in that file, one after the other. These files, which are known as *do-files* because they use the extension .do, have the same function as "syntax files" in SPSS or "batch files" in other statistics packages. For more serious or complex work, we *always* use do-files because they make it easier to redo the analysis with small modifications later and because they provide an exact record of what has been done.

To get an idea of how do-files work, consider the file `example.do` saved in the working directory:

```
log using example, replace text
use binlfp2, clear
tabulate hc wc, row nolabel
log close
```

To execute a do-file, you execute the command

```
do dofilename
```

from the Command window. For example, `do example` tells Stata to run each of the commands in `example.do`. (Note: If the do-file is not in the working directory, you need to specify the directory, such as `do d:\spostdata\example`.) Executing `example.do` begins by opening the log `example.log`, then loads `binlfp2.dta`, and finally constructs a table with `hc` and `wc`. Here is what the output looks like:

```
-------------------------------------------------------------------------------
       log:  f:\spostdata\example.log
  log type:  text
 opened on:  26 May 2003, 15:44:45
. use http://www.stata-press.com/data/lfr/binlfp2, clear
(Data from 1976 PSID-T Mroz)
```

```
. tabulate hc wc, row nolabel
+----------------+
| Key            |
|----------------|
|    frequency   |
|  row percentage |
+----------------+

   Husband |  Wife College: 1=yes
  College: |        0=no
 1=yes 0=no |        0          1 |    Total
-----------+----------------------+----------
         0 |      417         41 |      458
           |    91.05        8.95 |   100.00
-----------+----------------------+----------
         1 |      124        171 |      295
           |    42.03       57.97 |   100.00
-----------+----------------------+----------
     Total |      541        212 |      753
           |    71.85       28.15 |   100.00
. log close
       log:  f:\spostdata\example.log
  log type:  text
 closed on:  26 May 2003, 15:44:45
--------------------------------------------------------------------------
```

2.9.1 Adding comments

Stata has several different methods for denoting comments. We will make extensive use of two methods. First, on any given line, Stata treats everything that comes after // or after * as comments that are simply echoed to the output. Second, on any given line, Stata ignores whatever comes after /// and treats the next line as a continuation of the current line. For example, the following do-file executes the same commands as the one above but includes comments:

```
//
// ==> short simple do-file
// ==> for didactic purposes
//
log using example, replace  // this comment is ignored
// next we load the data
use binlfp2, clear
// tabulate husband´s and wife´s education
tabulate hc wc,    /// the next line is treated as a continuation of this one
      row nolabel
// close up
log close
// make sure there is a cr at the end!
```

If you look at the do-files on our web site that reproduce the examples in this book, you will see that we use many comments. They are extremely helpful if others will be using your do-files or log files, or if there is a chance that you will use them again at a later time.

2.9.2   Long lines

Sometimes you need to execute a command that is longer than the text that can fit onto a screen. If you are entering the command interactively, the Command window simply pushes the left part of the command off the screen as space is needed. Before entering a long command line in a do-file, however, you can use `#delimit ;` to tell Stata to interpret ";" as the end of a command. After the long command is entered, you can enter `#delimit cr` to return to using the carriage return as the end-of-line delimiter. For example,

```
#delimit ;
recode income91 1=500 2=1500 3=3500 4=4500 5=5500 6=6500 7=7500 8=9000
9=11250 10=13750 11=16250 12=18750 13=21250 14=23750 15=27500 16=32500
17=37500 18=45000 19=55000 20=67500 21=75000 *=. ;
#delimit cr
```

Instead of the `#delimit` command, we could have used `///`. For example,

```
recode income91 1=500 2=1500 3=3500 4=4500 5=5500 6=6500 7=7500 8=9000     ///
     9=11250 10=13750 11=16250 12=18750 13=21250 14=23750 15=27500 16=32500 ///
     17=37500 18=45000 19=55000 20=67500 21=75000 *=.
```

2.9.3   Stopping a do-file while it is running

If you are running a command or a do-file that you want to stop before it completes execution, click on [X] or press Ctrl-Break.

2.9.4   Creating do-files

Using Stata's do-file editor

do-files can be created with Stata's built-in do-file editor. To use the editor, enter the command `doedit` to create a file to be named later or `doedit` *filename* to create or edit a file named *filename*.do. Alternatively, you can click on [icon]. The do-file editor is easy to use and works like most text editors (see *Getting Started* for further details).

After you finish your do-file, select Tools→Do to execute the file or click on [icon].

Using other editors to create do-files

Since do-files are plain text files, you can create do-files with any program that creates text files. Specialized text editors work much better than word processors such as WordPerfect or Microsoft Word. Among other things, with word processors it is easy to forget to save the file as plain text. Our own preference for creating do-files is TextPad (*http://www.textpad.com*), which runs in Windows. This program has many features that make it faster to create do-files. For example, you can create a "clip library" that contains frequently entered material, and you can obtain a syntax file from our web site that provides color coding of reserved words for Stata. TextPad also allows you to have several different files open at once, which is often handy for complex data analyses.

If you use an editor other than Stata's built-in editor, you cannot run the do-file by clicking on an icon or selecting from a menu. Instead, you must switch from your editor and then enter the command do *filename*.

Warning Stata executes commands when it encounters a carriage return (i.e., the Enter
key). If you do not include a carriage return after the last line in a do-file, that
last line will not be executed. TextPad has a feature to enter that final, pesky
carriage return automatically. To set this option in TextPad 4, select the option
"Automatically terminate the last line of the file" in the preferences for the editor.

2.9.5   A recommended structure for do-files

This is the basic structure that we recommend for do-files:

```
// including version number ensures compatibility with later Stata releases
version 8
// if a log file is open, close it
capture log close
// don´t pause when output scrolls off the page
set more off
// log results to file myfile.log
log using myfile, replace text
// * myfile.do - written 19 jan 2001 to illustrate do-files
//
// * your commands go here
//
// close the log file.
log close
```

While the comments (which you can remove) should explain most of the file, there are a few points that we need to explain.

- The **version 8** command indicates that the program was written for use in Stata 8. This command tells any future version of Stata that you want the commands that follow to work just as they did when you ran them in Stata 8. This prevents the problem of old do-files not running correctly in newer releases of the program.

- The command **capture log close** is very useful. Suppose you have a do-file that starts with **log using mylog, replace**. You run the file and it "crashes" before reaching **log close**, which means that the log file remains open. If you revise the do-file and run it again, an error is generated when it tries to open the log file because the file is already open. The prefix **capture** tells Stata not to stop the do-file if the command that follows produces an error. Accordingly, `capture log close` closes the log file if it is open. If it is not open, the error generated by trying to close an already closed file is ignored.

Tip: The command `cmdlog` is very much like the `log` command, except that it creates a text file with extension `.txt` that saves all subsequent commands that are entered in the Command window (it does not save commands that are executed within a do-file). This is handy because it allows you to use Stata interactively and then make a do-file based on what you have done. You simply load the cmdlog that you saved, rename it to *newname*`.do`, delete commands you no longer want, and execute the new do-file. Your interactive session is now documented as a do-file. The syntax for opening and closing cmdlog files is the same as the syntax for `log` (i.e., `cmdlog using` to open and `cmdlog close` to close), and you can have log and cmdlog files open simultaneously.

2.10   Using Stata for serious data analysis

Voltaire is said to have written *Candide* in three days. Creative work often rewards such inspired, seat-of-the-pants, get-the-details-later activity. *Data management does not.* Instead, effective data management rewards forethought, carefulness, double- and triple-checking of details, and meticulous, albeit tedious, documentation. Errors in data management are astonishingly (and painfully) easy to make. Moreover, tiny errors can have disastrous implications that can cost hours and even weeks of work. The extra time it takes to conduct data management carefully is rewarded many times over by the reduced risk of errors. Put another way, it helps prevent you from getting incorrect results that you do not know are incorrect. With this in mind, we begin with some broad, perhaps irritatingly practical, suggestions for doing data analysis efficiently and effectively.

1. *Ensure replicability by using do-files and log files for everything.* For data analysis to be credible, you must be able to reproduce *entirely and exactly* the trail from the original data to the tables in your paper. Thus, any permanent changes you make to the data should be made by running do-files rather than by using the interactive mode. If you work interactively, be sure that the first thing you do is to open a log or cmdlog file. Then when you are done, you can use these files to create a do-file to reproduce your interactive results.

2. *Document your do-files.* The reasoning that is obvious today can be baffling in six months. We use comments extensively in our do-files, which are invaluable for remembering what we did and why we did it.

3. *Keep a research log.* For serious work, you should keep a diary that includes a description of *every* program you run, the research decisions that are being made (e.g., the reasons for recoding a variable in a particular way), and the files that are created. A good research log allows you to reproduce everything you have done starting only with the original data. We cannot overemphasize how helpful such notes are when you return to a project that was put on hold, when you are responding to reviewers, or when you moving on to the next stage of your research.

4. *Develop a system for naming files.* Usually it makes the most sense to have each do-file generate one log file with the same prefix (e.g., clean_data.do, clean_data.log). Names are easiest to organize when brief, but they should be long enough and logically related enough to make sense of the task the file does.[4] Scott prefers to keep the names short and organized by major task (e.g., recode01.do), while Jeremy likes longer names (e.g., make_income_vars.do). Either is fine as long as it works for you.

5. *Use new names for new variables and files.* Never change a dataset and save it with the original name. If you drop three variables from pcoms1.dta and create two new variables, call the new file pcoms2.dta. When you transform a variable, give it a new name rather than simply replacing or recoding the old variable. For example, if you have a variable workmom with a five-point attitude scale, and you want to create a binary variable indicating positive and negative attitudes, create a new variable called workmom2.

6. *Use labels and notes.* When you create a new variable, give it a variable label. If it is a categorical variable, assign value labels. You can add a note about the new variable using the notes command (described below). When you create a new dataset, you can also use notes to document what it is.

7. *Double-check every new variable.* Cross-tabulating or graphing the old variable and the new variable are often effective for verifying new variables. As we describe below, using list with a subset of cases is similarly effective for checking transformations. At the very least, be sure to look carefully at the frequency distributions and summary statistics of variables in your analysis. You would not believe how many times puzzling regression results turn out to involve miscodings of variables that would have been immediately apparent by looking at the descriptive statistics.

8. *Practice good archiving.* If you want to retain hard copies of all your analyses, develop a system of binders for doing so rather than a set of intermingling piles on your desk. Back up everything. Make off-site backups or keep any on-site backups in a fireproof box. Should cataclysm strike, you will have enough other things to worry about without also having lost months or years of work.

---

[4]Students sometimes find it amusing to use names like dumbproject.do or joanieloveschachi.do. The fun ends when one needs to reconstruct something but can no longer recall which file does what.

2.11   The syntax of Stata commands

Think about the syntax of commands in everyday, spoken English. They usually begin with a verb telling the other person what they are supposed to do. Sometimes the verb is the entire command: "Help!" or "Stop!" Sometimes the verb needs to be followed by an object that indicates who or what the verb is to be performed on: "Help Dave!" or "Stop the car!" In some cases, the verb is followed by a qualifier that gives specific conditions under which the command should or should not be performed: "Give me a piece of pizza *if it doesn't have mushrooms*" or "Call me *if you get home before nine*". Verbs can also be followed by adverbs that specify that the action should be performed in some way that is different from how it might normally be, such as when a teacher commands her students to "Talk *clearly*" or "Walk *single file*".

Stata follows an analogous logic, albeit with some additional wrinkles that we will introduce later. The basic syntax of a command has four parts:

1. *Command*: What action do you want performed?

2. *Names of variables, files, or other objects*: On what things is the command to be performed?

3. *Qualifier on observations*: On which observations should the command be performed?

4. *Options*: What special things should be done in executing the command?

All commands in Stata require the first of these parts, just as it is hard to issue spoken commands without a verb. Each of the other three parts can be required, optional, or not allowed, depending on the particular command and circumstances. Here is an example of a command that features all four parts and uses `binlfp2.dta`, which we loaded earlier:

```
. tabulate hc wc if age>=40, row
```

Key
frequency *row percentage*

Husband College: 1=yes 0=no	Wife College: 1=yes 0=no NoCol College		Total
NoCol	275 92.28	23 7.72	298 100.00
College	61 38.85	96 61.15	157 100.00
Total	336 73.85	119 26.15	455 100.00

If you want to suppress the key, you can add the option `nokey`. For example, `tabulate hc wc, row nokey`.

`tabulate` is a command for making one- or two-way tables of frequencies. In this example, we want a two-way table of the frequencies of variables `hc` by `wc`. By putting `hc` first, we make this the row variable and `wc` the column variable. By specifying `if age>40`, we specify that the frequencies should only include observations for those older than 40. The option `row` indicates that row percentages should be printed as well as frequencies. These allow us to see that in 61% of the cases in which the husband had attended college, the wife had also done so, while wives had attended college only in 8% of cases in which the husband had not. Notice the comma preceding `row`: *whenever options are specified, they are at the end of the command with a single comma to indicate where the list of options begins.* The precise ordering of multiple options after the comma is never important.

Next, we provide more information on each of the four components.

2.11.1   Commands

Commands define the tasks that Stata is to perform. A great thing about Stata is that the set of commands is deliciously open ended. It expands not just with new releases of Stata but also when users add their own commands, such as our SPost commands. Each new command is stored in its own file, ending with the extension `.ado`. Whenever Stata encounters a command that is not in its built-in library, it searches various directories for the appropriate ado-file. The list of the directories it searches (and the order that it searches them) can be obtained by typing `adopath`.

2.11.2   Variable lists

Variable names are case sensitive. For example, you could have three different variables named `income`, `Income`, and `inCome`. Of course, this is not a good idea because it leads to confusion. To keep life simple, we stick exclusively to lowercase names. Starting with Stata 7, Stata allows variable names up to 32 characters long, compared with the 8-character maximum imposed by earlier versions of Stata and many other statistics packages. In practice, we try not to give variables names more than 8 characters, as this makes it easier to share data with people who use other packages. Additionally, we recommend using short names because longer variable names become unwieldy to type. (Although variable names can be abbreviated to whatever initial set of characters identifies the variable uniquely, we worry that too much reliance on this feature might cause one to make mistakes.)

If you do not list any variables, many commands assume that you want to perform the operation on every variable in the dataset. For example, the `summarize` command provides summary statistics on the listed variables:

```
. sum age inc k5
    Variable |        Obs        Mean    Std. Dev.        Min         Max
-------------+--------------------------------------------------------
         age |        753    42.53785     8.072574         30          60
         inc |        753    20.12897      11.6348  -.0290001          96
          k5 |        753    .2377158      .523959          0           3
```

Alternatively, we could get summary statistics on every variable in our dataset by just typing

```
. sum
    Variable |        Obs        Mean    Std. Dev.        Min         Max
-------------+--------------------------------------------------------
         lfp |        753    .5683931     .4956295          0           1
          k5 |        753    .2377158      .523959          0           3
        k618 |        753    1.353254     1.319874          0           8
         age |        753    42.53785     8.072574         30          60
          wc |        753    .2815405     .4500494          0           1
-------------+--------------------------------------------------------
          hc |        753    .3917663     .4884694          0           1
         lwg |        753    1.097115     .5875564  -2.054124    3.218876
         inc |        753    20.12897      11.6348  -.0290001          96
```

You can also select all variables that begin or end with the same letter(s) by using the wildcard operator *. For example,

```
. sum k*
    Variable |        Obs        Mean    Std. Dev.        Min         Max
-------------+--------------------------------------------------------
          k5 |        753    .2377158      .523959          0           3
        k618 |        753    1.353254     1.319874          0           8
```

2.11.3   if and in qualifiers

Stata has two qualifiers that restrict the sample that is analyzed: if and in. in performs operations on a range of consecutive observations. Typing sum in 20/100 gives summary statistics based only on the 20th through 100th observations. in restrictions are dependent on the current sort order of the data, meaning that if you resort your data, the 81 observations selected by the restriction sum in 20/100 might be different.[5]

In practice, if conditions are used much more often than in conditions. if restricts the observations to those that fulfill a specified condition. For example, sum if age<50 provides summary statistics for only those observations where age is less than 50. Here is a list of the elements that can be used to construct logical statements for selecting observations with if:

[5]In Stata 6 and earlier, some official Stata commands changed the sort order of the data, but fortunately this quirk was removed in Stata 7. As of Stata 7, no properly written Stata command should change the sort order of the data, although readers should beware that user-written programs may not always follow proper Stata programming practice.

Operator	Definition	Example
==	equal to	if female==1
!=	not equal to	if female!=1
>	greater than	if age>20
>=	greater than or equal to	if age>=21
<	less than	if age<66
<=	less than or equal to	if age<=65
&	and	if age==21 & female==1
|	or	if age==21|educ>16

There are two important things to note about the if qualifier:

1. To specify a condition to test, use a double equal sign (e.g., sum if female==1). When assigning a value to something, such as when creating a new variable, use a single equal sign (e.g., gen newvar=1). Putting these examples together results in gen newvar=1 if female==1.

2. **The missing value codes are the largest positive numbers**. This implies that Stata treats missing cases as positive infinity when evaluating if expressions. In other words, if you type sum ed if age>50, the summary statistics for ed are calculated on all observations where age is greater than 50, including cases where the value of age is missing. You must be careful of this when using if with > or >= expressions. If you type sum ed if age<., Stata gives summary statistics for cases where age is not missing (Note: . is the smallest of the 27 missing value codes. See section 2.12.3 for more on missing values.). Entering sum ed if age>50 & age<. provides summary statistics for those cases where age is greater than 50 and is not missing.

Examples of if qualifier

If we wanted summary statistics on income for only those respondents who were between the ages of 25 and 65, we would type

```
. sum income if age>=25 & age<=65
```

If we wanted summary statistics on income for only female respondents who were between the ages of 25 and 65, we would type

```
. sum income if age>=25 & age<=65 & female==1
```

If we wanted summary statistics on income for the remaining female respondents—that is, those who are younger than 25 or older than 65—we would type

```
. sum income if (age<25 | age>65) & age<. & female==1
```

Notice that we need to include `& age<.` because Stata treats missing codes as positive infinity. The condition (`age<25 | age>65`) would otherwise include those cases for which `age` is missing.

Tip: Removing the separator If you do not like the horizontal separator that appears after every five variables in the output for `summarize`, you can remove the lines with the option `sep(0)`.

2.11.4   Options

Options are set off from the rest of the command by a comma. Options can often be abbreviated, although whether and how they can be abbreviated varies across commands. In this book, we rarely cover all of the available options available for any given command, but you can check the manual or use `help` for further options that might be useful for your analyses.

2.12   Managing data

2.12.1   Looking at your data

There are two easy ways to look at your data.

`browse` opens a spreadsheet in which you can scroll to look at the data, but you cannot change the data. You can look and change data with the `edit` command, but this is risky. We much prefer making changes to our data using do files, even when we are only changing the value of one variable for one observation. The browser is also available by clicking on , while the data editor is available by clicking on .

`list` creates a list of values of specified variables and observations. `if` and `in` qualifiers can be used to look at just a portion of the data, which is sometimes useful for checking that transformations of variables are correct. For example, if you want to confirm that the variable `lninc` has been correctly constructed as the natural log of `inc`, typing `list inc lninc in 1/20` lets you see the values of `inc` and `lninc` for the first 20 observations.

2.12.2   Getting information about variables

There are several methods for obtaining basic information about your variables. Here are five commands that we find useful. Which one you use depends in large part on the kind and level of detail you need.

describe provides information on the size of the dataset and the names, labels, and types of variables. For example,

```
. use http://www.stata-press.com/data/lfr/binlfp2, clear
. describe
Contains data from http://www.stata-press.com/data/lfr/binlfp2.dta
  obs:           753                          Data from 1976 PSID-T Mroz
  vars:            8                          30 Apr 2001 16:17
  size:        13,554 (98.7% of memory free)  (_dta has notes)
```

	storage	display	value	
variable name	type	format	label	variable label
lfp	byte	%9.0g	lfplbl	Paid Labor Force: 1=yes 0=no
k5	byte	%9.0g		# kids < 6
k618	byte	%9.0g		# kids 6-18
age	byte	%9.0g		Wife's age in years
wc	byte	%9.0g	collbl	Wife College: 1=yes 0=no
hc	byte	%9.0g	collbl	Husband College: 1=yes 0=no
lwg	float	%9.0g		Log of wife's estimated wages
inc	float	%9.0g		Family income excluding wife's

```
Sorted by:  lfp
```

summarize provides summary statistics. By default, summarize presents the number of nonmissing observations, the mean, the standard deviation, the minimum values, and the maximum. Adding the detail option includes additional information. For example,

```
. sum age, detail
                        Wife's age in years
        Percentiles      Smallest
  1%         30              30
  5%         30              30
 10%         32              30       Obs                753
 25%         36              30       Sum of Wgt.        753
 50%         43                       Mean          42.53785
                           Largest    Std. Dev.     8.072574
 75%         49              60
 90%         54              60       Variance      65.16645
 95%         56              60       Skewness       .150879
 99%         59              60       Kurtosis      1.981077
```

tabulate creates the frequency distribution for a variable. For example,

```
. tab hc
   Husband
   College:
 1=yes 0=no |      Freq.     Percent        Cum.
------------+-----------------------------------
      NoCol |        458       60.82       60.82
    College |        295       39.18      100.00
------------+-----------------------------------
      Total |        753      100.00
```

If you do not want the value labels included, type

```
. tab hc, nolabel
  Husband |
 College: |
 1=yes 0=no |      Freq.      Percent        Cum.
----------+---------------------------------------
        0 |        458        60.82       60.82
        1 |        295        39.18      100.00
----------+---------------------------------------
    Total |        753       100.00
```

If you want a two-way table, type

```
. tab hc wc
  Husband | Wife College: 1=yes
 College: |         0=no
 1=yes 0=no |   NoCol     College  |    Total
----------+------------------------+----------
    NoCol |     417          41    |     458
  College |     124         171    |     295
----------+------------------------+----------
    Total |     541         212    |     753
```

By default, **tabulate** does not tell you the number of missing values for either variable. Specifying the **missing** option includes missing values. We recommend this option whenever you are generating a frequency distribution to check that some transformation was done correctly. The options **row**, **col**, and **cell** request row, column, and cell percentages along with the frequency counts. The option **chi2** reports the chi-square for a test that the rows and columns are independent.

tab1 presents univariate frequency distributions for each variable listed. For example,

```
. tab1 hc wc
-> tabulation of hc
  Husband |
 College: |
 1=yes 0=no |      Freq.      Percent        Cum.
----------+---------------------------------------
    NoCol |        458        60.82       60.82
  College |        295        39.18      100.00
----------+---------------------------------------
    Total |        753       100.00
-> tabulation of wc
     Wife |
 College: |
 1=yes 0=no |      Freq.      Percent        Cum.
----------+---------------------------------------
    NoCol |        541        71.85       71.85
  College |        212        28.15      100.00
----------+---------------------------------------
    Total |        753       100.00
```

dotplot generates a quick graphical summary of a variable, which is very useful for quickly checking your data. For example, the command `dotplot age` leads to the following graph:

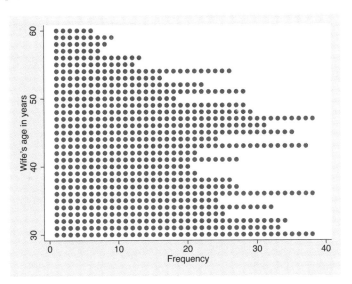

This graph will appear in a new window called the Graph window. Details on saving, printing, and enhancing graphs are given in Section 2.16.

codebook summarizes a variable in a format designed for printing a codebook. For example, `codebook age` produces

```
. codebook age
```

age				Wife's age in years

type:	numeric (byte)			
range:	[30,60]		units:	1
unique values:	31		missing .:	0/753
mean:	42.5378			
std. dev:	8.07257			

percentiles:	10%	25%	50%	75%	90%
	32	36	43	49	54

2.12.3 Missing values

While numeric missing values are automatically excluded when Stata fits models, they are stored as the largest positive values. Twenty-seven missing values are available, with the ordering

$$\text{all numbers} < \,.\, < .a < .b < \cdots < .z$$

This way of handling missing values can have unexpected consequences when determining samples. For instance, the expression `if age>65` is true when `age` has a value greater than 65 and when `age` is missing. Similarly, the expression `occupation!=1` is true if `occupation` is not equal to 1 or `occupation` is missing. When expressions such as these are required, be sure to explicitly exclude any unwanted missing values. For instance, `age>65 & age<.` would be true only for those people whose age is not missing and who are over 65. Similarly, `occupation!=1 & occupation <.` would be true only when the `occupation` is not missing and not equal to 1.

The different missing values can be used to record the distinct reasons why a variable is missing. For instance, consider a survey that asked people about their driving records. The variable that records whether someone received a ticket after being involved in an accident could be missing because the respondent had not been involved in any accidents or because the person refused to answer the question.

2.12.4   Selecting observations

As previously mentioned, you can select cases using with the `if` and `in` options. For example, `summarize age if wc==1` provides summary statistics on `age` for only those observations where `wc` equals 1. In some cases it is simpler to remove the cases with either the `drop` or `keep` commands. `drop` removes observations from memory (not from the .dta file) based on an `if` or `in` specification. The syntax is

drop [in *range*] [if *exp*]

Only observations that do *not* meet those conditions are left in memory. For example, `drop if wc==1` keeps only those cases where `wc` is not equal to 1, including observations with missing values on `wc`.

`keep` has the same syntax as `drop` and deletes all cases *except* those that meet the condition. For example, `keep if wc==1` keeps only those cases where `wc` is 1; all other observations, including those with missing values for `wc`, are dropped from memory. After selecting the observations that you want, you can save the remaining variables to a new dataset with the `save` command.

2.12.5   Selecting variables

You can also select which variables you want to keep. The syntax is

drop *variable_list*

keep *variable_list*

With **drop**, all variables are kept except those that are explicitly listed. With **keep**, only those variables that are explicitly listed are kept. After selecting the variables that you want, you can save the remaining variables to a new dataset with the **save** command.

2.13   Creating new variables

The variables that you analyze are often constructed differently than the variables in the original dataset. In this section we consider basic methods for creating new variables. Our examples always create a new variable from an old variable rather than transforming an existing variable. Even though it is possible to simply transform an existing variable, we find that this leads to mistakes.

2.13.1   generate command

generate creates new variables. For example, to create **age2** as an exact copy of **age**, type

```
. generate age2 = age
. summarize age2 age
```

Variable	Obs	Mean	Std. Dev.	Min	Max
age2	753	42.53785	8.072574	30	60
age	753	42.53785	8.072574	30	60

The results of **summarize** show that the two variables are identical. Note that we used a single equal sign since we are making a variable equal to some value.

Observations excluded by **if** or **in** qualifiers in the **generate** command are coded as missing. For example, to generate **age3** that equals **age** for those over 40 but is otherwise missing, type

```
. gen age3 = age if age>40
(318 missing values generated)
. sum age3 age
```

Variable	Obs	Mean	Std. Dev.	Min	Max
age3	435	48.3977	4.936509	41	60
age	753	42.53785	8.072574	30	60

Whenever **generate** (or **gen** as it can be abbreviated) produces missing values, it tells you how many cases are missing.

generate can also create variables that are mathematical functions of existing variables. For example, we can create **agesq** that is the square of **age** and **lnage** that is the natural log of **age**:

```
. gen agesq = age^2
. gen lnage = ln(age)
```

For a complete list of the mathematical functions in Stata, type `help functions`. For quick reference, here is a list of particularly useful functions:

Function	Definition	Example
+	addition	`gen y = a+b`
−	subtraction	`gen y = a-b`
/	division	`gen density = pop/area`
*	multiplication	`gen y = a*b`
^	take to a power	`gen y = a^3`
ln	natural log	`gen lnwage = ln(wage)`
exp	exponential	`gen y = exp(a)`
sqrt	square root	`gen agesqrt = sqrt(age)`

2.13.2   replace command

`replace` has the same syntax as `generate` but is used to change values of a variable that already exists. For example, say we want to make a new variable, `age4`, that equals `age` if `age` is over 40, but equals 40 for all persons aged 40 and under. First, we create `age4` equal to `age`. Then, we replace those values we want to change:

```
. gen age4 = age
. replace age4 = 40 if age<40
(298 real changes made)
. sum age4 age
    Variable |       Obs        Mean    Std. Dev.        Min         Max
        age4 |       753    44.85126     5.593896         40          60
         age |       753    42.53785     8.072574         30          60
```

Note that `replace` reports how many values were changed. This is useful in verifying that the command did what you intended. Also, `summarize` confirms that the minimum value of `age` is 30 and that `age4` now has a minimum of 40 as intended.

Warning Of course, we could have simply changed the original variable: `replace age = 40 if age<40`. But, if we did this and saved the data, there would be no way to return to the original values for `age` if we later needed them.

2.13.3   recode command

The values of *existing* variables can also be changed using the `recode` command. With `recode` you specify a set of correspondences between old values and new ones. For example, you might want old values of 1 and 2 to correspond to new values of 1, old values of 3 and 4 to correspond to new values of 2, and so on. This is particularly useful for combining categories. To use this command, we recommend that you start by making a copy of an existing variable. Then, recode the copy. Or, to be more efficient, you can use the **generate**(*newvariablename*) option with **recode**. With this option, Stata creates a new variable instead of overwriting the old one. **recode** is best explained by example, several of which we include below (for more, type **help recode**).

To change 1 to 2 and 3 to 4 but leave all other values unchanged, type

```
. recode origvar (1=2) (3=4), generate(myvar1)
(23 differences between origvar and mvar1)
```

To change 2 to 1 and change all other values (including missing) to 0:

```
. recode origvar (2=1) (*=0), gen(myvar2)
(100 differences between origvar and myvar2)
```

where the asterisk indicates all values, including missing values, that have not been explicitly recoded.

To change 2 to 1 and change all other values *except missing* to 0:

```
. recode origvar (2=1) (nonmissing=0), gen(myvar3)
(89 differences between origvar and myvar3)
```

To change values from 1 to 4 inclusive to 2 and keep other values unchanged:

```
. recode origvar (1/4=2), gen(myvar4)
(40 differences between origvar and myvar4)
```

To change values 1, 3, 4 and 5 to 7 and keep other values unchanged:

```
. recode origvar (1 3 4 5=7), gen(myvar5)
(55 differences between origvar and myvar5)
```

To change all values from the minimum through 5 to the minimum:

```
. recode origvar (min/5=min), gen(myvar6)
(56 differences between origvar and myvar6)
```

To change missing values to 9:

```
. recode origvar (missing=9), gen(myvar7)
(11 differences between origvar and myvar7)
```

To change values of −999 to missing:

```
. recode origvar (-999=.), gen(myvar8)
(56 differences between origvar and myvar8)
```

Note that **recode** can be used to recode several variables at once if they are all to be recoded the same way. Just include all the variable names before the instructions on how they are to be recoded, and include all the names for new variables (if you do not want the old variables to be overwritten) within the parentheses of the `generate()` option.

2.13.4   Common transformations for RHS variables

For the models we discuss in later chapters, you can use many of the tricks you learned for coding right-hand side (i.e., independent) variables in the linear regression model. Here are some useful examples. Details on how to interpret such variables in regression models are given in Chapter 8.

Breaking a categorical variable into a set of binary variables

To use a j-category nominal variable as an independent variable in a regression model, you need to create a set of $j-1$ binary variables, also known as dummy variables or indicator variables. To illustrate how to do this, we use educational attainment (`degree`), which is coded as $0 =$ no diploma, $1 =$ high school diploma, $2 =$ associate's degree, $3 =$ bachelor's degree, and $4 =$ postgraduate degree, with some missing data. We want to make four binary variables with the "no diploma" category serving as our reference category. We also want observations that have missing values for `degree` to have missing values in each of the dummy variables that we create. The simplest way to do this is to use the `generate` option with `tabulate`:

```
. tab degree, gen(edlevel)
```

rs highest degree	Freq.	Percent	Cum.
lt high school	801	17.47	17.47
high school	2,426	52.92	70.40
junior college	273	5.96	76.35
bachelor	750	16.36	92.71
graduate	334	7.29	100.00
Total	4,584	100.00	

The `generate(`*name*`)` option creates a new binary variable for each category of the specified variable. In our example, `degree` has 5 categories, so five new variables are created. These variables all begin with `edlevel`, the root that we specified with the `generate(edlevel)` option. We can check the five new variables by typing `sum edlevel*`:

```
. sum edlevel*
```

Variable	Obs	Mean	Std. Dev.	Min	Max
edlevel1	4584	.1747382	.3797845	0	1
edlevel2	4584	.5292321	.4991992	0	1
edlevel3	4584	.059555	.2366863	0	1
edlevel4	4584	.1636126	.369964	0	1
edlevel5	4584	.0728621	.2599384	0	1

By cross-tabulating the new `edlevel1` by the original `degree`, we can see that `edlevel1` equals 1 for individuals with no high school diploma and equals 0 for everyone else except the 14 observations with missing values for `degree`:

```
. tab degree edlevel1, missing
```

rs highest degree	degree==lt high school 0	1	.	Total
lt high school	0	801	0	801
high school	2,426	0	0	2,426
junior college	273	0	0	273
bachelor	750	0	0	750
graduate	334	0	0	334
.	0	0	14	14
Total	3,783	801	14	4,598

One limitation of using the `generate(`*name*`)` option of `tab` is that it only works when there is a one-to-one correspondence between the original categories and the dummy variables that we wish to create. So, let's suppose that we want to combine high school graduates and those with associate's degrees when creating our new binary variables. Say also that we want to treat those without high school diplomas as the omitted category. The following is one way to create the three binary variables that we need:

```
. gen hsdeg = (degree==1 | degree==2) if degree<.
(14 missing values generated)
. gen coldeg = (degree==3) if degree<.
(14 missing values generated)
. gen graddeg = (degree==4) if degree<.
(14 missing values generated)
. tab degree coldeg, missing
```

rs highest degree	coldeg 0	1	.	Total
lt high school	801	0	0	801
high school	2,426	0	0	2,426
junior college	273	0	0	273
bachelor	0	750	0	750
graduate	334	0	0	334
.	0	0	14	14
Total	3,834	750	14	4,598

To understand how this works, you need to know that when Stata is presented with an expression (e.g., `degree==3`) where it expects a value, it evaluates the expression and assigns it a value of 1 if true and 0 if false. Consequently, `gen coldeg = (degree==3)` creates the variable `coldeg` that equals 1 whenever `degree` equals 3 and 0 otherwise. By adding `if degree<.` to the end of the command, we assign these values *only* to observations in which the value of `degree` is not missing. If an observation has a missing value for `degree`, these cases are given a missing value.

More examples of creating binary variables

Binary variables are used so often in regression models that it is worth providing more examples of generating them. In the dataset that we use in Chapter 5, the independent variable for respondent's education (`ed`) is measured in years. We can create a dummy variable that equals 1 if the respondent has at least 12 years of education and 0 otherwise:

```
. gen ed12plus = (ed>=12) if ed<.
```

Alternatively, we might want to create a set of variables that indicates whether an individual has less than 12 years of education, between 13 and 16 years of education, or 17 or more years of education. This is done as follows:

```
. gen edlt13 = (ed<=12) if ed<.
. gen ed1316 = (ed>=13 & ed<=16) if ed<.
. gen ed17plus = (ed>17) if ed<.
```

Tip: Naming dummy variables Whenever possible, we name dummy variables so that 1 corresponds to "yes" and 0 to "no". With this convention, a dummy variable called `female` is coded 1 for women (i.e., yes, the person is female) and 0 for men. If the dummy variable was named `sex`, there would be no immediate way to know what 0 and 1 mean.

The `recode` command can also be used to create binary variables. The variable `warm` contains responses to the question of whether working women can have as warm a relationship with their children as women who do not work: 1=strongly disagree, 2=disagree, 3=agree, and 4=strongly agree. To create a dummy indicating agreement as opposed to disagreement, type

```
. gen wrmagree = warm
. recode wrmagree 1=0 2=0 3=1 4=1
(wrmagree: 2293 changes made)
. tab wrmagree warm
```

wrmagree	Mom can have warm relations with child				Total
	SD	D	A	SA	
0	297	723	0	0	1,020
1	0	0	856	417	1,273
Total	297	723	856	417	2,293

Nonlinear transformations

Nonlinear transformations of the independent variables are commonly used in regression models. For example, researchers often include both age and age^2 as explanatory variables to allow the effect of a one-year increase in age to change as one gets older. We can create a squared term as

```
. gen agesq = age*age
```

Likewise, income is often logged so that the impact of each additional dollar decreases as income increases. The new variable can be created as

```
. gen lnincome = ln(income)
(495 missing values generated)
```

We can use the minimum and maximum values reported by summarize as a check on our transformations:

```
. sum age agesq income lnincome
```

Variable	Obs	Mean	Std. Dev.	Min	Max
age	4598	46.12375	17.33162	18	99
agesq	4598	2427.72	1798.477	324	9801
income	4103	34790.7	22387.45	1000	75000
lnincome	4103	10.16331	.8852605	6.907755	11.22524

Interaction terms

In regression models, you can include interactions by taking the product of two independent variables. For example, we might think that the effect of family income differs for men and women. If gender is measured as the dummy variable female, we can construct an interaction term as follows:

```
. gen feminc = female * income
(495 missing values generated)
```

Tip: The xi command can be used to create interaction variables automatically. While this is a very powerful command that can save time, it that can be confusing unless you use the command frequently. Accordingly, we do not recommend it. Constructing interactions with generate is a good way to make sure you understand what the interactions mean.

2.14 Labeling variables and values

Variable labels provide descriptive information about what a variable measures. For example, the variable agesq might be given the label "age squared", or warm could have the label "Mother has a warm relationship". *Value* labels provide descriptive information about the different values of a categorical variable. For example, value labels might indicate that the values 1–4 correspond to survey responses of strongly agree, agree, disagree, and strongly disagree. Adding labels to variables and values is not much fun, but in the long run it can save a great deal of time and prevent misunderstandings. Also, many of the commands in SPost produce output that is more easily understood if the dependent variable has value labels.

2.14.1   Variable labels

The `label variable` command attaches a label of up to 80 characters to a variable. For example,

```
. label variable agesq "Age squared"
. describe agesq
```

variable name	storage type	display format	value label	variable label
agesq	float	%9.0g		Age squared

If no label is specified, any existing variable label is removed. For example,

```
. label variable agesq
. describe agesq
```

variable name	storage type	display format	value label	variable label
agesq	float	%9.0g		

Tip: Use short labels While variable labels of up to 80 characters are allowed, output often does not show all 80 characters. We find it works best to keep variable labels short, with the most important information in the front of the label. That way, if the label is truncated, you will see the critical information.

2.14.2   Value labels

Beginners often find value labels in Stata confusing. The key thing to keep in mind is that Stata splits the process of labeling values into two steps: creating labels and then attaching the labels to variables.

Step 1 defines a set of labels *without* reference to a variable. Here are some examples of value labels:

```
. label define yesno 1 yes 0 no
. label define posneg4 1 veryN 2 negative 3 positive 4 veryP
. label define agree4 1 StrongA 2 Agree 3 Disagree 4 StrongD
. label define agree5 1 StrongA 2 Agree 3 Neutral 4 Disagree 5 StrongD
```

Notice several things. First, each *set* of labels is given a unique name (e.g., yesno, agree4). Second, individual labels are associated with a specific value. Third, none of our labels have spaces in them (e.g., we use StrongA not Strong A). While you can have spaces if you place the label within quotes, some commands crash when they encounter blanks in value labels. So, it is easier not to do it. We have also found that the characters ., :, and { in value labels can cause similar problems. Fourth, our labels are 8 letters or shorter in length because some programs have trouble with value labels longer than 8 letters.

Step 2 assigns the value labels to variables. Let's say that variables `female`, `black`, and `anykids` all imply yes/no categories with 1 as yes and 0 as no. To assign labels to the values, we would use the following commands:

```
. label values female yesno
. label values black yesno
. label values anykids yesno
. describe female black anykids
```

variable name	storage type	display format	value label	variable label
female	byte	%9.0g	yesno	Female
black	byte	%9.0g	yesno	Black
anykids	byte	%9.0g	yesno	R have any children?

The output for `describe` shows which value labels were assigned to which variables. The new value labels are reflected in the output from `tabulate`:

```
. tab anykids
```

R have any children?	Freq.	Percent	Cum.
no	1,267	27.64	27.64
yes	3,317	72.36	100.00
Total	4,584	100.00	

For the `degree` variable that we looked at earlier, we assign labels with

```
. label define degree 0 "no_hs" 1 "hs" 2 "jun_col" 3 "bachelor" 4 "graduate"
. label values degree degree
. tab degree
```

rs highest degree	Freq.	Percent	Cum.
no_hs	801	17.47	17.47
hs	2,426	52.92	70.40
jun_col	273	5.96	76.35
bachelor	750	16.36	92.71
graduate	334	7.29	100.00
Total	4,584	100.00	

Notice that we used _s (underscores) instead of spaces.

2.14.3   notes command

The `notes` command allows you to add notes to the dataset as a whole or to specific variables. Because the notes are saved in the dataset, the information is always available when you use the data. In the following example, we add one note describing the dataset and two that describe the `income` variable:

```
. notes: General Social Survey extract for Stata book
. notes income: self-reported family income, measured in dollars
. notes income: refusals coded as missing
```

We can review the notes by typing `notes`:

```
. notes
_dta:
  1.  General Social Survey extract for Stata book
income:
  1.  self-reported family income, measured in dollars
  2.  refusals coded as missing
```

If we save the dataset after adding notes, the notes become a permanent part of the dataset.

2.15   Global and local macros

While macros are most often used when writing ado-files, they are also very useful in do-files. Later in the book, and especially in Chapter 8, we use macros extensively. Accordingly, we will discuss them briefly here. Readers with less familiarity with Stata might want to skip this section for now and read it later when macros are used in our examples.

In Stata, you can assign values or strings to *macros*. Whenever Stata encounters the macro name, it automatically substitutes the contents of the macro. For example, pretend that you want to generate a series of two-by-two tables where you want cell percentages, requiring the `cell` option; missing values, requiring the `missing` option; values printed instead of value labels, requiring the `nolabel` option; and the chi-squared test statistic, requiring the `chi2` option. Even if you use the shortest abbreviations, this would require typing "`, ce m nol ch`" at the end of each `tab` command. Alternatively, you could use the following command to define a global macro called `myopt`:

```
. global myopt = ", cell miss nolabel chi2 nokey"
```

Then, whenever you type `$myopt` (the `$` tells Stata that `myopt` is a global macro), Stata substitutes `, cell miss nolabel chi2 nokey`. If you type

```
. tab lfp wc $myopt
```

Stata interprets this as if you had typed

```
. tab lfp wc, cell miss nolabel chi2
```

Global macros are "global" because, once they are set, they can be accessed by any do (or ado) program in the current session. The flip side is that the global macros that you are using can be reset by any of the do- or ado-files that you use along the way. In contrast, "local" macros can only be accessed within the do- or ado-file in which they are defined. When the do- or ado-file program terminates, the local macro disappears. We prefer using local macros whenever possible because you do not have to worry about conflicts with other programs or do-files that try to use the same macro name for a different purpose. Local macros are defined using the `local` command, and they are referenced by placing the name of the local macro in single quotes; for example, `` `myopt' ``. Notice that the two single quote marks use different symbols (on many keyboards, the left single quote is in the upper left-hand corner, while the right single quote is next to the Enter key). If the operations we just performed were in a do-file, we could have produced the same output with the following lines:

```
. local opt = ", cell miss nolabel chi2 nokey"
. tab lfp wc `opt'
  (output omitted)
```

You can also define macros to equal the result of computations. After entering `global four = 2+2`, the value 4 will be substituted for `$four`. In addition, Stata contains many *macro functions* in which items retrieved from memory are assigned to macros. For example, to display the variable label that you have assigned to the variable `wc`, you can type

```
. global wclabel : variable label wc
. display "$wclabel"
Wife College: 1=yes 0=no
```

We have only scratched the surface of the potential of macros. Macros are immensely flexible and are indispensable for a variety of advanced tasks that Stata can perform. Perhaps most importantly, macros are essential for doing any meaningful Stata programming. If you look at the ado-files for the commands we have written for this book, you will see many instances of macros, and even of macros within macros. For users interested in advanced applications, the `macro` entry in the *Programming Reference Manual* should be read closely.

2.16   Graphics

Version 8 introduced huge changes to Stata's graphics. Stata comes with an entire manual dedicated to graphics and is capable of making many more kinds of graphs than those used in this book. In this section, we provide only a brief introduction to graphics in Stata, focusing on the type of graph that we use most often in later chapters. Namely, we consider plots of one or more outcomes against an independent variable using the command `graph twoway`. The syntax of `graph twoway` has the form

<u>gr</u>aph <u>tw</u>oway *plottype* ···

The *Stata Graphics Reference Manual* lists 29 different *plottypes* for the `graph twoway` command. Since we only discuss two (`scatter` and `connected`) *plottypes* here, interested readers are encouraged to consult the *Graphics Reference Manual* or to type `help graph` for more information.[6]

Graphs that you create in Stata are drawn in their own window, which should appear on top of the four windows we discussed above. If the Graph window is hidden, you can bring it to the front by clicking on ⬛. You can make the Graph window larger or smaller by clicking and dragging the borders.

2.16.1   The graph command

The type of graph that we use most often shows how the predicted probability of observing a given outcome changes as a continuous variable changes over a specified range. For example, in Chapter 4 we show you how to compute the predicted probability of a woman being in the labor force according to the number of children she has and the family's income. In later chapters, we show you how to compute these predictions, but for now you can simply load them into memory with the command `use lfpgraph2, clear`. The variable `income` is family income measured in thousands of dollars, excluding any contribution made by the woman of the household, while the next three variables show the predicted probabilities of working for a woman who has no children under six (`kid0p1`), one child under six (`kid1p1`), or two children under six (`kid2p1`). Since there are only eleven values, we can easily list them:

```
. use http://www.stata-press.com/data/lfr/lfpgraph2, clear
(Sample predictions to plot.)
. list income kid0p1 kid1p1
```

	income	kid0p1	kid1p1
1.	10	.7330963	.3887608
2.	18	.6758616	.3256128
3.	26	.6128353	.2682211
4.	34	.54579	.2176799
5.	42	.477042	.1743927
6.	50	.409153	.1381929
7.	58	.3445598	.1085196
8.	66	.285241	.0845925
9.	74	.2325117	.065553
10.	82	.18698	.0505621
11.	90	.1486378	.0388569

---

[6]In Chapter 6, we discuss a third plottype called `rarea`. Because we only use that plottype in one example, we do not discuss it here.

We see that as annual income increases the predicted probability of being in the labor force decreases. Also, by looking across any row, we see that for a given level of income the probability of being in the labor force decreases as the number of young children increases. We want to display these patterns graphically.

graph twoway scatter can be used to draw a scatterplot in which the values of one or more y-variables are plotted against values of an x-variable. In our case, income is the x-variable, and the predicted probabilities kid0p1, kid1p1, and kid2p1 are the y-variables. Thus, for each value of x, we have three values of y. In making scatterplots with graph twoway scatter, the y-variables are listed first, and the x-variable is listed last. If we type,

```
. graph twoway scatter kid0p1 kid1p1 kid2p1 income, ytitle(Probability)
```

we obtain the following graph:

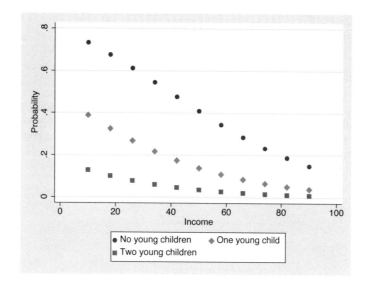

Our simple scatterplot shows the pattern of decreasing probabilities as income or number of children increases. While this simple command produces a reasonable first graph, we can make it a more effective graph by using some of the many options that are available for graph twoway.

While there are always exceptions, four points are usually true about the graphics system that has been available since Stata 8.

1. The defaults usually produce a good starting graph.

2. You will frequently want to change the title for the y-axis and set the overall title.

3. You sometimes may want to change the way the axes are labeled.

4. If you want to change some aspect of the graph, you can almost certainly get what you want by specifying some set of options.

Points 1, 2, and 3 imply that we should discuss setting titles and axis labels while we can safely ignore many of the finer points of the graphs. While point 4 extols the flexibility of the graph system, it also warns us that any attempt at being comprehensive would lead us to write something similar to the over 500 hundred pages in the *Graphics Reference Manual*. Still, in order to use the graphics system, it is necessary to have basic understanding of how it works.

A slightly more detailed syntax[7] for **graph twoway** is

graph twoway $plot_1$ $[plot_2]...[plot_N]$ $[$**if** $exp]$ $[$**in** $range]$ $[$, $options]$

where $plot_i$ is defined to be

$[($ *plottype varlist*, $[$ **title**("*string*") **subtitle**("*string*") **ytitle**("*string*")
xtitle("*string*") **caption**("*string*") **xlabel**(*values*) **ylabel**(*values*)
other_options$]$ $[)]$

This syntax highlights the fact that it is possible to put multiple plots on the same graph.[8] The plots can be of different plottypes. For instance, suppose that we wanted the symbols in the plot corresponding to "No young children" to be connected. This plottype is called **connected**. If we type,

```
. graph twoway (connected kid0p1 income)
        (scatter kid1p1 kid2p1 income), ytitle(Probability)
```

then we obtain the graph

---

[7]The syntax presented here is far from complete. We only wish to explain the elements that we have found ourselves using in presenting analyses like those in this book. See the *Stata Graphics Reference Manual* for more information.

[8]The parentheses are used to separate the different plots when there are multiple plots. When there is only one plot, the parentheses are not required.

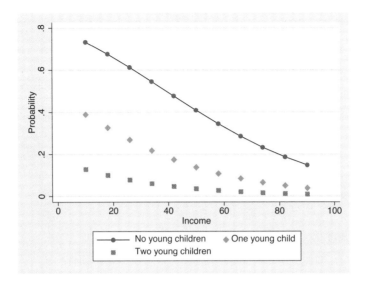

With the exception of the title on the y-axis, the default choices for the symbols, line styles, etcetera, all are all quite nice. Stata made these choices within the context of an overall look or scheme. For example, because our book is published in monochrome, we wanted our graphs to be drawn in monochrome. The *Graphics Reference Manual* describes how they could be changed. At the time of this writing, there are nine different schemes from which to choose. (Type `help schemes` in Stata for the latest information about the available schemes.) Users can choose the overall look of their graphs by setting the scheme. In writing this book, we simply include the line

```
. set scheme s2manual
```

at the top of our do-files.

Adding titles

Now, we provide a quick introduction that illustrates how to set the five titles that we frequently wish to change. We have found that we routinely set the (1) overall title, (2) overall subtitle, (3) y-axis title, (4) x-axis title, and (5) graph caption. The options for setting each of these five titles are in the syntax diagram above. The command and graph below illustrate how we might use each of these titles

```
. graph twoway (connected kid0p1 kid1p1 kid2p1 income),
    ytitle("Probability")
    title("Predicted Probability of Female LFP")
    subtitle("(as predicted by logit model)")
    xtitle("Family income, excluding wife's")
    caption("Data from 1976 PSID-T Mroz")
```

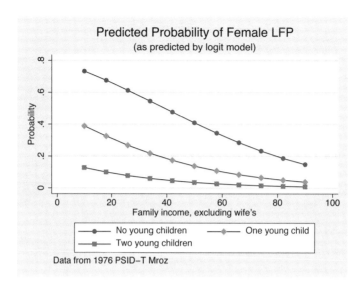

This graph is much more effective in illustrating that the probability of a woman being in the labor force declines as family income increases, and that the differences in predicted probabilities between women with no young children and those with one or two young children are greatest at the lowest levels of income.

Labeling the axes

Even though the defaults are nice, it is common to want to change the labeling of the ticks on the x-axis or y-axis. The `ylabel()` and `xlabel()` options allow users to specify either a rule or a set of values for the tick marks. A rule is simply a compact way to specify a list of values.

Let's consider specifying a list of values first. A common change is to alter the frequency or range of tick marks. This change can also be made with the `xlabel()` and `ylabel()` options. For instance, suppose that we liked the frequency of the ticks on the x-axis but wanted to restrict the range to $[10, 90]$. We make this change in the command.

```
. graph twoway (connected kid0p1 kid1p1 kid2p1 income),
    ytitle("Probability")
    title("Predicted Probability of Female LFP")
    subtitle("(as predicted by logit model)")
    xtitle("Family income, excluding wife's")
    caption("Data from 1976 PSID-T Mroz")
    xlabel(10 20 30 40 50 60 70 80 90)
```

produces the graph

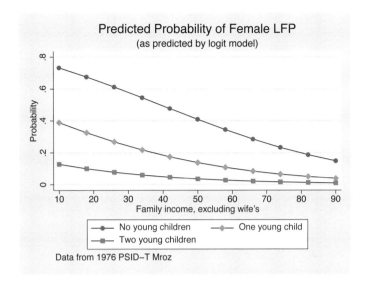

We could have obtained the same graph by specifying a rule for a new set of x-axis values. While there are several ways to specify a rule,[9] we find the form $\#_1(\#_2)\#_3$ most useful. In this form, the user specifies three numbers: $\#_1$ specifies the beginning of the sequence of values, $\#_2$ specifies the increment between each of the values, and $\#_3$ specifies the maximum value. For instance, instead of specifying the option

```
xlabel(10 20 30 40 50 60 70 80 90)
```

in the previous graph, we could have specified

```
xlabel(10(10)90)
```

to obtain the same graph.

Saving graphs

Graphs can be saved in either a file or in memory. When a graph is saved to a file, it remains there until the file is erased. When a graph is saved in memory, it remains there until you exit Stata or drop the graph from memory. Specifying `saving`(*filename,* `replace`) saves the graph to a file in Stata's proprietary format (indicated by the suffix `.gph`) in the working directory. Including `replace` tells Stata to overwrite a file with that name if it exists. Specifying `name`(*name,* `replace`) saves the graph in memory. The `replace` option tells Stata to replace any existing graphs saved under that name.

Graphs must be saved, either to memory or files when you want to combine graphs for display, as discussed below. For example, if we were to need the graph we just created later on, we could save it in memory under the name **graph1** with the command

---

[9]Type `help axis_label_options` for other ways to specify a rule.

```
. graph twoway (connected kid0p1 kid1p1 kid2p1 income),
        ytitle("Probability")
        title("Predicted Probability of Female LFP")
        subtitle("(as predicted by logit model)")
        xtitle("Family income, excluding wife's")
        caption("Data from 1976 PSID-T Mroz")
        xlabel(10(10)90) name(graph1, replace)
```

Tip: Exporting graphs to other programs If you are using Windows and want to export graphs to another program, such as a word processor, we find that it works best to save them as a Windows Metafile, which has the `.wmf` extension. This can be done using the `graph export` command. If the graph is currently in the graph window, you can save it in `.wmf` format with the command `graph export` *filename*`.wmf`. If the file is already saved in `.gph` format, you can export it to `.wmf` format in two steps. First, redisplay the graph with the command `graph use` *filename*. Then export the graph with the command `graph export` *filename*`.wmf`. The `replace` option can be used with `graph export` to automatically overwrite a graph of the same name, which is useful in do-files.

2.16.2 Displaying previously drawn graphs

There are several commands used for manipulating graphs that have been previously drawn and saved to memory or disk. `graph dir` lists graphs previously saved in memory or to a file in the current working directory. `graph use` copies a graph stored in a file into memory and displays it. `graph display` redisplays a graph stored in memory.

2.16.3 Printing graphs

It is easiest to print a graph once it is in the Graph window. When a graph is in the Graph window, you may print it by selecting File→Print Graph from the menus or by clicking on [printer icon]. You can also print a graph in the Graph window with the command `graph print`. To print a graph saved to memory or disk, first use `graph use` or `graph display` to redisplay it, and then print it with the command `graph print`.

2.16.4 Combining graphs

Multiple graphs that have been saved can be combined into a single graph. This is useful, for example, when you want to place two graphs side-by-side or one above the other. In Chapter 5, we will find it useful to combine two graphs into a single graph. Here we use two of the graphs that we discuss in detail in section 5.8.6 to illustrate `graph combine`. When we originally drew the graphs, we saved them in memory under the names `graph1` and `graph2`. Now, we use `graph combine` to put the two graphs side-by-side in a single graph window.

> . graph combine graph1 graph2, imargin(small)

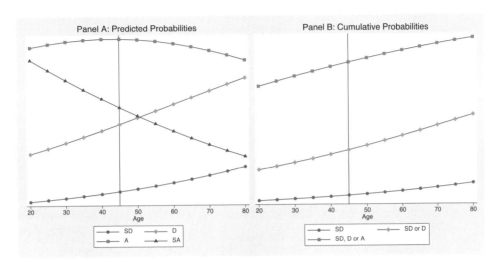

This combined graph is not as effective as one in which the graphs are stacked. The trick is to understand that when multiple graphs are combined, Stata divides the Graphics window into an array. The `rows()` and `cols()` options can be used to set the number of rows and columns in the array. Of course, as with most aspects of a graph, the *Graphics Reference Manual* describes how almost any part of the combined graph can be changed.[10] By default, the individual graphs are allocated over the rows in the order in which the filenames are listed in the `graph combine` command.

To display the graphs vertically, one over the other, specify the `col()` option:

> . graph combine graph1 graph2, iscale(*.9) imargin(small)
>   ysize(3.9) xsize(3.5405) col(1)

(Continued on next page)

---

[10]In particular, see the **Advanced use** section of the entry for `graph combine` for a rather impressive example.

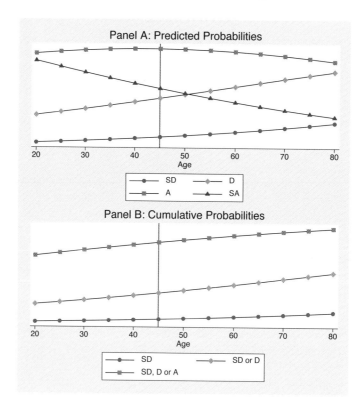

As we described earlier, `graph export` can be used to save the graph as a Windows Metafile that can be imported to a word processor or other program. More details on combining graphs can be found in the *Stata Graphics Reference Manual*.

2.17   A brief tutorial

This tutorial uses the `science2.dta` data that is available from the book's web site. You can use your own dataset as you work through this tutorial, but you will need to change some of the commands to correspond to the variables in your data. In addition to our tutorial, the *User's Guide* provides a wealth of information for new users.

Opening a log

The first step is to open a log file for recording your results. Remember that all commands are case sensitive. The commands are listed with a period in front, but you do *not* type the period:

```
. capture log close
. log using tutorial, text
```

```
  log:  d:\spostdata\tutorial.smcl
  log type:  text
opened on:    26 May 2003, 11:18:15
```

Loading the data

We assume that `science2.dta` is in your working directory. `clear` tells Stata to "clear out" any existing data from memory before loading the new dataset:

```
. use http://www.stata-press.com/data/lfr/science2, clear
(Note that some of the variables have been artificially constructed.)
```

The message after loading the data reflects that this dataset was created for teaching. While most of the variables contain real information, some variables have been artificially constructed.

Examining the dataset

`describe` gives information about the dataset.

```
. describe
Contains data from science2.dta
  obs:            308                          Note that some of the variables
                                                 have been artificially
                                                 constructed.
  vars:            35                          10 Mar 2001 05:51
  size:        17,556 (98.3% of memory free)   (_dta has notes)
```

variable name	storage type	display format	value label	variable label
id	float	%9.0g		ID Number.
cit1	int	%9.0g		Citations: PhD yr -1 to 1.
cit3	int	%9.0g		Citations: PhD yr 1 to 3.
cit6	int	%9.0g		Citations: PhD yr 4 to 6.
cit9	int	%9.0g		Citations: PhD yr 7 to 9.
enrol	byte	%9.0g		Years from BA to PhD.
fel	float	%9.0g		Fellow or PhD prestige.
felclass	byte	%9.0g	prstlb	* Fellow or PhD prestige class.
fellow	byte	%9.0g	fellbl	Postdoctoral fellow: 1=y,0=n.
female	byte	%9.0g	femlbl	Female: 1=female,0=male.
job	float	%9.0g		Prestige of 1st univ job.
jobclass	byte	%9.0g	prstlb	* Prestige class of 1st job.
mcit3	int	%9.0g		Mentor's 3 yr citation.
mcitt	int	%9.0g		Mentor's total citations.
mmale	byte	%9.0g	malelb	Mentor male: 1=male,0=female.
mnas	byte	%9.0g	naslb	Mentor NAS: 1=yes,0=no.
mpub3	byte	%9.0g		Mentor's 3 year publications.
nopub1	byte	%9.0g	nopublb	1=No pubs PhD yr -1 to 1.
nopub3	byte	%9.0g	nopublb	1=No pubs PhD yr 1 to 3.

nopub6	byte	%9.0g	nopublb	1=No pubs PhD yr 4 to 6.
nopub9	byte	%9.0g	nopublb	1=No pubs PhD yr 7 to 9.
phd	float	%9.0g		Prestige of Ph.D. department.
phdclass	byte	%9.0g	prstlb	* Prestige class of Ph.D. dept.
pub1	byte	%9.0g		Publications: PhD yr -1 to 1.
pub3	byte	%9.0g		Publications: PhD yr 1 to 3.
pub6	byte	%9.0g		Publications: PhD yr 4 to 6.
pub9	byte	%9.0g		Publications: PhD yr 7 to 9.
work	byte	%9.0g	worklbl	Type of first job.
workadmn	byte	%9.0g	wadmnlb	Admin: 1=yes; 0=no.
worktch	byte	%9.0g	wtchlb	* Teaching: 1=yes; 0=no.
workuniv	byte	%9.0g	wunivlb	* Univ Work: 1=yes; 0=no.
wt	byte	%9.0g		
faculty	byte	%9.0g	faclbl	1=Faculty in University
jobrank	byte	%9.0g	joblbl	Rankings of University Job.
totpub	byte	%9.0g		Total Pubs in 9 Yrs post-Ph.D.
				* indicated variables have notes

Sorted by:

Examining individual variables

A series of commands gives us information about individual variables. You can use whichever command you prefer, or all of them.

```
. sum work
     Variable |      Obs        Mean    Std. Dev.       Min        Max
-------------+--------------------------------------------------------
         work |      302    2.062914     1.37829         1          5
. tab work, missing
```

Type of first job.	Freq.	Percent	Cum.
FacUniv	160	51.95	51.95
ResUniv	53	17.21	69.16
ColTch	26	8.44	77.60
IndRes	36	11.69	89.29
Admin	27	8.77	98.05
.	6	1.95	100.00
Total	308	100.00	

```
. codebook work
```

work				Type of first job.

```
            type:  numeric (byte)
           label:  worklbl

           range:  [1,5]                       units:  1
   unique values:  5                        missing .:  6/308

      tabulation:  Freq.   Numeric   Label
                     160         1   FacUniv
                      53         2   ResUniv
                      26         3   ColTch
                      36         4   IndRes
                      27         5   Admin
                       6         .
```

Graphing variables

Graphs are also useful for examining data. The command

```
. dotplot work
```

creates the following graph:

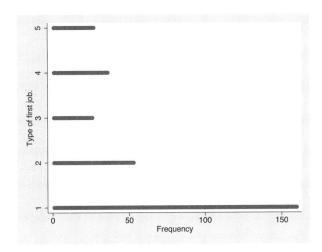

Saving graphs

To save the above graph as a Windows Metafile, type

```
. graph export myname.wmf, replace
(file d:\spostdata\myname.wmf written in Windows Metafile format)
```

Adding comments

To add comments to your output, which allows you to document your command files, type * at the beginning of each comment. The comments are listed in the log file:

```
. * saved graph as work.wmf
```

Creating a dummy variable

Now, let's make a dummy variable with faculty in universities coded 1 and all others coded 0. The command `gen isfac = (work==1) if work<.` generates `isfac` as a dummy variable where `isfac` equals 1 if `work` is 1, else 0. The statement `if work<.` makes sure that missing values are kept as missing in the new variable.

```
. gen isfac = (work==1) if work<.
(6 missing values generated)
```

Six missing values were generated because **work** contained six missing observations.

Checking transformations

One way to check transformations is with a table. In general, it is best to look at the missing values, which requires the **missing** option:

```
. tab isfac work, missing
```

isfac	FacUniv	ResUniv	Type of first job. ColTch	IndRes	Admin	Total
0	0	53	26	36	27	142
1	160	0	0	0	0	160
.	0	0	0	0	0	6
Total	160	53	26	36	27	308

isfac	Type of first job. .	Total
0	0	142
1	0	160
.	6	6
Total	6	308

Labeling variables and values

For many of the regression commands, value labels for the dependent variable are essential. We start by creating a variable label, then create **isfac** to store the value labels, and finally assign the value labels to the variable **isfac**:

```
. label variable isfac "1=Faculty in University"
. label define isfac 0 "NotFac" 1 "Faculty"
. label values isfac isfac
```

Then, we can get labeled output:

```
. tab isfac
```

1=Faculty in University	Freq.	Percent	Cum.
NotFac	142	47.02	47.02
Faculty	160	52.98	100.00
Total	302	100.00	

Creating an ordinal variable

The prestige of graduate programs is often referred to using the categories of adequate, good, strong, and distinguished. Here, we create such an ordinal variable from the continuous variable for the prestige of the first job. `missing` tells Stata to show cases with missing values.

```
. tab job, missing
```

Prestige of 1st univ job.	Freq.	Percent	Cum.
1.01	1	0.32	0.32
1.2	1	0.32	0.65
1.22	1	0.32	0.97
1.32	1	0.32	1.30
1.37	1	0.32	1.62
(output omitted)			
3.97	6	1.95	48.38
4.18	2	0.65	49.03
4.42	1	0.32	49.35
4.5	6	1.95	51.30
4.69	5	1.62	52.92
.	145	47.08	100.00
Total	308	100.00	

The `recode` command makes it easy to group the categories from `job`. Of course, we then label the variable:

```
. gen jobprst=job
(145 missing values generated)
. recode jobprst .=. 1/1.99=1 2/2.99=2 3/3.99=3 4/5=4
(jobprst: 162 changes made)
. label variable jobprst "Rankings of University Job"
. label define prstlbl 1 "Adeq" 2 "Good" 3 "Strong" 4 "Dist"
. label values jobprst prstlbl
```

Here is the new variable (note that we use the `missing` option so that missing values are included in the tabulation):

```
. tab jobprst, missing
```

Rankings of University Job	Freq.	Percent	Cum.
Adeq	31	10.06	10.06
Good	47	15.26	25.32
Strong	71	23.05	48.38
Dist	14	4.55	52.92
.	145	47.08	100.00
Total	308	100.00	

Combining variables

Now, we create a new variable by summing existing variables. If we add pub3, pub6, and pub9, we can obtain the scientist's total number of publications over the nine years following receipt of the Ph.D.

```
. gen pubsum=pub3+pub6+pub9
. label variable pubsum "Total Pubs in 9 Yrs post-Ph.D."
. sum pub3 pub6 pub9 pubsum
```

Variable	Obs	Mean	Std. Dev.	Min	Max
pub3	308	3.185065	3.908752	0	31
pub6	308	4.165584	4.780714	0	29
pub9	308	4.512987	5.315134	0	33
pubsum	308	11.86364	12.77623	0	84

A scatterplot matrix graph can be used to plot all pairs of variables simultaneously:

```
. graph matrix pubsum pub3 pub6 pub9,  half
```

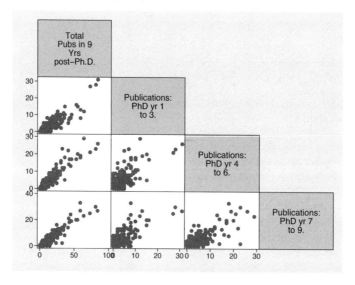

Saving the new data

After you make changes to your dataset, it is a good idea to save the data with a new filename:

```
. save sciwork, replace
file sciwork.dta saved
```

Closing the log file

Last, we need to close the log file so that we can refer to it in the future.

```
. log close
  log:   d:\spostdata\tutorial.smcl
  log type:  smcl
  closed on:   26 May 2003, 11:18:27
```

A batch version

If you have read Section 2.9, you know that a better idea is to create a batch (do-) file, perhaps called `tutorial.do`:[11]

```
// batch version of tutorial do file
// 26 May 2003
// version 1.0.0
//
version 8
set scheme s2manual
set more off
capture log close
log using ch2tutorial, replace
// * load and describe the data
use science2, clear
describe
// * check variable work
sum work
tab work, missing
codebook work
dotplot work
graph export 02dotplot2.wmf, replace
// * saved graph as work.wmf
// * dummy variable indicating faculty
gen isfac = (work==1) if work<.
tab isfac work, missing
label variable isfac "1=Faculty in University"
label define isfac 0 "NotFac" 1 "Faculty"
label values isfac isfac
tab isfac
// * clean and recode job variable
tab job, missing
gen jobprst=job
recode jobprst .=. 1/1.99=1 2/2.99=2 3/3.99=3 4/5=4
label variable jobprst "Rankings of University Job"
label define prstlbl 1 "Adeq" 2 "Good" 3 "Strong" 4 "Dist"
label values jobprst prstlbl
tab jobprst, missing
// * create total publications variable
gen pubsum=pub3+pub6+pub9
label variable pubsum "Total Pubs in 9 Yrs post-Ph.D."
sum pub3 pub6 pub9 pubsum
graph matrix pubsum pub3 pub6 pub9,  half
// * save the new data
save sciwork, replace
// * close the log
log close
```

Then type `do tutorial` in the Command window or select File→Do... from the menu.

---

[11]If you download this file from our web site, it is called `st8ch2tutorial.do`.

3   Estimation, Testing, Fit, and Interpretation

Our book deals with what we think are the most fundamental and useful cross-sectional regression models for categorical and count outcomes: binary logit and probit, ordinal logit and probit, multinomial and conditional logit, Poisson regression, negative binomial regression, and zero-inflated models for counts.[1] While these models differ in many respects, they share common features:

1. Each model is fitted by maximum likelihood.

2. The estimates can be tested with Wald and LR tests.

3. Measures of fit can be computed.

4. Models can be interpreted by examining predicted values of the outcome.

As a consequence of these similarities, the same principles and commands can be applied to each model. Coefficients can be listed with `listcoef`. Wald and likelihood-ratio tests can be computed with `test` and `lrtest`. Measures of fit can be computed with `fitstat`, and our SPost suite of post-estimation commands for interpretation can be used to interpret the predictions.

Building on the overview that this chapter provides, later chapters focus on the application of these principles and commands to exploit the unique features of each model. Additionally, this chapter serves as a reference for the syntax and options for the SPost commands that we introduce here. Accordingly, we encourage you to read this chapter before proceeding to the chapter that covers the models of greatest interest to you.

(Continued on next page)

---

[1] While many of the principles and procedures discussed in this book apply to panel models, such as fitted by Stata's `xt` commands, or the multiple-equation systems, such as `biprobit` or `treatreg`, these models are not considered here.

3.1   Estimation

Each of the models that we consider is fitted by maximum likelihood (ML).[2] ML estimates are the values of the parameters that have the greatest likelihood (i.e., the *maximum likelihood*) of generating the observed sample of data if the assumptions of the model are true. To obtain the maximum likelihood estimates, a *likelihood function* calculates how likely it is that we would observe the data we actually observed if a given set of parameter estimates were the true parameters. For example, in linear regression with a single independent variable, we need to estimate both the slope β and the intercept α (for simplicity, we are ignoring the parameter σ^2). For any combination of possible values for α and β, the likelihood function tells us how likely it is that we would have observed the data that we did observe if these values were the true population parameters. If we imagine a surface in which the range of possible values of α comprises one axis and the range of β comprises another axis, the resulting graph of the likelihood function would look like a hill, and the ML estimates would be the parameter values corresponding to the top of this hill. The variance of the estimates corresponds roughly to how quickly the slope is changing near the top of the hill.

For all but the simplest models, the only way to find the maximum of the likelihood function is by numerical methods. *Numerical methods* are the mathematical equivalent of how you would find the top of a hill if you were blindfolded and only knew the slope of the hill at the spot where you are standing and how the slope at that spot is changing (which you could figure out by poking your foot in each direction). The search begins with start values corresponding to your location as you start your climb. From the start position, the slope of the likelihood function and the rate of change in the slope determine the next guess for the parameters. The process continues to *iterate* until the maximum of the likelihood function is found, called *convergence*, and the resulting estimates are reported. Advances in numerical methods and computing hardware have made estimation by numerical methods routine.

3.1.1   Stata's output for ML estimation

The process of iteration is reflected in the initial lines of Stata's output. Consider the first lines of the output from the logit model of labor force participation that we use as an example in Chapter 4:

```
. logit lfp k5 k618 age wc hc lwg inc
Iteration 0:   log likelihood =  -514.8732
Iteration 1:   log likelihood = -454.32339
Iteration 2:   log likelihood = -452.64187
Iteration 3:   log likelihood = -452.63296
Iteration 4:   log likelihood = -452.63296
Logit estimates                              Number of obs   =        753
    (output omitted )
```

---

[2]In many situations, there are convincing reasons for using Bayesian or exact methods for the estimation of these models. However, these methods are not generally available and hence are not considered here.

The output begins with the iteration log, where the first line reports the value of the *log* likelihood at the start values, reported as iteration 0. While earlier we talked about maximizing the likelihood equation, in practice, programs maximize the log of the likelihood, which simplifies the computations and yields the same ultimate result. For the probability models considered in this book, the log likelihood is always negative, because the likelihood itself is always between 0 and 1. In our example, the log likelihood at the start is -514.8732. The next four lines in this example show the progress in maximizing the log likelihood, converging to the value of -452.63296. The rest of the output is discussed later in this section.

3.1.2   ML and sample size

Under the usual assumptions (see Cramer 1986 or Eliason 1993 for specific details), the ML estimator is consistent, efficient, and asymptotically normal. These properties hold as the sample size approaches infinity. While ML estimators are not necessarily bad estimators in small samples, the small sample behavior of ML estimators for the models we consider is largely unknown. With the exception of the logit and Poisson regression, which can be estimated using exact permutation methods with LogXact (Cytel Corporation 2000), alternative estimators with known small sample properties are generally not available. With this in mind, Long (1997, 54) proposed the following guidelines for the use of ML in small samples:

> It is risky to use ML with samples smaller than 100, while samples over 500 seem adequate. These values should be raised depending on characteristics of the model and the data. First, if there are many parameters, more observations are needed .... A rule of at least 10 observations per parameter seems reasonable .... This does not imply that a minimum of 100 is not needed if you have only two parameters. Second, if the data are ill-conditioned (e.g., independent variables are highly collinear) or if there is little variation in the dependent variable (e.g., nearly all of the outcomes are 1), a larger sample is required. Third, some models seem to require more observations [such as the ordinal regression model or the zero-inflated count models].

3.1.3   Problems in obtaining ML estimates

While the numerical methods used by Stata to compute ML estimates are highly refined and generally work extremely well, you can encounter problems. If your sample size is adequate, but you cannot get a solution or appear to get the wrong solution (i.e., the estimates do not make substantive sense), our experience suggests that the most common cause is that the data have not been properly "cleaned". In addition to mistakes in constructing variables and selecting observations, the scaling of variables can cause problems. The larger the ratio between the largest and smallest standard deviations among variables in the model, the more problems you are likely to encounter with numerical methods due to rounding. For example, if income is measured in units of $1, income is likely to have a very large standard deviation relative to other variables. Recoding income to units of $1,000 can solve the problem.

Overall, however, numerical methods for ML estimation work well when your model is appropriate for your data. Still, Cramer's (1986, 10) advice about the need for care in estimation should be taken very seriously:

> Check the data, check their transfer into the computer, check the actual computations (preferably by repeating at least a sample by a rival program), and always remain suspicious of the results, regardless of the appeal.

3.1.4 The syntax of estimation commands

All single-equation estimation commands have the same syntax:[3]

command depvar $\big[$*indepvars*$\big]$ $\big[$*weight*$\big]$ $\big[$**if** *exp*$\big]$ $\big[$**in** *range*$\big]$ $\big[$, *option(s)* $\big]$

Elements in brackets [ ] are optional. Here are a few examples for a `logit` model with `lfp` as the dependent variable:

```
. logit lfp k5 k618 age wc lwg
(output omitted)
. logit lfp k5 k618 age wc lwg if hc == 1
(output omitted)
. logit lfp k5 k618 age wc lwg [pweight=wgtvar]
(output omitted)
. logit lfp k5 k618 age wc lwg if hc == 1, level(90)
(output omitted)
```

You can review the output from the last estimation by typing the command name again. For example, if the most recent model that you fitted was a logit model, you could have Stata replay the results by simply typing `logit`.

Variable lists

depvar is the dependent variable. *indepvars* is a list of the independent variables. If no independent variables are given, a model with only the intercept(s) is fitted. Stata automatically corrects some mistakes in specifying independent variables. For example, if you include `wc` as an independent variable when the sample is restricted based on `wc` (e.g., `logit lfp wc k5 k618 age hc if wc==1`), Stata drops `wc` from the list of variables. Or, suppose that you recode a k-category variable into a set of k dummy variables.

---

[3]`mlogit` is a multiple-equation estimation command, but the syntax is the same as single-equation commands because the independent variables included in the model are the same for all equations. The zero-inflated count models `zip` and `zinb` are the only multiple-equation commands considered in our book where different sets of independent variables can be used in each equation. Details on the syntax for these models are given in Chapter 7.

Recall that one of the dummy variables must be excluded to avoid perfect collinearity. If you included all *k* dummy variables in *indepvars*, Stata automatically excludes one of them.

Specifying the estimation sample

`if` and `in` restrictions can be used to define the estimation sample (i.e., the sample used to fit the model), where the syntax for `if` and `in` conditions follows the guidelines in Chapter 2. For example, if you want to fit a logit model only for women who went to college, you could specify `logit lfp k5 k618 age hc lwg if wc==1`.

Missing data

Estimation commands use *listwise deletion* to exclude cases in which there are missing values for any of the variables in the model. Accordingly, if two models are fitted using the same dataset but have different sets of independent variables, it is possible to have different samples. The easiest way to understand this is with a simple example.[4] Suppose that among the 753 cases in the sample, 23 have missing data for at least one variable. If we fitted a model using all variables, we would get

```
. logit lfp k5 k618 age wc hc lwg inc, nolog
Logit estimates                         Number of obs   =        730
  (output omitted)
```

Suppose that seven of the missing cases were missing only for `k618` and that we fit a second model that excludes this variable:

```
. logit lfp k5 age wc hc lwg inc, nolog
Logit estimates                         Number of obs   =        737
  (output omitted)
```

The estimation sample for the second model has increased by seven cases. Importantly, this means that you cannot compute a likelihood-ratio test comparing the two models (see Section 3.3) and that any changes in the estimates could be due either to changes in the model specification or to the use of different samples to fit the models. When *you compare coefficients across models, you want the samples to be exactly the same.* If they are not, you cannot compute likelihood-ratio tests, and any interpretations of why the coefficients have changed must take into account differences between the samples.

While Stata uses listwise deletion when fitting models, *this does not mean that this is the only or the best way to handle missing data.* While the complex issues related to missing data are beyond the scope of our discussion (see Little and Rubin 1987; Schafer 1997; Allison 2001), we recommend that you make explicit decisions about which cases to include in your analyses, rather than let cases be dropped implicitly. Personally, we wish that Stata would issue an error rather than automatically dropping cases.

---

[4]This example uses `binlfp2.dta`, which does not have any missing data. We have artificially created missing data. Remember that all of our examples are available from *http://www.indiana.edu/~jslsoc/spost.htm* or can be obtained by typing `net search spost`.

The `mark` and `markout` commands make it simple to explicitly exclude missing data. `mark` *markvar* generates a new variable *markvar* that equals 1 for all cases. `markout` *markvar varlist* changes the values of *markvar* from 1 to 0 for any cases in which values of any of the variables in *varlist* are missing. The following example illustrates how this works (missing data were artificially created):

```
. mark nomiss
. markout nomiss lfp k5 k618 age wc hc lwg inc
. tab nomiss
```

nomiss	Freq.	Percent	Cum.
0	23	3.05	3.05
1	730	96.95	100.00
Total	753	100.00	

```
. logit lfp k5 k618 age wc hc lwg inc if nomiss==1, nolog
Logit estimates                              Number of obs   =       730
    (output omitted)
. logit lfp k5 age wc hc lwg inc if nomiss==1, nolog
Logit estimates                              Number of obs   =       730
    (output omitted)
```

Because the `if` condition excludes the same cases from both equations, the sample size is the same for both models. Alternatively, after using `mark` and `markout`, we could have used `drop if nomiss==0` to delete observations with missing values.

Post-estimation commands and the estimation sample

Excepting `predict`, the post-estimation commands for testing, assessing fit, and making predictions that are discussed below use only observations from the estimation sample, unless you specify otherwise. Accordingly, you do not need to worry about `if` and `in` conditions or cases deleted due to missing data when you use these commands. Further details are given below.

Weights

Weights indicate that some observations should be given more weight than others when computing estimates. The syntax for specifying weights is [*type=varname*], where the brackets [ ] are part of the command, *type* is the abbreviation for the type of weight to be used, and *varname* is the weighting variable. Stata recognizes four types of weights:

1. `fweight`s, or frequency weights, indicate that an observation represents multiple observations with *identical* values. For example, if an observation has an `fweight` of 5, this is equivalent to having 5 identical, duplicate observations. In very large datasets, `fweight`s can substantially reduce the size of the data file. If you do not include a weight option in your estimation command, this is equivalent to specifying `fweight=1`.

2. `pweights`, or sampling weights, denote the inverse of the probability that the observation is included due to the sampling design. For example, if a case has a `pweight` of 1200, that case represents 1200 observations in the population.

3. `aweights`, or analytic weights, are inversely proportional to the variance of an observation. The variance of the jth observation is assumed to be σ^2/w_j, where w_j is the analytic weight. Analytic weights are used most often when observations are averages and the weights are the number of elements that gave rise to the average. For example, if each observation is the cell mean from a larger dataset, the data are heteroskedastic, because we would expect the variance of these means to decrease as the number of observations used to calculate them increases. For some estimation problems, analytic weights can be used to transform the data to reinstate the homoskedasticity assumption.

4. `iweights`, or importance weights, have no formal statistical definition. They are used by programmers to facilitate certain types of computations under specific conditions.

The use of weights is a complex topic, and it is easy to apply weights incorrectly. If you need to use weights, we encourage you to read the detailed discussion in the *Stata User's Guide* ([U] **14.1.6 weight** and [U] **23.16 Weighted estimation**). Winship and Radbill (1994) also provide a useful introduction to weights in the linear regression model.

`svy` **Commands** For more complex sampling designs that include sampling weights, strata, and PSU identifier variables, Stata provides a set of `svy` commands. For example, `svylogit` fits a binary logit model with corrections for a complex sampling design. While the interpretation of the estimated coefficients is the same for these commands as their non`svy` counterparts, we do not consider these commands further here. For further details, type `help svy` or see Hosmer and Lemeshow (2000, Chapter 6), the *Stata Survey Data Reference Manual*, or [U] **30 Overview of survey estimation**.

Options

The following options apply to most regression models. Unique options for specific models are considered in later chapters.

`noconstant` constrains the intercept(s) to equal 0. For example, in a linear regression the command `regress y x1 x2, noconstant` would fit the model $y = \beta_1 x_1 + \beta_2 x_2 + \varepsilon$.

`nolog` suppresses the iteration history. While this option shortens the output, the iteration history might contain information that indicates problems with your model. If you use this option and you have problems in obtaining estimates, it is a good idea to refit the model without this option and with the `trace` option.

`trace` lets you see the values of the parameters for each step of the iteration. This can be useful for determining which variables may be causing a problem if your model has difficulty converging.

`level(#)` specifies the level of the confidence interval. By default, Stata provides 95% confidence intervals for estimated coefficients, meaning that the interval around the estimated $\widehat{\beta}$ would capture the true value of β 95 percent of the time if repeated samples were drawn. `level` allows you to specify other intervals. For example, `level(90)` specifies a 90% interval. You can also change the default level with the command `set level 90` (for 90% confidence intervals).

`score(newvar)` creates *newvar* containing $u_j = \partial \ln L_j / \partial (\mathbf{x}_j \mathbf{b})$ for each observation j in the sample. The score vector is $\sum \partial \ln L_j / \partial \mathbf{b} = \sum u_j \mathbf{x}_j$; i.e., the product of *newvar* with each covariate summed over observations. See [U] **23.15 Obtaining scores**.

`cluster(varname)` specifies that the observations are independent across the clusters that are defined by unique values of *varname*, but are not necessarily independent within clusters. Specifying this option leads to robust standard errors, as discussed below, with an additional correction for the effects of clustered data. See Hosmer and Lemeshow (2000, Section 8.3) for a detailed discussion of logit models with clustered data.

In some cases, observations share similarities that violate the assumption of independent observations. For example, the same person might provide information at more than one point in time. Or, several members of the same family might be in the sample, again violating independence. In these examples, it is reasonable to assume that the observations within the groups, which are known as clusters, are not independent. With clustering, the usual standard errors will be incorrect.

`robust` replaces the traditional standard errors with robust standard errors, which are also known as Huber, White, or sandwich standard errors. These estimates are considered robust in the sense that they provide correct standard errors in the presence of violations of the assumptions of the model. For example, if the correct model is a binary logit model and a binary probit model is used, the model has been misspecified. The estimates obtained by fitting a logit model cannot be maximum likelihood estimates since an incorrect likelihood function is being used (i.e., a logistic probability density is used instead of the correct normal density). In this situation, the estimator is referred to by White (1982) as a *minimum ignorance estimator* since the estimators provide the best possible approximation to the true probability density function. When a model is misspecified in this way, the usual standard errors are incorrect. Arminger (1995) makes a compelling case for why robust standard errors should be used. He writes: "If one keeps in mind that most researchers misspecify the model..., it is obvious that their estimated parameters can usually be interpreted only as minimum ignorance estimators and that the standard errors and test statistics may be far away from the correct asymptotic values, depending on the discrepancy between the assumed density and the actual density that generated the data". However, we have not seen any information on the small sample properties of robust standard errors for nonlinear models (i.e., how well these standard errors work in finite samples). Long and Ervin (2000) consider this problem in the context of the linear regression model, where they found that two small sample versions of the robust standard error work quite well, while the asymptotic version often does *worse* than the usual standard errors.[5]

Robust estimators are automatically used with the `svy` commands and with the `cluster()` option. See the section [U] **23.14 Obtaining robust variance estimates** in the *Stata User's Guide*, and see Gould Pitblado, and Sribney (2003, **1.2.5 Robust variance estimator**) for a detailed discussion of how robust standard errors are computed in Stata; see Arminger (1995, 111–113) for a more mathematical treatment.

3.1.5   Reading the output

We have already discussed the iteration log, so in the following example we suppress it with the `nolog` option. Here, we consider other parts of the output from estimation commands. While the sample output is from `logit`, our discussion generally applies to other regression models.

```
. use http://www.stata-press.com/data/lfr/binlfp2, clear
(Data from 1976 PSID-T Mroz)

. logit lfp k5 k618 age wc hc lwg inc, nolog
Logit estimates                                Number of obs   =       753
                                               LR chi2(7)      =    124.48
                                               Prob > chi2     =    0.0000
Log likelihood = -452.63296                    Pseudo R2       =    0.1209
```

lfp	Coef.	Std. Err.	z	P>|z|	[95% Conf. Interval]	
k5	-1.462913	.1970006	-7.43	0.000	-1.849027	-1.076799
k618	-.0645707	.0680008	-0.95	0.342	-.1978499	.0687085
age	-.0628706	.0127831	-4.92	0.000	-.0879249	-.0378162
wc	.8072738	.2299799	3.51	0.000	.3565215	1.258026
hc	.1117336	.2060397	0.54	0.588	-.2920969	.515564
lwg	.6046931	.1508176	4.01	0.000	.3090961	.9002901
inc	-.0344464	.0082084	-4.20	0.000	-.0505346	-.0183583
_cons	3.18214	.6443751	4.94	0.000	1.919188	4.445092

Header

1. `Log likelihood = -452.63296` corresponds to the value of the log likelihood at convergence.

2. `Number of obs` is the number of observations, excluding those with missing values and after any `if` or `in` conditions have been applied.

---
[5]These versions can be computed by using the `hc2` or `hc3` options for `regress`. Long and Ervin (2000) recommend using `hc3`.

3. LR chi2(7) is the value of a likelihood-ratio chi-squared for the test of the null hypothesis that all of the coefficients associated with independent variables are simultaneously equal to zero. The *p*-value is indicated by Prob > chi2. The number in parentheses is the number of coefficients being tested.

4. Pseudo R2 is the measure of fit also known as McFadden's R^2. Details on how this measure is computed are given below, along with a discussion of alternative measures of fit.

Estimates and standard errors

1. The left column lists the variables in the model, with the dependent variable listed at the top. The independent variables are always listed in the same order as they were entered on the command line. The constant, labeled _cons, is last.

2. Column Coef. contains the ML estimates.

3. Column Std. Err. is the standard error of the estimates.

4. The resulting *z*-test, equal to the estimate divided by its standard error, is labeled z with the two-tailed significance level listed as P > | z |. A significance level listed as 0.000 means that $P < .0005$ (for example, .0006 is rounded to .001, while .00049 is rounded to .000).

5. The start and end points for the confidence interval for each estimate are listed under [95% Conf. Interval].

3.1.6   Reformatting output with estimates table

estimates table can be used to reformat from an estimation command to look more like the tables that are seen in articles that use regression models. estimates table also makes it easier to move estimation results into a word processor or spreadsheet to make a presentation-quality table there. We strongly recommend using this command or some other automated procedure (such as outreg discussed in the next section) rather than retyping the results. Not only is it much less tedious, but it also diminishes the possibility of errors. The syntax for estimates table is

estimates table [*modelname* [*modelname* ...]] [, *other options*]

where *modelname* is the name of a model whose results have been saved using estimates store, or is the previously fitted model if none was specified. As we will show later, the command can be used to place the results from multiple models side by side.

After fitting the logit model we presented above, we could run estimates table as follows:

```
. logit lfp k5 k618 age wc hc lwg, nolog
  (output omitted )
```

```
. estimates table, b(%9.3f) t label varwidth(30)
```

Variable	active
# kids < 6	-1.439
	-7.44
# kids 6-18	-0.087
	-1.31
Wife's age in years	-0.069
	-5.49
Wife College: 1=yes 0=no	0.693
	3.10
Husband College: 1=yes 0=no	-0.142
	-0.73
Log of wife's estimated wages	0.561
	3.77
Constant	2.939
	4.67

```
                         legend: b/t
```

estimates table provides a good deal of flexibility for what you include in your table. While you should check the *Stata Base Reference Manual* or type help est_table for complete information, here are some of the most helpful options:

se, t, p, and star specify whether and how standard errors are to be included in the table. se tells Stata to print standard errors along with the coefficients, t specifies t or z statistics, and p tells Stata to print the p values. In contrast, star tells Stata to print one star by the coefficient if the p value is $< .05$, two if $< .01$, and three if $< .001$. At least as of this writing, the star option cannot be used in conjunction with the se (or t or p) option.

b(*format*) specifies the format in which the coefficients are printed; e.g., the number of decimal places shown. Also, formats can be specified in parentheses after se, t, or p. We use the format %9.3f for many of our tables; the 3 in this format means three decimal places. For more information on formats, see help format or the *Stata User's Guide*.

keep(*varlist*) or drop(*varlist*) can be used to specify which of the independent variables from the regression you wish to include in the table, if you do not wish to include them all.

label indicates that variable labels should be used instead of variable names in the rows of the table.

varwidth(#) specifies the width of the column that includes variable names and labels, which is useful when these are long.

stats(*list*) indicates that the scalar statistic(s) included in the *list* should also be included in the model. N is one such statistic that can be specified here. Others, including several goodness-of-fit statistics are also available. Type help est_table for more information.

3.1.7   Reformatting output with outreg

As an alternative to `estimates table`, you can use `outreg`, a user-written command (Gallup 2001). To install `outreg`, type `findit outreg` and follow the links. `outreg` can be used to reformat the output from an estimation command to look more like the tables that are seen in articles that use regression models. `outreg` also makes it easier to move estimation results into a word processor or spreadsheet to make a presentation-quality table there. The syntax of `outreg` is

outreg $[varlist]$ using *filename* $\left[, options \right]$

where *varlist* contains the names of the independent variables in the order you want them presented in the table and where *filename* is the name of the file to contain the reformatted results. When you run `outreg`, the results are not posted in the Results window but are only written to this file.

After fitting the logit model that we presented above, we could run `outreg` as follows:

```
. logit lfp k5 k618 age wc hc lwg
  (output omitted)
. outreg using model1, replace
```

File `model1.out` is saved to the working directory. We needed to tinker with this file in a text editor to get the spacing just right, but our end result looks like this:

```
Dependent variable: In paid labor force.
-----------------------------------------
# kids <= 5.                    -1.439
                                (7.44)**
# kids 6-18.                    -0.087
                                (1.31)
Wife's age in years.            -0.069
                                (5.49)**
Wife College: 1=yes,0=no         0.693
                                (3.10)**
Husband College: 1=yes,0=no     -0.142
                                (0.73)
Log of wife´s estimated wages    0.561
                                (3.77)**
Constant                         2.939
                                (4.67)**
-----------------------------------------
Observations     753
Absolute value of z-statistics in parentheses
* significant at 5%; ** significant at 1%
```

`outreg` is a very flexible command, with too many options to consider all of them here. Some of the most commonly used options are the following:

`replace` specifies that *filename*`.out` should be overwritten if it already exists.

append indicates that the estimation output should be appended to that of an existing file. This allows you to construct tables with the results from several regressions, as we illustrate in Chapter 4.

se reports standard errors instead of t/z statistics.

pvalue reports p-values instead of t/z statistics.

title(*text*) adds a title to be printed at the top of the output.

addnote(*text*) adds a note to be printed at the bottom of the output.

nolabel specifies that variable names rather than variable labels should be printed in the output.

A full list of options and further information about using the command can be obtained by typing help outreg.

3.1.8 Alternative output with listcoef

The interpretation of regression models often involves transformations of the usually estimated parameters. For some official Stata estimation commands, there are options to list transformations of the parameters, such as the or option to list odds ratios in logit or the beta option to list standardized coefficients for regress. While Stata is commendably clear in explaining the meaning of the estimated parameters, in practice it is easy to be confused about proper interpretations. For example, the zip model (discussed in Chapter 8) simultaneously estimates a binary and count model, and it is easy to be confused regarding the direction of the effects.

For the estimation commands considered in this book (plus some not considered here), our command listcoef lists estimated coefficients in ways that facilitate interpretation. You can list coefficients by name or significance level, list transformations of the coefficients, and request help on proper interpretation. In fact, in many cases you won't even need the normal output from the estimation. You could suppress this output with the prefix quietly (e.g., quietly logit lfp k5 wc hc) and then use the listcoef command. The syntax is

listcoef [*varlist*] [, pvalue(#) [factor | percent | std] constant help]

where *varlist* indicates that coefficients for only these variables are to be listed. If no *varlist* is given, coefficients for all variables are listed.

Options for types of coefficients

Depending on the model estimated and the specified options, listcoef computes standardized coefficients, factor changes in the odds or expected counts, or percent changes in the odds or expected counts. More information on these different types of coefficients is provided below, as well as in the chapters that deal with specific types of outcomes.

The table below lists which options (details on these options are given below) are available for each estimation command. If an option is the default, it does not need to be specified.

	Option		
	std	factor	percent
Type 1: `regress, probit, cloglog, oprobit,` `tobit, cnreg, intreg`	Default	No	No
Type 2: `logit, logistic ologit`	Yes	Default	Yes
Type 3: `clogit, mlogit, poisson, nbreg, zip, zinb`	No	Default	Yes

`factor` requests factor change coefficients.

`percent` requests percent change coefficients instead of factor change coefficients.

`std` indicates that coefficients are to be standardized to a unit variance for the independent and dependent variables. For models with a latent dependent variable, the variance of the latent outcome is estimated.

Other options

`pvalue(#)` specifies that only coefficients significant at the # significance level or smaller will be printed. For example, `pvalue(.05)` specifies that only coefficients significant at the .05 level of less should be listed. If `pvalue` is not given, all coefficients are listed.

`constant` includes the constant(s) in the output. By default, they are not listed.

`help` gives details for interpreting each coefficient.

Standardized coefficients

`std` requests coefficients after some or all of the variables have been standardized to a unit variance. Standardized coefficients are computed as follows:

x-standardized coefficients

The linear regression model can be expressed as

$$y = \beta_0 + \beta_1 x_1 + \beta_2 x_2 + \varepsilon \tag{3.1}$$

The independent variables can be standardized with simple algebra. Let σ_k be the standard deviation of x_k. Then, dividing each x_k by σ_k and multiplying the corresponding β_k by σ_k

$$y = \beta_0 + (\sigma_1 \beta_1) \frac{x_1}{\sigma_1} + (\sigma_2 \beta_2) \frac{x_2}{\sigma_2} + \varepsilon$$

$\beta_k^{S_x} = \sigma_k \beta_k$ is an x-standardized coefficient. For a continuous variable, $\beta_k^{S_x}$ can be interpreted as

> For a standard deviation increase in x_k, y is expected to change by $\beta_k^{S_x}$ units, holding all other variables constant.

The same method of standardization can be used in all of the other models we consider in this book.

y and y^*-standardized coefficients

To standardize for the dependent variable, let σ_y be the standard deviation of y. We can standardize y by dividing (3.1) by σ_y:

$$\frac{y}{\sigma_y} = \frac{\beta_0}{\sigma_y} + \frac{\beta_1}{\sigma_y} x_1 + \frac{\beta_2}{\sigma_y} x_2 + \frac{\varepsilon}{\sigma_y}$$

Then $\beta_k^{S_y} = \beta_k / \sigma_y$ is a y-standardized coefficient that can be interpreted as

> For a unit increase in x_k, y is expected to change by $\beta_k^{S_y}$ standard deviations, holding all other variables constant.

For a dummy variable,

> Having characteristic x_k (as opposed to not having the characteristic) results in an expected change in y of $\beta_k^{S_y}$ standard deviations, holding all other variables constant.

In models with a latent dependent variable, the equation $y^* = \beta_0 + \beta_1 x_1 + \beta_2 x_2 + \varepsilon$ can be divided by $\widehat{\sigma}_{y^*}$. To estimate the variance of the latent variable, the quadratic form is used:

$$\widehat{\mathrm{Var}}(y^*) = \widehat{\beta}' \widehat{\mathrm{Var}}(\mathbf{x}) \, \widehat{\beta} + \mathrm{Var}(\varepsilon)$$

where $\widehat{\beta}$ is a vector of estimated coefficients and $\widehat{\mathrm{Var}}(\mathbf{x})$ is the covariance matrix for the xs computed from the observed data. By assumption, $\mathrm{Var}(\varepsilon) = 1$ in probit models, and $\mathrm{Var}(\varepsilon) = \pi^2/3$ in logit models.

Fully standardized coefficients

In the linear regression model, it is possible to standardize both y and the xs:

$$\frac{y}{\sigma_y} = \frac{\beta_0}{\sigma_y} + \left(\frac{\sigma_1 \beta_1}{\sigma_y}\right) \frac{x_1}{\sigma_1} + \left(\frac{\sigma_2 \beta_2}{\sigma_y}\right) \frac{x_2}{\sigma_2} + \frac{\varepsilon}{\sigma_y}$$

Then, $\beta_k^S = (\sigma_k \beta_k)/\sigma_y$ is a *fully standardized coefficient* that can be interpreted as follows:

For a standard deviation increase in x_k, y is expected to change by β_k^S standard deviations, holding all other variables constant.

The same approach can be used in models with a latent dependent variable y^*.

Example of listcoef for standardized coefficients

Here, we illustrate the computation of standardized coefficients for the regression model. Examples for other models are given in later chapters. The standard output from `regress` is

```
. use science2, clear
(Note that some of the variables have been artificially constructed.)
. regress job female phd mcit3 fellow pub1 cit1
```

Source	SS	df	MS
Model	28.8930452	6	4.81550754
Residual	95.7559074	154	.621791607
Total	124.648953	160	.779055954

Number of obs = 161
F(6, 154) = 7.74
Prob > F = 0.0000
R-squared = 0.2318
Adj R-squared = 0.2019
Root MSE = .78854

job	Coef.	Std. Err.	t	P>|t|	[95% Conf. Interval]	
female	-.1243218	.1573559	-0.79	0.431	-.4351765	.1865329
phd	.2898888	.0732633	3.96	0.000	.145158	.4346196
mcit3	.0021852	.0023485	0.93	0.354	-.0024542	.0068247
fellow	.1839757	.133502	1.38	0.170	-.0797559	.4477073
pub1	-.0068635	.0255761	-0.27	0.789	-.0573889	.0436618
cit1	.0080916	.0041173	1.97	0.051	-.0000421	.0162253
_cons	1.763224	.2361352	7.47	0.000	1.296741	2.229706

Now, we use `listcoef`:

```
. listcoef female cit1, help
regress (N=161): Unstandardized and Standardized Estimates
  Observed SD: .88264146
  SD of Error: .78853764
```

job	b	t	P>|t|	bStdX	bStdY	bStdXY	SDofX
female	-0.12432	-0.790	0.431	-0.0534	-0.1409	-0.0605	0.4298
cit1	0.00809	1.965	0.051	0.1719	0.0092	0.1947	21.2422

```
       b = raw coefficient
       t = t-score for test of b=0
   P>|t| = p-value for t-test
   bStdX = x-standardized coefficient
   bStdY = y-standardized coefficient
  bStdXY = fully standardized coefficient
   SDofX = standard deviation of X
```

By default for `regress`, `listcoef` lists the standardized coefficients. Notice that we only requested information on two of the variables.

Factor and percent change

In logit-based models and models for counts, coefficients can be expressed as (1) a factor or multiplicative change in the odds or the expected count (requested in `listcoef` by the `factor` option), or (2) the percent change in the odds or expected count (requested with the `percent` option). While these can be computed with options to some estimation commands, `listcoef` provides a single method to compute these. Details on these coefficients are given in later chapters for each specific model.

3.1.9 Storing estimation results

Stata's `estimates` command provides the facility to store estimation results in memory under a given name. Storing the estimation results allows post-estimation commands to use them as input. For instance, to perform a likelihood-ratio test, the Stata command `lrtest` needs information from both the constrained and the unconstrained estimation results. In this section, we discuss how to use the `estimates` command to store results from a given model and sample.[6] In the next section, we will discuss how to perform post-estimation analysis on stored estimation results.

After running any estimation command in Stata, the syntax of the `estimates` command for storing results is

`estimates store` *name*

As an example, suppose that we know that we will want to perform post-estimation analysis on the results from the above regression. To store the estimation results in memory under the name `reg1`, we could type

```
. regress job female phd mcit3 fellow pub1 cit1
  (output omitted)
. estimates store reg1
```

We could then perform post-estimation analysis on the estimation results stored as `reg1` as described in the next section.

3.2 Post-estimation analysis

There are three types of post-estimation analysis that we consider in the remainder of this chapter. The first is statistical testing that goes beyond routine tests of a single coefficient. This is done with Stata's powerful `test` and `lrtest` commands. In later chapters, we present other tests of interest for a given model (e.g., tests of the parallel regression assumption for ordered regression models). The second post-estimation task

---

[6]The `estimates` command can do more than just store estimation results. The command actually provides an environment for storing and manipulating estimation results. Type `help estimates` for more information.

is assessing the fit of a model using scalar measures computed by our command `fitstat`. Examining outliers and residuals for binary models is considered in Chapter 4. The third task, and the focus of much of this book, is interpreting the predictions from nonlinear models. We begin by discussing general issues that apply to all nonlinear models. We then discuss our SPost commands that implement these methods of interpretation.

3.3   Testing

Coefficients estimated by ML can be tested with Wald tests using `test` and likelihood ratio (LR) tests using `lrtest`. For both types of tests, there is a null hypothesis H_0 that implies constraints on the model's parameters. For example, H_0: $\beta_{wc} = \beta_{hc} = 0$ hypothesizes that two of the parameters are zero in the population.

The Wald test assesses H_0 by considering two pieces of information. First, all else being equal, the greater the distance between the estimated coefficients and the hypothesized values, the less support we have for H_0. Second, the greater the curvature of the log-likelihood function, the more certainty we have about our estimates. This means that smaller differences between the estimates and hypothesized values are required to reject H_0. The LR test assesses a hypothesis by comparing the log likelihood from the full model (i.e., the model that does not include the constraints implied by H_0) and a restricted model that imposes the constraints. If the constraints significantly reduce the log likelihood, then H_0 is rejected. Thus, the LR test requires fitting two models. Even though the LR and Wald tests are asymptotically equivalent, in finite samples they give different answers, particularly for small samples. In general, it is unclear which test is to be preferred. Rothenberg (1984) and Greene (2003) suggest that neither test is uniformly superior, although many statisticians prefer the LR.

3.3.1   Wald tests

`test` computes Wald tests for linear hypotheses about parameters from the last model fitted. Here, we consider the most useful features of this command for regression models. Information on features for multiple-equation models, such as `mlogit`, `zip`, and `zinb`, are discussed in Chapters 6 and 7. Use `help test` for additional features and `help testnl` for testing nonlinear hypotheses.

The first syntax for `test` allows you to specify that one or more coefficients from the last estimation are simultaneously equal to 0:

<u>test</u> *varlist* [, <u>a</u>ccumulate]

where *varlist* contains names of one or more independent variable from the last estimation. The `accumulate` option will be discussed shortly.

Some examples should make this first syntax clear. With a single variable listed, `k5` in this case, we are testing H_0: $\beta_{k5} = 0$.

```
. use http://www.stata-press.com/data/lfr/binlfp2, clear
(Data from 1976 PSID-T Mroz)
. logit lfp k5 k618 age wc hc lwg inc, nolog
  (output omitted)
. test k5
 ( 1)  k5 = 0
            chi2( 1) =    55.14
          Prob > chi2 =    0.0000
```

The resulting chi-squared test with 1 degree of freedom equals the square of the z-test in the output from the estimation command, and we can reject the null hypothesis.

With two variables listed, we are testing $H_0\colon \beta_{k5} = \beta_{k618} = 0$:

```
. test k5 k618
 ( 1)  k5 = 0
 ( 2)  k618 = 0
            chi2( 2) =    55.16
          Prob > chi2 =    0.0000
```

We can reject the hypothesis that the effects of young and older children are simultaneously zero.

In our last example, we include all the independent variables.

```
. test k5 k618 age wc hc lwg inc
 ( 1)  k5 = 0
 ( 2)  k618 = 0
 ( 3)  age = 0
 ( 4)  wc = 0
 ( 5)  hc = 0
 ( 6)  lwg = 0
 ( 7)  inc = 0
            chi2( 7) =    94.98
          Prob > chi2 =    0.0000
```

This is a test of the hypothesis that all the coefficients except the intercept are simultaneously equal to zero. As noted above, a likelihood-ratio test of this same hypothesis is part of the standard output of estimation commands (e.g., `LR chi2(7)=124.48` from the earlier `logit` output).

The second syntax for `test` allows you to test hypotheses about linear combinations of coefficients:

<u>test</u> $\big[exp{=}exp\big]$ $\big[$, <u>a</u>ccumulate $\big]$

For example, to test that two coefficients are equal, for example $H_0\colon \beta_{k5} = \beta_{k618}$:

```
. test k5=k618
 ( 1)  k5 - k618 = 0
            chi2( 1) =    49.48
          Prob > chi2 =    0.0000
```

Because the test statistic is significant, we can reject the null hypothesis that the effect of having young children on labor force participation is equal to the effect of having older children.

The accumulate option

The `accumulate` option allows you to build more complex hypotheses based on the prior use of the `test` command. For example, you might begin with a test of H_0: $\beta_{k5} = \beta_{k618}$:

```
. test k5=k618
 ( 1)   k5 - k618 = 0
           chi2(  1) =    49.48
         Prob > chi2 =    0.0000
```

Then, add the constraint that $\beta_{wc} = \beta_{hc}$:

```
. test wc=hc, accumulate
 ( 1)   k5 - k618 = 0
 ( 2)   wc - hc = 0
           chi2(  2) =    52.16
         Prob > chi2 =    0.0000
```

This results in a test of H_0: $\beta_{k5} = \beta_{k618}$, $\beta_{wc} = \beta_{hc}$.

3.3.2   LR tests

`lrtest` compares nested models using an LR test. The syntax is

`lrtest` $model_1$ $\left[model_2 \right]$

where $model_1$, and $model_2$ if specified, are the names of estimation results saved using `estimates store`. When $model_2$ is not specified, the most recent estimation results are used in its place.

We prefer to save the results of both models before running `lrtest`. Typically, we begin by fitting the full or unconstrained model and then save the results using `estimates store`. For example,

```
. logit lfp k5 k618 age wc hc lwg inc, nolog
  (output omitted)
. estimates store fmodel
```

where `fmodel` is the name of the estimation results from the the full model. While any name up to 27 characters can be used, we recommend keeping the names short but informative. After you save the results, you fit a model that is *nested* in the full model. A nested model is one that can be created by imposing constraints on the coefficients in the prior model. Most commonly, one excludes some of the variables that were included in the first model, which in effect constrains the coefficients on these variables to be zero. For example, if we drop `k5` and `k618` from the last model

```
. logit lfp age wc hc lwg inc, nolog
  (output omitted)
. estimates store nmodel
. lrtest fmodel nmodel
likelihood-ratio test                           LR chi2(2)   =      66.49
(Assumption: nmodel nested in fmodel)           Prob > chi2 =     0.0000
```

We stored the results for the nested models as `nmodel`. The output indicates that the test assumes that `nmodel` is nested in `fmodel`. It is up to the user to ensure that the models are nested. In this case, the models are nested. The results is an LR test of the hypothesis H_0: $\beta_{k5} = \beta_{k618} = 0$. The significant chi-squared statistic means that we reject the null hypothesis that these two coefficients are simultaneously equal to zero.

The syntax of `lrtest` allows us to fit the constrained model first, as in the following example.

```
. logit lfp age wc hc lwg inc, nolog
  (output omitted)
. estimates store nmodel
. logit lfp k5 k618 age wc hc lwg inc, nolog
  (output omitted)
. estimates store fmodel
. lrtest nmodel fmodel
likelihood-ratio test                           LR chi2(2)   =      66.49
(Assumption: nmodel nested in fmodel)           Prob > chi2 =     0.0000
```

The output from `lrtest` states that, in this case, the `nmodel` estimation results are assumed to be based on a model that nests in `fmodel`.

Avoiding invalid LR tests

`lrtest` does *not* always prevent you from computing an invalid test. There are two things that you must check. First, the two models must be nested. Second, the two models must be fitted on exactly the same sample. While `lrtest` will exit with an error message if the number of observations differs over the models, this check does not catch those cases in which the number of observations is the same but the samples are different. If either of these conditions are violated, the results of `lrtest` are meaningless. For details on ensuring the same sample size, see our discussion of `mark` and `markout` in Section 3.1.4.

3.4   Measures of fit

Assessing fit involves both the analysis of the fit of individual observations and the evaluation of scalar measures of fit for the model as a whole. Regarding the former, Pregibon (1981) extended methods of residual and outlier analysis from the linear regression model to the case of binary logit and probit (see also Cook and Weisberg 1999, Part IV). These measures are considered in Chapter 4. Measures for many count models are also available (Cameron and Trivedi 1998). Unfortunately, similar methods for ordinal and nominal outcomes are not available. Many scalar measures have been developed to summarize the overall goodness of fit for regression models of continuous, count, or categorical dependent variables. A scalar measure can be useful in comparing competing models and ultimately in selecting a final model. Within a substantive area, measures of fit can provide a *rough* index of whether a model is adequate. However, *there is no convincing evidence that selecting a model that maximizes the value of a given measure results in a model that is optimal in any sense other than the model having a larger (or, in some instances, smaller) value of that measure.* While measures of fit provide some information, it is only partial information that must be assessed within the context of the theory motivating the analysis, past research, and the estimated parameters of the model being considered.

Syntax of fitstat

Our command `fitstat` calculates a large number of fit statistics for the estimation commands we consider in this book. With its `saving()` and `using()` options, the command also allows the comparison of fit measures across two models. While `fitstat` duplicates some measures computed by other commands (e.g., the pseudo-R^2 in standard Stata output or `lfit`), `fitstat` adds many more measures and makes it convenient to compare measures across models. The syntax is

fitstat [, <u>s</u>aving(*name*) <u>u</u>sing(*name*) <u>b</u>ic force save <u>d</u>if]

While many of the calculated measures are based on values returned by the estimation command, for some measures it is necessary to compute additional statistics from the estimation sample. This is done automatically using the estimation sample from the last estimation command. `fitstat` can also be used when models are fitted with weighted data, with two limitations. First, some measures cannot be computed with some types of weights. Second, with `pweights`, we use the "pseudo-likelihoods" rather than the likelihood to compute our measures of fit. Given the heuristic nature of the various measures of fit, we see no reason why the resulting measures would be inappropriate.

`fitstat` terminates with an error if the last estimation command does not return a value for the log-likelihood equation with only an intercept (i.e., if `e(ll_0)=.`). This occurs, for example, if the `noconstant` option is used to fit a model.

Options

saving(*name*) saves the computed measures in a matrix for subsequent comparisons. *name* must be four characters or shorter.

using(*name*) compares the fit measures for the current model with those of the model saved as *name*. *name* cannot be longer than four characters.

bic presents only BIC and other information measures. When comparing two models, fitstat reports Raftery's (1996) guidelines for assessing the strength of one model over another, which are detailed at the end of this section.

force is required to compare two models when the number of observations or the estimation method varies between the two models.

save and dif are equivalent to saving(0) and using(0).

Models and measures

Details on the measures of fit are given below. Here, we only summarize which measures are computed for which models. ■ indicates that a measure is computed, and □ indicates that the measure is not computed.

	regress	logit probit	cloglog	ologit oprobit	clogit mlogit	cnreg intreg tobit	gologit nbreg poisson zinb zip
Log likelihood	■	■	■[1]	■	■	■	■[2]
Deviance & LR chi-squared	■	■	■	■	■	■	■
AIC, AIC*n, BIC, BIC′	■	■	■	■	■	■	■
R^2 & Adjusted R^2	■	□	□	□	□	□	□
Efron's R^2	□	■	■	□	□	□	□
McFadden's, ML, C&U's R^2	□	■	■	■	■	□	□
Count & Adjusted Count R^2	□	■	■	■	■[3]	□	□
Var(e), Var(y^*) and M&Z's R^2	□	■	□	■	□	■	□

1: For cloglog the log likelihood for the intercept-only model does not correspond to the first step in the iterations.
2: For zip and zinb, the log likelihood for the intercepts-only model is calculated by estimating
   zip | zinb *depvar* , inf(_cons).
3: The adjusted count R^2 is not defined for clogit.

Example of fitstat

To compute fit statistics for a single model, we first fit the model and then run fitstat:

```
. logit lfp k5 k618 age wc hc lwg inc, nolog
(output omitted)

. fitstat

Measures of Fit for logit of lfp
Log-Lik Intercept Only:      -514.873    Log-Lik Full Model:      -452.633
D(745):                       905.266    LR(7):                    124.480
                                         Prob > LR:                  0.000
McFadden's R2:                  0.121    McFadden's Adj R2:          0.105
Maximum Likelihood R2:          0.152    Cragg & Uhler's R2:         0.204
McKelvey and Zavoina's R2:      0.217    Efron's R2:                 0.155
Variance of y*:                 4.203    Variance of error:          3.290
Count R2:                       0.693    Adj Count R2:               0.289
AIC:                            1.223    AIC*n:                    921.266
BIC:                        -4029.663    BIC':                     -78.112
```

`fitstat` is particularly useful for comparing two models. To do this, you fit a model and then save the results from `fitstat`. Here, we use `quietly` to suppress the output from `fitstat` because we list those results in the next step:

```
. logit lfp k5 k618 age wc hc lwg inc, nolog
  (output omitted )
. quietly fitstat, saving(mod1)
```

Next, we generate `agesq` which is the square of `age`. The new model adds `agesq` and drops `k618`, `hc`, and `lwg`. To compare the saved model with the current model, type

```
. generate agesq = age*age
. logit lfp k5 age agesq wc inc, nolog
  (output omitted )
. fitstat, using(mod1)
```

Measures of Fit for logit of lfp

	Current	Saved	Difference
Model:	logit	logit	
N:	753	753	0
Log-Lik Intercept Only:	-514.873	-514.873	0.000
Log-Lik Full Model:	-461.653	-452.633	-9.020
D:	923.306(747)	905.266(745)	18.040(2)
LR:	106.441(5)	124.480(7)	18.040(2)
Prob > LR:	0.000	0.000	0.000
McFadden's R2:	0.103	0.121	-0.018
McFadden's Adj R2:	0.092	0.105	-0.014
Maximum Likelihood R2:	0.132	0.152	-0.021
Cragg & Uhler's R2:	0.177	0.204	-0.028
McKelvey and Zavoina's R2:	0.182	0.217	-0.035
Efron's R2:	0.135	0.155	-0.020
Variance of y*:	4.023	4.203	-0.180
Variance of error:	3.290	3.290	0.000
Count R2:	0.677	0.693	-0.016
Adj Count R2:	0.252	0.289	-0.037
AIC:	1.242	1.223	0.019
AIC*n:	935.306	921.266	14.040
BIC:	-4024.871	-4029.663	4.791
BIC':	-73.321	-78.112	4.791

Difference of   4.791 in BIC' provides positive support for saved model.

Note: p-value for difference in LR is only valid if models are nested.

Methods and formulas for fitstat

This section provides brief descriptions of each measure computed by `fitstat`. Full details along with citations to original sources are found in Long (1997). The measures are listed in the same order as the output above.

Log-likelihood based measures

Stata begins maximum likelihood iterations by computing the log likelihood of the model with all parameters but the intercept(s) constrained to zero, referred to as

$L\left(M_{\text{Intercept}}\right)$. The log likelihood upon convergence, referred to as M_{Full}, is also listed. This information is usually presented as the first step of the iteration log and in the header for the estimation results.[7]

Chi-squared test of all coefficients   An LR test of the hypothesis that all coefficients except the intercept(s) are zero can be computed by comparing the log likelihoods: $LR = 2 \ln L(M_{\text{Full}}) - 2 \ln L(M_{\text{Intercept}})$. This statistic is sometimes designated as G^2. LR is reported by Stata as LR chi2(7) = 124.48, where the degrees of freedom, (7), are the number of constrained parameters. fitstat reports this statistic as LR(7): 124.48. For zip and zinb, LR tests that the coefficients in the count portion (not the binary portion) of the model are zero.

Deviance   The *deviance* compares a given model with a model that has one parameter for each observation so that the model reproduces perfectly the observed data. The deviance is defined as $D = -2 \ln L(M_{\text{Full}})$, where the degrees of freedom equal N minus the number of parameters. Note that D does not have a chi-squared distribution.

R^2 in the LRM

For regress, fitstat reports the standard coefficient of determination, which can be defined variously as

$$R^2 = 1 - \frac{\sum_{i=1}^{N}(y_i - \widehat{y}_i)^2}{\sum_{i=1}^{N}(y_i - \overline{y})^2} = \frac{\widehat{\text{Var}}(\widehat{y})}{\widehat{\text{Var}}(\widehat{y}) + \widehat{\text{Var}}(\widehat{\varepsilon})} = 1 - \left\{\frac{L(M_{\text{Intercept}})}{L(M_{\text{Full}})}\right\}^{2/N} \quad (3.2)$$

The adjusted R^2 is defined as

$$\overline{R}^2 = \left(R^2 - \frac{K}{N-1}\right)\left(\frac{N-1}{N-K-1}\right)$$

where K is the number of independent variables.

Pseudo-R^2s

While each of the definitions of R^2 in (3.2) give the same numeric value in the LRM, they give different answers and thus provide different measures of fit when applied to the other models evaluated by fitstat.

---

[7]In cloglog, the value at iteration 0 is not the log likelihood with only the intercept. For zip and zinb, the "intercept-only" model can be defined in different ways. These commands return as e(ll_0), the value of the log likelihood with the binary portion of the model unrestricted while only the intercept is free for the Poisson or negative binomial portion of the model. fitstat returns the value of the log likelihood from the model with only an intercept in both the binary and the count portion of the model.

McFadden's R^2 McFadden's R^2, also known as the "likelihood-ratio index", compares a model with just the intercept to a model with all parameters. It is defined as

$$R^2_{\text{McF}} = 1 - \frac{\ln \widehat{L}(M_{\text{Full}})}{\ln \widehat{L}(M_{\text{Intercept}})}$$

If model $M_{\text{Intercept}} = M_{\text{Full}}$, R^2_{McF} equals 0, but R^2_{McF} can never exactly equal 1. This measure, which is computed by Stata as `Pseudo R2 = 0.1209`, is listed in `fitstat` as `McFadden's R2: 0.121` Because R^2_{McF} always increases as new variables are added, an adjusted version is also available:

$$\overline{R}^2_{\text{McF}} = 1 - \frac{\ln \widehat{L}(M_{\text{Full}}) - K^*}{\ln \widehat{L}(M_{\text{Intercept}})}$$

where K^* is the number of parameters (not independent variables).

Maximum likelihood R^2 Another analogy to R^2 in the LRM was suggested by Maddala:

$$R^2_{\text{ML}} = 1 - \left\{ \frac{L(M_{\text{Intercept}})}{L(M_{\text{Full}})} \right\}^{2/N} = 1 - \exp(-G^2/N)$$

Cragg & Uhler's R^2 Since R^2_{ML} only reaches a maximum of $1 - L(M_{\text{Intercept}})^{2/N}$, Cragg and Uhler suggested a normed measure:

$$R^2_{\text{C\&U}} = \frac{R^2_{\text{ML}}}{\max R^2_{\text{ML}}} = \frac{1 - \left\{ L(M_{\text{Intercept}})/L(M_{\text{Full}}) \right\}^{2/N}}{1 - L(M_{\text{Intercept}})^{2/N}}$$

Efron's R^2 For binary outcomes, Efron's pseudo-R^2 defines $\widehat{y} = \widehat{\pi} = \widehat{\Pr}(y = 1 \mid \mathbf{x})$ and equals

$$R^2_{\text{Efron}} = 1 - \frac{\sum_{i=1}^{N} (y_i - \widehat{\pi}_i)^2}{\sum_{i=1}^{N} (y_i - \overline{y})^2}$$

$V(y^*)$, $V(\varepsilon)$ and McKelvey and Zavoina's R^2 Some models can be defined in terms of a latent variable y^*. This includes the models for binary or ordinal outcomes: `logit`, `probit`, `ologit` and `oprobit`, as well as some models with censoring: `tobit`, `cnreg`, and `intreg`. Each model is defined in terms of a regression on a latent variable y^*:

$$y^* = \mathbf{x}\beta + \varepsilon$$

Using $\widehat{\text{Var}}(\widehat{y}^*) = \widehat{\beta}'\widehat{\text{Var}}(\mathbf{x})\,\widehat{\beta}$, McKelvey and Zavoina proposed

$$R^2_{M\&Z} = \frac{\widehat{\text{Var}}(\widehat{y}^*)}{\widehat{\text{Var}}(y^*)} = \frac{\widehat{\text{Var}}(\widehat{y}^*)}{\widehat{\text{Var}}(\widehat{y}^*) + \text{Var}(\varepsilon)}$$

In models for categorical outcomes, $\text{Var}(\varepsilon)$ is assumed to identify the model.

Count and adjusted count R^2 Observed and predicted values can be used in models with categorical outcomes to compute what is known as the count R^2. Consider the binary case where the observed y is 0 or 1 and $\pi_i = \widehat{\Pr}(y = 1 \mid \mathbf{x}_i)$. Define the expected outcome as

$$\widehat{y}_i = \begin{cases} 0 & \text{if } \widehat{\pi}_i \leq 0.5 \\ 1 & \text{if } \widehat{\pi}_i > 0.5 \end{cases}$$

This allows us to construct a table of observed and predicted values, such as that produced for the logit model by the Stata command `lstat`:

```
. lstat
Logistic model for lfp
                     ------- True -------
Classified |       D             ~D    |      Total
-----------+------------------------------+----------
     +     |      342           145     |        487
     -     |       86           180     |        266
-----------+------------------------------+----------
   Total   |      428           325     |        753

Classified + if predicted Pr(D) >= .5
True D defined as lfp != 0
--------------------------------------------------
Sensitivity                    Pr( +| D)    79.91%
Specificity                    Pr( -|~D)    55.38%
Positive predictive value      Pr( D| +)    70.23%
Negative predictive value      Pr(~D| -)    67.67%
--------------------------------------------------
False + rate for true ~D       Pr( +|~D)    44.62%
False - rate for true D        Pr( -| D)    20.09%
False + rate for classified +  Pr(~D| +)    29.77%
False - rate for classified -  Pr( D| -)    32.33%
--------------------------------------------------
Correctly classified                        69.32%
```

From this output, we can see that positive responses were predicted for 487 observations, of which 342 of these were correctly classified because the observed response was positive $(y = 1)$, while the other 145 were incorrectly classified because the observed response was negative $(y = 0)$. Likewise, of the 266 observations for which a negative response was predicted, 180 were correctly classified, and 86 were incorrectly classified.

A seemingly appealing measure is the proportion of correct predictions, referred to as the *count* R^2,

$$R^2_{\text{Count}} = \frac{1}{N} \sum_j n_{jj}$$

where the n_{jj}s are the number of correct predictions for outcome j. The count R^2 can give the faulty impression that the model is predicting very well. In a binary model *without* knowledge about the independent variables, it is possible to correctly predict at least 50 percent of the cases by choosing the outcome category with the largest percentage of observed cases. To adjust for the largest row marginal,

$$R^2_{\text{AdjCount}} = \frac{\sum_j n_{jj} - \max_r (n_{r+})}{N - \max_r (n_{r+})}$$

where n_{r+} is the marginal for row r. The *adjusted count R^2* is the proportion of correct guesses beyond the number that would be correctly guessed by choosing the largest marginal.

Information measures

This class of measures can be used to compare models across different samples or to compare non-nested models.

AIC   Akaike's (1973) information criterion is defined as AIC $= \{-2 \ln \widehat{L}(M_k) + 2P\}/N$, where $\widehat{L}(M_k)$ is the likelihood of the model and P is the number of parameters in the model (e.g., $K+1$ in the binary regression model where K is the number of regressors). All else being equal, the model with the smaller AIC is considered the better-fitting model. Some authors define AIC as being N times the value we report; see, e.g., the `mlfit` add-on command by Tobias and Campbell (1998). We report this quantity as `AIC*n`.

BIC **and** BIC′   The Bayesian information criterion has been proposed by Raftery (1996 and the literature cited therein) as a measure of overall fit and a means to compare nested and non-nested models. Consider the model M_k with deviance $D(M_k)$. BIC is defined as

$$\text{BIC}_k = D(M_k) - df_k \ln N$$

where df_k is the degrees of freedom associated with the deviance. The more negative the BIC_k, the better the fit. A second version of BIC is based on the LR chi-squared with df'_k equal to the number of regressors (not parameters) in the model. Then,

$$\text{BIC}'_k = -G^2(M_k) + df'_k \ln N$$

The more negative the BIC'_k, the better the fit. The difference in the BICs from two models indicates which model is more likely to have generated the observed data. Since $\text{BIC}_1 - \text{BIC}_2 = \text{BIC}'_1 - \text{BIC}'_2$, the choice of which BIC measure to use is a matter of convenience. If $\text{BIC}_1 - \text{BIC}_2 < 0$, then the first model is preferred. If $\text{BIC}_1 - \text{BIC}_2 > 0$, then the second model is preferred. Raftery (1996) suggested guidelines for the strength of evidence favoring M_2 against M_1 based on a difference in BIC or BIC′:

Absolute Difference	Evidence
0-2	Weak
2-6	Positive
6-10	Strong
>10	Very Strong

3.5 Interpretation

Models for categorical outcomes are nonlinear. Understanding the implications of nonlinearity is fundamental to the proper interpretation of these models. In this section we begin with a heuristic discussion of the idea of nonlinearity and the implications of nonlinearity for the proper interpretation of these models. We then introduce a set of commands that facilitate proper interpretation. Later chapters contain the details for specific models.

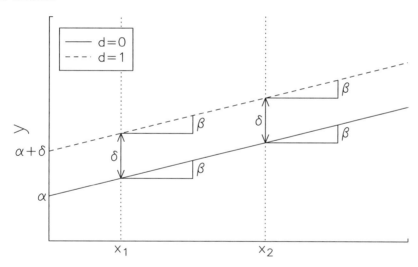

Figure 3.1: A simple linear model.

Linear models

Figure 3.1 shows a simple, linear regression model, where y is the dependent variable, x is a continuous independent variable, and d is a binary independent variable. The model being fitted is

$$y = \alpha + \beta x + \delta d$$

where for simplicity we assume that there is no error term. The solid line plots y as x changes holding $d = 0$; that is, $y = \alpha + \beta x$. The dashed line plots y as x changes when $d = 1$, which has the effect of changing the intercept: $y = \alpha + \beta x + \delta 1 = (\alpha + \delta) + \beta x$.

The effect of x on y can be computed as the partial derivative or slope of the line relating x to y, often called the *marginal effect* or *marginal change*:

$$\frac{\partial y}{\partial x} = \frac{\partial \left(\alpha + \beta x + \delta d\right)}{\partial x} = \beta$$

This equation is the ratio of the change in y to the change in x, when the change in x is infinitely small, holding d constant. In a linear model, the marginal is the same at *all* values of x and d. Consequently, when x increases by one unit, y increases by β units regardless of the current values for x and d. This is shown by the four small triangles with bases of length 1 and heights of β.

The effect of d cannot be computed with a partial derivative because d is discrete. Instead, we measure the *discrete change* in y as d changes from 0 to 1, holding x constant:

$$\frac{\Delta y}{\Delta d} = (\alpha + \beta x + \delta\, 1) - (\alpha + \beta x + \delta\, 0) = \delta$$

When d changes from 0 to 1, y changes by δ units regardless of the level of x. This is shown by the two arrows marking the distance between the solid and dashed lines. As a consequence of the linearity of the model, the discrete change equals the partial change in linear models.

The distinguishing feature of interpretation in the LRM *is that the effect of a given change in an independent variable is the same regardless of the value of that variable at the start of its change and regardless of the level of the other variables in the model.* That is, interpretation only needs to specify which variable is changing and by how much, and specify that all other variables are being held constant.

Given the simple structure of linear models, such as `regress`, most interpretations only require reporting the estimates. In some cases, it is useful to standardize the coefficients, which can be obtained with `listcoef` as discussed earlier.

(Continued on next page)

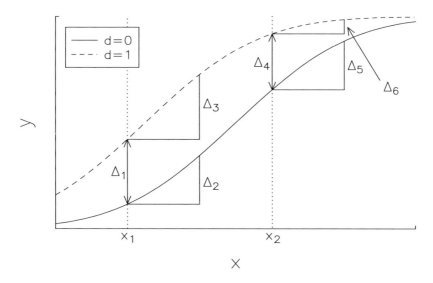

Figure 3.2: A simple nonlinear model.

Nonlinear models

Figure 3.2 plots a logit model where $y = 1$ if the outcome event occurs, say, if a person is in the labor force, else $y = 0$. The curves are from the logit equation[8]:

$$\Pr\left(y = 1\right) = \frac{\exp\left(\alpha + \beta x + \delta d\right)}{1 + \exp\left(\alpha + \beta x + \delta d\right)} \tag{3.3}$$

Once again, x is continuous, and d is binary.

The nonlinearity of the model makes it more difficult to interpret the effects of x and d on the probability of an event occurring. For example, neither the marginal nor the discrete change with respect to x is constant:

$$\frac{\partial \Pr\left(y = 1\right)}{\partial x} \neq \beta$$
$$\frac{\Delta \Pr\left(y = 1\right)}{\Delta d} \neq \delta$$

This is illustrated by the triangles. Because the solid curve for $d = 0$ and the dashed curve for $d = 1$ are not parallel, $\Delta_1 \neq \Delta_4$. And, the effect of a unit change in x differs according to the level of both d and x: $\Delta_2 \neq \Delta_3 \neq \Delta_5 \neq \Delta_6$. *In nonlinear models the effect of a change in a variable depends on the values of all variables in the model and is no longer simply equal to one of the parameters of the model.*

---

[8]The α, β, and δ parameters in this equation are unrelated to those in Figure 3.1.

3.5.1   Approaches to interpretation

There are several general approaches for interpreting nonlinear models:

1. Predictions can be computed for each observation in the sample using `predict`.

2. The marginal or discrete change in the outcome can be computed at a representative value of the independent variables using `prchange`.

3. Predicted values for substantively meaningful "profiles" of the independent variables can be compared using `prvalue`, `prtab`, or `prgen`.

4. The nonlinear model can be transformed to a model that is linear in some other outcome. As we discuss in Chapter 4, the logit model in (3.3) can be written as

$$\ln\left\{\frac{\Pr(y=1)}{1-\Pr(y=1)}\right\} = \alpha + \beta x + \delta d$$

which can then be interpreted with methods for linear models or the exponential of the coefficients can be interpreted in terms of factor changes in the odds.

The first three of these methods are now considered. Details on using these approaches for specific models are given in later chapters.

3.5.2   Predictions using predict

`predict` can be used to compute predicted values for each observation in the current dataset. While `predict` is a powerful command with many options, we consider only the simplest form of the command that provides all the details that we need. For additional options, you can enter `help predict`. The simplest syntax for `predict` is

predict *newvarname*

where *newvarname* is the name or names of the new variables that are being generated. The quantity computed for *newvarname* depends on the model that was fitted, and the number of new variables created depends on the model. The defaults are listed in the following table.

Estimation Quantity	Command Computed
regress	Predicted value $\widehat{y} = \mathbf{x}\widehat{\beta}$
logit, logistic, probit, cloglog	Predicted probability $\widehat{\Pr}(y=1)$
ologit, oprobit, clogit, mlogit	Predicted probabilities $\widehat{\Pr}(y=k)$
poisson, nbreg, zip, zinb	Predicted count or rate

In the following example, we generate predicted probabilities for a logit model of women's labor force participation. The values of `pr1` generated by `predict` are the probabilities of a woman being in the labor force for each of the observations in the dataset:

```
. logit lfp k5 k618 age wc hc lwg inc
(output omitted)

. predict pr1
(option p assumed; Pr(lfp))

. sum pr1
```

Variable	Obs	Mean	Std. Dev.	Min	Max
pr1	753	.5683931	.1944213	.0139875	.9621198

The summary statistics show that the predicted probabilities in the sample range from .013 to .962, with an average probability of .568. Further discussion of predicted probabilities for the logit model is provided in Chapter 4.

For models for ordinal or nominal categories, `predict` computes the predicted probability of an observation falling into each of the outcome categories. So, instead of specifying one variable for predictions, you specify as many names as there are categories. For example, after fitting a model for a nominal dependent variable with four categories, you can type `predict pr1 pr2 pr3 pr4`. These new variables contain the predicted probabilities of being in the first, second, third, and fourth categories, as ordered from the lowest value of the dependent variable to the highest.

For count models, `predict` computes predicted counts for each observation. Our command `prcounts` computes the predicted probabilities of observing specific counts, e.g., $\Pr(y=3)$ and the cumulative probabilities, e.g., $\Pr(y \leq 3)$. Further details are given in Chapter 7.

3.5.3 Overview of prchange, prgen, prtab, and prvalue

We have written the post-estimation commands `prchange`, `prgen`, `prtab`, and `prvalue` to make it simple to compute specific predictions that are useful for interpreting models for categorical and count outcomes. Details on installing these programs are given in Chapter 1.

`prchange` computes discrete and marginal changes in the predicted outcomes.

`prgen` computes predicted outcomes as a single independent variable changes over a specified range, holding other variables constant. New variables containing these values are generated and can then be plotted. `prgen` is limited in that it cannot handle complex model specifications in which a change in the value of the key independent variable implies a change in another independent variable, such as in models that include terms for both age and age squared. For these models, we have created the more general, but more complex, command `praccum`, which we describe in Chapter 8.

`prtab` creates a table of predicted outcomes for a cross-classification of up to four categorical independent variables, while other independent variables are held at specified values.

`prvalue` computes predicted values of the outcomes for specified values of the independent variables, and can compute differences in predictions for two sets of values. `prvalue` is the most basic command. Indeed, it can be used to compute all of the quantities except marginal change from the next three commands.

The most effective interpretation involves using all of these commands in order to discover the most convincing way to convey the predictions from the model. This section is intended to give you an overview of the syntax and options for these commands. Many of the specific details might only be clear after you read the more detailed discussions in later chapters.

Specifying the levels of variables

Each command computes predicted values for the last regression model that was fitted. To compute predicted values, you must specify values for all of the independent variables in the regression. By default, all variables are set to their means in the estimation sample.[9] Using the `x()` and `rest()` options, variables can be assigned to specific values or to a sample statistic computed from the data in memory.

`x(`*variable1*`=`*value1* [...]`)` assigns *variable1* to *value1*, *variable2* to *value2*, and so on. While equal signs are optional, they make the commands easier to read. You can assign values to as many or as few variables as you want. The assigned value is either a specific *number* (e.g., `female=1`) or a *mnemonic* specifying the descriptive statistic (e.g., `phd=mean` to set variable `phd` to the mean; `pub3=max` to assign `pub3` to the maximum value). Details on the mnemonics that can be used are given below.

`rest(`*stat*`)` sets the values of all variables not specified in `x()` to the sample statistic indicated by *stat*. For example, `rest(mean)` sets all variables to their mean. If `x()` is not specified, all variables are set to *stat*. The value of *stat* can be calculated for the whole sample or can be conditional based on the values specified by `x()`. For example, if `x(female=1)` is specified, `rest(grmean)` specifies that all other variables should equal their mean in the sample defined by `female=1`. This is referred to as a group statistic (i.e., statistics that begin with *gr*). If you specify a group statistic for `rest()`, only numeric values can be used for `x()`. For example, `x(female=mean) rest(grmean)` is not allowed. If `rest()` is not specified, it is assumed to be `rest(mean)`.

---

[9]The estimation sample includes only those cases that were used in fitting a model. Cases that were dropped due to missing values or `if` and `in` conditions are not part of the estimation sample.

The statistics that can be used with x() and rest() are

mean, median, min, and max specify the unconditional mean, median, minimum, and
maximum. By default, the estimation sample is used to compute these statistics.
If the option all is specified, all cases in memory are used for computing descriptive statistics, regardless of whether they were used in the estimation. if or in
conditions can also be used. For example, adding if female==1 to any of these
commands restricts the computations of descriptive statistics to only women, even
if the estimation sample included men and women.

previous sets values to what they were the last time the command was called; this
option can only be used if the set of independent variables is the same in both
cases. This can be useful if you want to change the value of only one variable from
the last time the command was used.

upper and lower set values to those that yield the maximum or minimum predicted
values, respectively. These options can only be used for binary models.

grmean, grmedian, grmin, and grmax compute statistics that are conditional on the
group specified in x(). For example, x(female=0) rest(grmean) sets female to
0 and all other variables to the means *of the subsample* in which female is 0 (i.e.,
the means of these other variables for male respondents).

Options controlling output

nobase suppresses printing of the base values of the independent variables.

brief provides only minimal output.

all specifies that any calculations of means, medians, etc., should use the entire sample
instead of the sample used to estimate the model.

3.5.4 Syntax for prchange

prchange computes marginal and discrete change coefficients. The syntax is

prchange [*varlist*] [if *exp*] [in *range*] [, x(*variable1=value1* [...]) rest(*stat*)
outcome(#) fromto brief nobase nolabel help all uncentered
delta(#)]

varlist specifies that changes are to be listed only for these variables. By default, changes
are listed for all variables.

Options

outcome(#) specifies that changes will be printed only for the outcome indicated. For example, if ologit was run with outcome categories 1, 2, and 3, outcome(1) requests that only changes in the probability of outcome 1 should be listed. For ologit, oprobit, and mlogit, the default is to provide results for all outcomes. For the count models, the default is to present results with respect to the predicted rate; specifying an outcome number will provide changes in the probability of that outcome.

fromto specifies that the starting and ending probabilities from which the discrete change is calculated for prchange should also be displayed.

nolabel uses values rather than value labels in the output.

help provides information explaining the output.

uncentered specifies that the uncentered discrete change, rather than the centered discrete change, is to be computed. By default, the change in an independent variable is centered on its value.

delta(#) specifies the amount of the discrete change in the independent variable. The default is a 1-unit change (i.e., delta(1)).

3.5.5   Syntax for prgen

prgen computes a variable containing predicted values as one variable changes over a range of values, which is useful for constructing plots. The syntax is

prgen *varname* [if *exp*] [in *range*] , generate(*prefix*) [from(#) to(#)
    ncases(#) x(*variable1=value1*[...]) rest(*stat*) maxcnt(#) brief all]

Options

varname is the name of the variable that changes while all other variables are held at specified values.

generate(*prefix*) sets the prefix for the new variables created by prgen. Choosing a prefix that is different from the beginning letters of any of the variables in your dataset makes it easier to examine the results. For example, if you choose the prefix abcd then you can use the command sum abcd* to examine all newly created variables.

from(#) and to(#) are the start and end values for *varname*. The default is for *varname* to range from the observed minimum to the observed maximum of *varname*.

ncases(#) specifies the number of predicted values prgen computes as *varname* varies from the start value to the end value. The default is 11.

maxcnt(*#*) is the maximum count value for which a predicted probability is computed for count models. The default is 9.

Variables generated

prgen constructs variables that can be graphed. The observations contain predicted values and probabilities for a range of values for the variable *varname*, holding the other variables at the specified values. *n* observations are created, where *n* is 11 by default or specified by ncases(). The new variables all start with the *prefix* specified by gen(). The variables created are

For which models	Name	Content
All models	*prefix*x	The values of *varname* from from(*#*) to to(*#*).
logit, probit	*prefix*p0	Predicted probability $\Pr(y = 0)$.
	*prefix*p1	Predicted probability $\Pr(y = 1)$.
ologit, oprobit	*prefix*pk	Predicted probability $\Pr(y = k)$ for all outcomes.
	*prefix*sk	Cumulative probability $\Pr(y \leq k)$ for all outcomes.
mlogit	*prefix*pk	Predicted probability $\Pr(y = k)$ for all outcomes.
poisson, nbreg, zip, zinb	*prefix*mu	Predicted rate μ.
	*prefix*pk	Predicted probability $\Pr(y = k)$, for $0 \leq k \leq$ maxcnt().
	*prefix*sk	Cumulative probability $\Pr(y \leq k)$, for $0 \leq k \leq$ maxcnt().
zip, zinb	*prefix*inf	Predicted probability $\Pr(\text{Always } 0= 1) = \Pr(inflate)$.
regress, tobit, cnreg, intreg	*prefix*xb	Predicted value of *y*.

3.5.6 Syntax for prtab

prtab constructs a table of predicted values for all combinations of up to three variables. The syntax is

prtab *rowvar* [*colvar* [*supercolvar*]] [if *exp*] [in *range*] [, by(*superrowvar*)

   x(*variable1=value1* [...]) <u>r</u>est(*stat*) <u>o</u>utcome(*#*) <u>no</u>base <u>nol</u>abel <u>nova</u>rlbl

   <u>b</u>rief all]

Options

rowvar, *colvar*, *supercolvar*, and *superrowvar* are independent variables from the previously fitted model. These define the table that is constructed.

by(*superrowvar*) specifies the categorical independent variable that is to be used to form the superrows of the table.

outcome(*#*) specifies that changes be printed only for the outcome indicated. The default for ologit, oprobit, and mlogit is to provide results for all outcomes. For the count models, the default is to present results with respect to the predicted rate; specifying an outcome number provides changes in the probability of that outcome.

nolabel uses a variable's numerical values rather than value labels in the output. Sometimes this is more readable.

novarlbl uses a variable's name rather than the variable label in the output. Sometimes this is more readable.

3.5.7 Syntax for prvalue

prvalue computes the predicted outcome for a single set of values of the independent variables. The syntax is

prvalue [if *exp*] [in *range*] [, x(*variable1=value1*[...]) rest(*stat*) level(#)
    maxcnt(#) save dif ystar nobase nolabel brief all]

Options

level(#) specifies the confidence level, as a percentage, for confidence intervals. The default is level(95) or as set by the Stata command set level.

maxcnt(#) is the maximum count value for which a predicted probability is computed for count models. The default is 9.

save preserves predicted values computed by prvalue for subsequent comparison.

dif compares predicted values computed by prvalue with those previously preserved with the save option.

ystar prints the predicted value of y^* for binary and ordinal models.

nolabel uses values rather than value labels in output.

3.5.8 Computing marginal effects using mfx compute

Stata 7 introduced the mfx command for calculating marginal effects. Recall from above that the marginal effect is the partial derivative of y with respect to x_k. For nonlinear models, the value of the marginal depends on the specific values of all the independent variables. After fitting a model, mfx compute will compute the marginal effects for all the independent variables, evaluated at values that are specified using the at() option. at() is similar to the x() and rest() syntax used in our commands. To compute the marginal effects while holding age at 40 and female at 0, the command is mfx compute, at(age=40 female=0). As with our commands for working with predicted values, unspecified independent variables are held at their mean by default.

   mfx has several features that make it worth exploring. For one, it works after many different estimation commands. For dummy independent variables, mfx computes the discrete change rather than the marginal effect. Of particular interest for economists, the command optionally computes elasticities instead of marginal effects. And, mfx also computes standard errors for the effects. The derivatives are calculated numerically, which means that the command can take a long time to execute when there are many independent variables and observations, especially when used with `mlogit`. While we do not provide further discussion of `mfx` in this book, readers who are interested in learning more about this command are encouraged to examine its entry in the *Base Reference Manual*.

3.6   Next steps

This concludes our discussion of the basic commands and options that are used for fitting, testing, assessing fit, and interpreting regression models. In the next four chapters we illustrate how each of the commands can be applied for models relevant to one particular type of outcome. While Chapter 4 has somewhat more detail than later chapters, readers should be able to proceed from here to any of the chapters that follow.

Part II

Models for Specific Kinds of Outcomes

In Part II, we provide information on the models appropriate for different kinds of dependent outcomes.

- **Chapter 4** considers *binary outcomes*. Models for binary outcomes are the most basic type that we consider, and to some extent, they provide a foundation for the models in later chapters. For this reason, Chapter 4 has more detailed explanations, and we recommend that all readers review this chapter, even if they are mainly interested other types of outcomes. We show how to fit the binary regression model, how to test hypotheses, how to compute residuals and influence statistics, and how to calculate scalar measures of model fit. Following this, we focus on interpretation, describing how these models can be interpreted using predicted probabilities, discrete and marginal change in these probabilities, and odds ratios.

Chapters 5, 6, and 7 can be read in any combination or order, depending on the reader's interests. Each chapter provides information on fitting the relevant models, testing hypotheses about the coefficients, and interpretation in terms of predicted probabilities. In addition,

- **Chapter 5** on *ordered outcomes* describes the parallel regression assumption that is made by the ordered logit and probit models and shows how this assumption can be tested. We also discuss interpretation in terms of the underlying latent variable and odds ratios.

- **Chapter 6** on *nominal outcomes* introduces the multinomial logit model. We show how to test the assumption of the independence of irrelevant alternatives and present two graphical methods of interpretation. We conclude by introducing the conditional logit model.

- **Chapter 7** on *count outcomes* presents the Poisson and negative binomial regression models. We show how to test the Poisson model's assumption of equidispersion and how to incorporate differences in exposure time into the models. We also describe versions of these models for data with a high frequency of zero counts.

- **Chapter 8** covers additional topics that extend material presented earlier. We discuss the use and interpretation of categorical independent variables, interactions, and nonlinear terms. We also provide tips on how to use Stata more efficiently and effectively.

4   Models for Binary Outcomes

Regression models for binary outcomes are the foundation from which more complex models for ordinal, nominal, and count models can be derived. Ordinal and nominal regression models are equivalent to the simultaneous estimation of a series of binary outcomes. While the link is less direct in count models, the Poisson distribution can be derived as the outcome of a large number of binary trials. More importantly for our purposes, the zero-inflated count models that we discuss in Chapter 7 merge a binary logit or probit with a standard Poisson or negative binomial model. Consequently, the principles of fitting, testing, and interpreting binary models provide tools that can be readily adapted to models in later chapters. Thus, while each chapter is largely self-contained, this chapter provides somewhat more detailed explanations than later chapters. As a result, even if your interests are in models for ordinal, nominal, or count outcomes, we think that you will benefit from reading this chapter.

Binary dependent variables have two values, typically coded as 0 for a negative outcome (i.e., the event did not occur) and 1 as a positive outcome (i.e., the event did occur). Binary outcomes are ubiquitous, and examples come easily to mind. Did a person vote? Is a manufacturing firm unionized? Does a person identify as a feminist or nonfeminist? Did a start-up company go bankrupt? Five years after a person was diagnosed with cancer, is he or she still alive? Was a purchased item returned to the store or kept?

Regression models for binary outcomes allow a researcher to explore how each explanatory variable affects the probability of the event occurring. We focus on the two most frequently used models, the binary logit and binary probit models, referred to jointly as the *binary regression model* (BRM). Because the model is nonlinear, the magnitude of the change in the outcome probability that is associated with a given change in one of the independent variables depends on the levels of all of the independent variables. The challenge of interpretation is to find a summary of the way in which changes in the independent variables are associated with changes in the outcome that best reflect the key substantive processes without overwhelming yourself or your readers with distracting detail.

The chapter begins by reviewing the mathematical structure of binary models. We then examine statistical testing and fit, and finally, methods of interpretation. These discussions are intended as a review for those who are familiar with the models. For a complete discussion, see Long (1997). You can obtain sample do-files and data files that reproduce the examples in this chapter by downloading the `spostst4` package (see Chapter 1 for details).

4.1   The statistical model

There are three ways to derive the BRM, with each method leading to the same mathematical model. First, an unobserved or latent variable can be hypothesized along with a measurement model relating the latent variable to the observed, binary outcome. Second, the model can be constructed as a probability model. Third, the model can be generated as a random utility or discrete choice model. This last approach is not considered in our review; see Long (1997, 155–156) for an introduction or Pudney (1989) for a detailed discussion.

4.1.1   A latent variable model

Assume a *latent* or unobserved variable y^* ranging from $-\infty$ to ∞ that is related to the observed independent variables by the structural equation,

$$y_i^* = \mathbf{x}_i \beta + \varepsilon_i$$

where i indicates the observation and ε is a random error. For a single independent variable, we can simplify the notation to

$$y_i^* = \alpha + \beta x_i + \varepsilon_i$$

These equations are identical to those for the linear regression model with the important difference that the dependent variable is unobserved.

The link between the observed binary y and the latent y^* is made with a simple measurement equation:

$$y_i = \begin{cases} 1 & \text{if } y_i^* > 0 \\ 0 & \text{if } y_i^* \leq 0 \end{cases}$$

Cases with positive values of y^* are observed as $y=1$, while cases with negative or zero values of y^* are observed as $y=0$.

Imagine a survey item that asks respondents if they agree or disagree with the proposition that "a working mother can establish just as warm and secure a relationship with her children as a mother who does not work". Obviously, respondents vary greatly in their opinions on this issue. Some people very adamantly agree with the proposition, some very adamantly disagree, and still others have only weak opinions one way or the other. We can imagine an underlying continuum of possible responses to this item, with every respondent having some value on this continuum (i.e., some value of y^*). Those respondents whose value of y^* is positive answer "agree" to the survey question ($y = 1$), and those whose value of y^* is 0 or negative answer "disagree" ($y = 0$). A shift in a respondent's opinion might move them from agreeing strongly with the position to agreeing weakly with the position, which would not change the response we observe. Or, the respondent might move from weakly agreeing to weakly disagreeing, in which case we would observe a change from $y = 1$ to $y = 0$.

Consider a second example, which we use throughout this chapter. Let $y = 1$ if a woman is in the paid labor force and $y = 0$ if she is not. The independent variables include variables such as number of children, education, and expected wages. Not all women in the labor force ($y = 1$) are there with the same certainty. One woman might be close to leaving the labor force, while another woman could be firm in her decision to work. In both cases, we observe $y = 1$. The idea of a latent y^* is that an underlying *propensity to work* generates the observed state. Again, while we cannot directly observe the propensity, at some point a change in y^* results in a change in what we observe, namely, whether the woman is in the labor force.

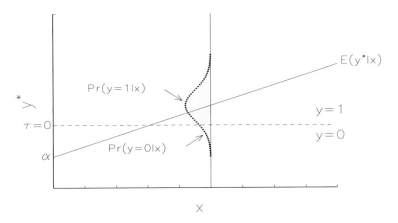

Figure 4.1: Relationship between latent variable y^* and $\Pr(y = 1)$ for the BRM.

The latent variable model for binary outcomes is illustrated in Figure 4.1 for a single independent variable. For a given value of x, we see that

$$\Pr(y = 1 \mid x) = \Pr(y^* > 0 \mid x)$$

Substituting the structural model and rearranging terms,

$$\Pr(y = 1 \mid x) = \Pr(\varepsilon > -[\alpha + \beta x] \mid x) \tag{4.1}$$

This equation shows that the probability depends on the distribution of the error ε.

Two distributions of ε are commonly assumed, both with an assumed mean of 0. First, ε is assumed to be distributed normally with $\mathrm{Var}(\varepsilon) = 1$. This leads to the binary probit model, in which (4.1) becomes

$$\Pr(y = 1 \mid x) = \int_{-\infty}^{\alpha + \beta x} \frac{1}{\sqrt{2\pi}} \exp\left(-\frac{t^2}{2}\right) dt$$

Alternatively, ε is assumed to be distributed logistically with $\mathrm{Var}(\varepsilon) = \pi^2/3$, leading to the binary logit model with the simpler equation

$$\Pr(y = 1 \mid x) = \frac{\exp\left(\alpha + \beta x\right)}{1 + \exp\left(\alpha + \beta x\right)} \tag{4.2}$$

The peculiar value assumed for $\text{Var}(\varepsilon)$ in the logit model illustrates a basic point about the identification of models with latent outcomes. In the LRM, $\text{Var}(\varepsilon)$ can be estimated because y is observed. For the BRM, the value of $\text{Var}(\varepsilon)$ must be assumed because the dependent variable is unobserved. The model is unidentified unless an assumption is made about the variance of the errors. For probit, we assume $\text{Var}(\varepsilon) = 1$ because this leads to a simple form of the model. If a different value was assumed, this would simply change the values of the structural coefficients in a uniform way. In the logit model, the variance is set to $\pi^2/3$ because this leads to the very simple form in (4.2). While the value assumed for $\text{Var}(\varepsilon)$ is arbitrary, the value chosen does *not* affect the computed value of the probability (see Long 1997, 49–50 for a simple proof). In effect, changing the assumed variance affects the spread of the distribution but not the proportion of the distribution above or below the threshold.

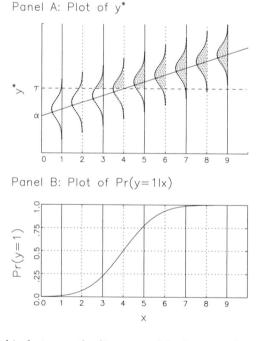

Figure 4.2: Relationship between the linear model $y^* = \alpha + \beta x + \varepsilon$ and the nonlinear probability model $\Pr(y = 1 \mid x) = F(\alpha + \beta x)$.

For both models, the probability of the event occurring is the cumulative density function (cdf) of ε evaluated at given values of the independent variables:

$$\Pr(y = 1 \mid \mathbf{x}) = F(\mathbf{x}\beta)$$

where F is the normal cdf Φ for the probit model and the logistic cdf Λ for the logit model. The relationship between the linear latent variable model and the resulting nonlinear probability model is shown in Figure 4.2 for a model with a single independent variable. Panel A shows the error distribution for nine values of x, which we have labeled 1, 2, ..., 9. The area where $y^* > 0$ corresponds to $\Pr(y = 1 \mid x)$ and has been shaded. Panel B plots $\Pr(y = 1 \mid x)$ corresponding to the shaded regions in Panel A. As we move from 1 to 2, only a portion of the thin tail crosses the threshold in Panel A, resulting in a small change in $\Pr(y = 1 \mid x)$ in Panel B. As we move from 2 to 3 to 4, thicker regions of the error distribution slide over the threshold, and the increase in $\Pr(y = 1 \mid x)$ becomes larger. The resulting curve is the well-known S-curve associated with the BRM.

4.1.2   A nonlinear probability model

Can all binary dependent variables be conceptualized as observed manifestations of some underlying latent propensity? While philosophically interesting, perhaps, the question is of little practical importance, as the BRM can also be derived without appealing to a latent variable. This is done by specifying a nonlinear model relating the xs to the probability of an event. Following Theil (1970), the logit model can be derived by constructing a model in which the predicted $\Pr(y = 1 \mid \mathbf{x})$ is forced to be within the range 0 to 1. For example, in the linear probability model,

$$\Pr(y = 1 \mid \mathbf{x}) = \mathbf{x}\beta + \varepsilon$$

the predicted probabilities can be greater than 1 and less than 0. To constrain the predictions to the range 0 to 1, we first transform the probability into the *odds*,

$$\Omega(\mathbf{x}) = \frac{\Pr(y = 1 \mid \mathbf{x})}{\Pr(y = 0 \mid \mathbf{x})} = \frac{\Pr(y = 1 \mid \mathbf{x})}{1 - \Pr(y = 1 \mid \mathbf{x})}$$

which indicate how often something happens ($y = 1$) relative to how often it does not happen ($y = 0$), and range from 0 when $\Pr(y = 1 \mid \mathbf{x}) = 0$ to ∞ when $\Pr(y = 1 \mid \mathbf{x}) = 1$. The log of the odds, or *logit*, ranges from $-\infty$ to ∞. This suggests a model that is *linear in the logit*:

$$\ln \Omega(\mathbf{x}) = \mathbf{x}\beta$$

This equation can be shown to be equivalent to the logit model from (4.2). Interpretation of this form of the logit model often focuses on factor changes in the odds, which are discussed below.

Other binary regression models are created by choosing functions of $\mathbf{x}\beta$ that range from 0 to 1. Cumulative distribution functions have this property and readily provide a number of examples. For example, the cdf for the standard normal distribution results in the probit model.

4.2   Estimation using logit and probit

Logit and probit can be estimated with the following commands:

logit *depvar* [*indepvars*] [*weight*] [if *exp*] [in *range*] [, level(#)
    noconstant or robust cluster(*varname*) score(*newvar*) nolog]

probit *depvar* [*indepvars*] [*weight*] [if *exp*] [in *range*] [, level(#)
    noconstant robust cluster(*varname*) score(*newvar*) nolog]

We have never had a problem with either of these models converging, even with small samples and data with wide variation in scaling.

Variable lists

depvar is the dependent variable. *indepvars* is a list of independent variables. If *indepvars* is not included, Stata fits a model with only an intercept.

Warning For binary models, Stata defines observations in which *depvar*= 0 as negative outcomes and observations in which *depvar* equals *any* other nonmissing value (including negative values) as positive outcomes. To avoid possible confusion, we urge you to explicitly create a 0/1 variable for use as *depvar*.

Specifying the estimation sample

if and in qualifiers can be used to restrict the estimation sample. For example, if you wanted to fit a logit model for only women who went to college (as indicated by the variable wc), you could specify logit lfp k5 k618 age hc lwg if wc==1.

Listwise deletion Stata excludes cases in which there are missing values for any of the variables in the model. Accordingly, if two models are fitted using the same dataset but have different sets of independent variables, it is possible to have different samples. We recommend that you use mark and markout (discussed in Chapter 3) to explicitly remove cases with missing data.

Weights

Both logit and probit can be used with fweights, pweights, and iweights. In Chapter 3, we provide a brief discussion of the different types of weights and how weighting variables are specified.

Options

level(*#*) specifies the level of the confidence interval. By default, Stata provides 95% confidence intervals for estimated coefficients. You can also change the default level, say to a 90% interval, with the command set level 90.

noconstant specifies that the model should not have a constant term. This would rarely be used for these models.

or reports the "odds ratios" defined as $\exp\left(\widehat{\beta}\right)$. Standard errors and confidence intervals are similarly transformed. Alternatively, our listcoef command can be used.

robust indicates that robust variance estimates are to be used. When cluster() is specified, robust standard errors are used automatically. We provide a brief general discussion of these options in Chapter 3.

cluster(*varname*) specifies that the observations are independent across the groups specified by unique values of *varname* but not necessarily within the groups.

score(*newvar*) creates *newvar* containing $u_j = \partial \ln L_j / \partial(\mathbf{x}_j \mathbf{b})$ for each observation j in the sample. The score vector is $\sum \partial \ln L_j / \partial \mathbf{b} = \sum u_j \mathbf{x}_j$; i.e., the product of *newvar* with each covariate summed over observations. See [U] **23.15 Obtaining scores**.

nolog suppresses the iteration history.

Example

Our example is from Mroz's (1987) study of the labor force participation of women, using data from the 1976 Panel Study of Income Dynamics.[1] The sample consists of 753 white, married women between the ages of 30 and 60. The dependent variable lfp equals 1 if a woman is employed and otherwise equals 0. Because we have assigned variable labels, a complete description of the data can be obtained using describe and summarize:

```
. use http://www.stata-press.com/data/lfr/binlfp2, clear
(Data from 1976 PSID-T Mroz)

. describe lfp k5 k618 age wc hc lwg inc

              storage  display   value
variable name  type    format    label      variable label
-------------------------------------------------------------------------
lfp            byte    %9.0g     lfplbl     Paid Labor Force: 1=yes 0=no
k5             byte    %9.0g                # kids < 6
k618           byte    %9.0g                # kids 6-18
age            byte    %9.0g                Wife's age in years
wc             byte    %9.0g     collbl     Wife College: 1=yes 0=no
hc             byte    %9.0g     collbl     Husband College: 1=yes 0=no
lwg            float   %9.0g                Log of wife's estimated wages
inc            float   %9.0g                Family income excluding wife's
```

---

[1]These data were generously made available by Thomas Mroz.

```
. summarize lfp k5 k618 age wc hc lwg inc
    Variable |       Obs        Mean    Std. Dev.        Min         Max
-------------+--------------------------------------------------------
         lfp |       753    .5683931    .4956295          0           1
          k5 |       753    .2377158     .523959          0           3
        k618 |       753    1.353254    1.319874          0           8
         age |       753    42.53785    8.072574         30          60
          wc |       753    .2815405    .4500494          0           1
-------------+--------------------------------------------------------
          hc |       753    .3917663    .4884694          0           1
         lwg |       753    1.097115    .5875564   -2.054124    3.218876
         inc |       753    20.12897     11.6348   -.0290001          96
```

Using these data, we fitted the model

$$\Pr\left(\mathtt{lfp}=1\right) = F(\beta_0 + \beta_\mathtt{k5}\mathtt{k5} + \beta_\mathtt{k618}\mathtt{k618} + \beta_\mathtt{age}\mathtt{age}$$
$$+ \beta_\mathtt{wc}\mathtt{wc} + \beta_\mathtt{hc}\mathtt{hc} + \beta_\mathtt{lwg}\mathtt{lwg} + \beta_\mathtt{inc}\mathtt{inc})$$

with both the `logit` and `probit` commands, and then we created a table of results with `estimates table`:

```
. logit lfp k5 k618 age wc hc lwg inc, nolog
Logit estimates                                 Number of obs   =        753
                                                LR chi2(7)      =     124.48
                                                Prob > chi2     =     0.0000
Log likelihood = -452.63296                     Pseudo R2       =     0.1209

-------------+--------------------------------------------------------------
         lfp |      Coef.   Std. Err.      z    P>|z|     [95% Conf. Interval]
-------------+--------------------------------------------------------------
          k5 |  -1.462913    .1970006    -7.43   0.000    -1.849027   -1.076799
        k618 |  -.0645707    .0680008    -0.95   0.342    -.1978499    .0687085
         age |  -.0628706    .0127831    -4.92   0.000    -.0879249   -.0378162
          wc |   .8072738    .2299799     3.51   0.000     .3565215    1.258026
          hc |   .1117336    .2060397     0.54   0.588    -.2920969     .515564
         lwg |   .6046931    .1508176     4.01   0.000     .3090961    .9002901
         inc |  -.0344464    .0082084    -4.20   0.000    -.0505346   -.0183583
       _cons |    3.18214    .6443751     4.94   0.000     1.919188    4.445092
-------------+--------------------------------------------------------------

. estimates store logit
. probit lfp k5 k618 age wc hc lwg inc, nolog
Probit estimates                                Number of obs   =        753
                                                LR chi2(7)      =     124.36
                                                Prob > chi2     =     0.0000
Log likelihood = -452.69496                     Pseudo R2       =     0.1208

-------------+--------------------------------------------------------------
         lfp |      Coef.   Std. Err.      z    P>|z|     [95% Conf. Interval]
-------------+--------------------------------------------------------------
          k5 |  -.8747112    .1135583    -7.70   0.000    -1.097281   -.6521411
        k618 |  -.0385945    .0404893    -0.95   0.340     -.117952    .0407631
         age |  -.0378235    .0076093    -4.97   0.000    -.0527375   -.0229095
          wc |   .4883144    .1354873     3.60   0.000     .2227642    .7538645
          hc |   .0571704    .1240052     0.46   0.645    -.1858754    .3002161
         lwg |   .3656287    .0877792     4.17   0.000     .1935847    .5376727
         inc |   -.020525    .0047769    -4.30   0.000    -.0298875   -.0111626
       _cons |   1.918422    .3806536     5.04   0.000     1.172355     2.66449
-------------+--------------------------------------------------------------

. estimates store probit
```

While the iteration log was suppressed by the `nolog` option, the value of the log likelihood at convergence is listed as `Log likelihood`. The information in the header and table of coefficients is in the same form as discussed in Chapter 3.

We can use `estimates table` to create a table that combines the results:

```
. estimates table logit probit, b(%9.3f) t label varwidth(30)
```

Variable	logit	probit
# kids < 6	-1.463	-0.875
	-7.43	-7.70
# kids 6-18	-0.065	-0.039
	-0.95	-0.95
Wife's age in years	-0.063	-0.038
	-4.92	-4.97
Wife College: 1=yes 0=no	0.807	0.488
	3.51	3.60
Husband College: 1=yes 0=no	0.112	0.057
	0.54	0.46
Log of wife's estimated wages	0.605	0.366
	4.01	4.17
Family income excluding wife's	-0.034	-0.021
	-4.20	-4.30
Constant	3.182	1.918
	4.94	5.04

legend: b/t

The estimated coefficients differ from logit to probit by a factor of about 1.7. For example, the ratio of the logit to probit coefficient for `k5` is 1.67 and for `inc` is 1.68. This illustrates how the magnitudes of the coefficients are affected by the assumed $\text{Var}(\varepsilon)$. The exception to the ratio of 1.7 is the coefficient for `hc`. This estimate has a great deal of sampling variability (i.e., a large standard error), and in such cases, the 1.7 rule often does not hold. Values of the z-tests are quite similar because they are not affected by the assumed $\text{Var}(\varepsilon)$. The z-test statistics are not exactly the same because the two models assume different distributions of the errors.

4.2.1   Observations predicted perfectly

ML estimation is not possible when the dependent variable does not vary within one of the categories of an independent variable. For example, say that you are fitting a logit model predicting whether a person voted in the last election, `vote`, and that one of the independent variables is whether the person is enrolled in college, `college`. If you had a small number of college students in your sample, it is possible that none of them voted in the last election. That is, `vote==0` every time `college==1`. The model cannot be fitted because the coefficient for `college` is effectively negative infinity. Stata's solution is to drop the variable `college` along with all observations where `college==1`. For example,

```
. logit vote college phd, nolog
Note: college!=0 predicts failure perfectly
      college dropped and 4 obs not used

Logit estimates                          Number of obs   =        299
   (output omitted)
```

4.3   Hypothesis testing with test and lrtest

Hypothesis tests of regression coefficients can be conducted with the z-statistics in the estimation output, with `test` for Wald tests of simple and complex hypotheses, and with `lrtest` for the corresponding likelihood-ratio tests. We consider the use of each of these to test hypotheses involving only one coefficient, and then we show you how both `test` and `lrtest` can be used to test hypotheses involving multiple coefficients.

4.3.1   Testing individual coefficients

If the assumptions of the model hold, the ML estimators (e.g., the estimates produced by `logit` or `probit`) are distributed asymptotically normally:

$$\widehat{\beta}_k \overset{a}{\sim} \mathcal{N}\left(\beta_k, \sigma^2_{\widehat{\beta}_k}\right)$$

The hypothesis $H_0\colon \beta_k = \beta^*$ can be tested with the z-statistic:

$$z = \frac{\widehat{\beta}_k - \beta^*}{\widehat{\sigma}_{\widehat{\beta}_k}}$$

z is included in the output from `logit` and `probit`. Under the assumptions justifying ML, if H_0 is true, then z is distributed approximately normally with a mean of zero and a variance of one for large samples. This is shown in the following figure, where the shading shows the rejection region for a two-tailed test at the .05 level:

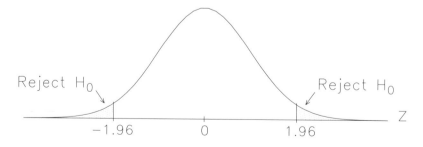

For example, consider the results for variable `k5` from the `logit` output generated in Section 4.2:

lfp	Coef.	Std. Err.	z	P>|z|	[95% Conf. Interval]
k5	-1.462913	.1970006	-7.43	0.000	-1.849027 -1.076799

(output omitted)

We conclude that

> having young children has a significant effect on the probability of working
> ($z = -7.43$, $p < 0.01$ for a two-tailed test).

One and two-tailed tests

The probability levels in the output for estimation commands are for two-tailed tests.
That is, the result corresponds to the area of the curve that is either greater than $|z|$
or less than $-|z|$. When past research or theory suggests the sign of the coefficient,
a one-tailed test can be used, and H_0 is only rejected when z is in the *expected* tail.
For example, assume that my theory proposes that having children can only have a
negative effect on labor force participation. For k618, $z = -0.950$ and P > | z | is
.342. This is the proportion of the sampling distribution for z that is less than -0.950
or greater than 0.950. Because we want a one-tailed test, and the coefficient is in the
expected direction, we only want the proportion of the distribution that is less than
-0.950, which is $.342/2 = .171$. We conclude that

> having older children does not significantly affect a woman's probability of
> working ($z = -0.95$, $p = .17$ for a one-tailed test).

You should only divide P > | z | by 2 when the estimated coefficient is in the
expected direction. For example, suppose I am testing a theory that having a husband
who went to college has a negative effect on labor force participation, but the estimated
coefficient is *positive* with $z = 0.542$ and P > | z | is .588. The one-tailed significance
level would be the percent of the distribution less than .542 (not the percent less than
$-.542$), which is equal to $1 - (.588/2) = .706$, not $.588/2 = .294$. We conclude that

> having a husband who attends college does not significantly affect a woman's
> probability of working ($z = 0.542$, $p = .71$ for a one-tailed test).

Testing single coefficients using test

The z-test included in the output of estimation commands is a Wald test, which can
also be computed using test. For example, to test $H_0: \beta_{k5} = 0$,

```
. test k5

 ( 1)   k5 = 0

           chi2(  1) =    55.14
         Prob > chi2 =    0.0000
```

We can conclude that

> The effect of having young children on the probability of entering the labor
> force is significant at the .01 level ($X^2 = 55.14$, $df = 1$, $p < .01$)

The value of a chi-squared test with 1 degree of freedom is identical to the square of
the corresponding z-test. For example, using Stata's `display` as a calculator

```
. display sqrt(55.14)
7.4256313
```

This corresponds to -7.426 from the `logit` output. Some packages, such as SAS, present
chi-squared tests rather than the corresponding z-test.

Testing single coefficients using lrtest

An LR test is computed by comparing the log likelihood from a full model with that of
a restricted model. To test a single coefficient, we begin by fitting the full model and
storing the results:

```
. logit lfp k5 k618 age wc hc lwg inc, nolog
Logit estimates                                Number of obs   =        753
                                               LR chi2(7)      =     124.48
                                               Prob > chi2     =     0.0000
Log likelihood = -452.63296                    Pseudo R2       =     0.1209
    (output omitted)
. estimates store fmodel
```

Then, we fit the model without k5 and run `lrtest`:

```
. logit lfp k618 age wc hc lwg inc, nolog
Logit estimates                                Number of obs   =        753
                                               LR chi2(6)      =      58.00
                                               Prob > chi2     =     0.0000
Log likelihood = -485.87503                    Pseudo R2       =     0.0563
    (output omitted)
. estimates store nmodel

. lrtest fmodel nmodel
likelihood-ratio test                          LR chi2(1)   =      66.48
(Assumption: nmodel nested in fmodel)          Prob > chi2 =     0.0000
```

The resulting LR test can be interpreted as indicating that

> the effect of having young children is significant at the .01 level
> ($LRX^2 = 66.48$, $df = 1$, $p < .01$).

4.3.2 Testing multiple coefficients

In many situations, you may wish to test complex hypotheses that involve more than one coefficient. For example, we have two variables that reflect education in the family, hc and wc. The conclusion that education has (or does not have) a significant effect on labor force participation cannot be based on a pair of tests of single coefficients. But, a joint hypothesis can be tested using either test or lrtest.

Testing multiple coefficients using test

To test that the effect of the wife attending college and of the husband attending college on labor force participation are both equal to 0, H_0: $\beta_{wc} = \beta_{hc} = 0$, we fit the full model and then

```
. test hc wc
 ( 1)   hc = 0
 ( 2)   wc = 0
           chi2(  2) =    17.66
         Prob > chi2 =     0.0001
```

We conclude that

> the hypothesis that the effects of the husband's and the wife's education are simultaneously equal to zero can be rejected at the .01 level ($X^2 = 17.66, df = 2, p < .01$).

This form of the test command can be readily extended to hypotheses regarding more than two independent variables by listing more variables; for example, test wc hc k5.

test can also be used to test the equality of coefficients. For example, to test that the effect of the wife attending college on labor force participation is equal to the effect of the husband attending college, H_0: $\beta_{wc} = \beta_{hc}$:

```
. test hc=wc
 ( 1)  - wc + hc = 0
           chi2(  1) =     3.54
         Prob > chi2 =     0.0600
```

Note that test has translated $\beta_{wc} = \beta_{hc}$ into the equivalent expression $-\beta_{wc} + \beta_{hc} = 0$. We conclude that

> the null hypothesis that the effects of husband's and wife's education are equal is marginally significant at the .05 level ($X^2 = 3.54, df = 1, p = .06$). This suggests that we have weak evidence that the effects are not equal.

Testing multiple coefficients using lrtest

To compute an LR test of multiple coefficients, we first fit the full model and then save the results using the command: `estimates store`. Then, to test the hypothesis that the effect of the wife attending college and of the husband attending college on labor force participation are both equal to zero, H_0: $\beta_{\mathtt{wc}} = \beta_{\mathtt{hc}} = 0$, we fit the model that excludes these two variables and then run `lrtest`:

```
. logit lfp k5 k618 age wc hc lwg inc, nolog
  (output omitted)
. estimates store fmodel
. logit lfp k5 k618 age lwg inc, nolog
  (output omitted)
. estimates store nmodel
. lrtest fmodel nmodel
likelihood-ratio test                                   LR chi2(2)   =      18.50
(Assumption: nmodel nested in fmodel)                   Prob > chi2  =     0.0001
```

We conclude that

the hypothesis that the effects of the husband's and the wife's education are simultaneously equal to zero can be rejected at the .01 level ($LRX^2 = 18.50, df = 2, p < .01$).

This logic can be extended to exclude other variables. Say that we wish to test the null hypothesis that all of the effects of the independent variables are simultaneously equal to zero. We do not need to fit the full model again because the results are still saved from our use of `estimates store fmodel` above. We fit the model with no independent variables and run `lrtest`:

```
. logit lfp, nolog
  (output omitted)
. estimates store intercept_only
. lrtest fmodel intercept_only
likelihood-ratio test                                   LR chi2(7)   =     124.48
(Assumption: intercept_only  nested in fmodel)          Prob > chi2  =     0.0000
```

We can reject the hypothesis that all coefficients except the intercept are zero at the .01 level ($LRX^2 = 124.48, df = 7, p < .01$).

Note that this test is identical to the test in the header of the `logit` output: `LR chi2(7) = 124.48`.

4.3.3   Comparing LR and Wald tests

While the LR and Wald tests are *asymptotically* equivalent, their values differ in finite samples. For example,

Hypothesis	df	LR Test		Wald Test	
		G^2	p	W	p
$\beta_{k5} = 0$	1	66.48	<0.01	55.14	<0.01
$\beta_{wc} = \beta_{hc} = 0$	2	18.50	<0.01	17.66	<0.01
All slopes $= 0$	7	124.48	<0.01	95.0	< 0.01

Statistical theory is unclear on whether the LR or Wald test is to be preferred in models for categorical outcomes, although many statisticians, ourselves included, prefer the LR test. The choice of which test to use is often determined by convenience, personal preference, and convention within an area of research.

4.4   Residuals and influence using predict

Examining residuals and outliers is an important way to assess the fit of a regression model. *Residuals* are the difference between a model's predicted and observed outcome for each observation in the sample. Cases that fit poorly (i.e., have large residuals) are known as *outliers*. When an observation has a large effect on the estimated parameters, it is said to be *influential*.

Not all outliers are influential, as Figure 4.3 illustrates. In the top panel Figure 4.3, we show a scatterplot of some simulated data, and we have drawn the line that results from the linear regression of y on x. The residual of any observation is its vertical distance from the regression line. The observation highlighted by the box has a very large residual and so is an outlier. Even so, it is not very influential on the slope of the regression line. In the bottom panel, the only observation whose value has changed is the highlighted one. Now, the magnitude of the residual for this observation is much smaller, but it is very influential; its presence is entirely responsible for the slope of the new regression line being positive instead of negative.

(*Continued on next page*)

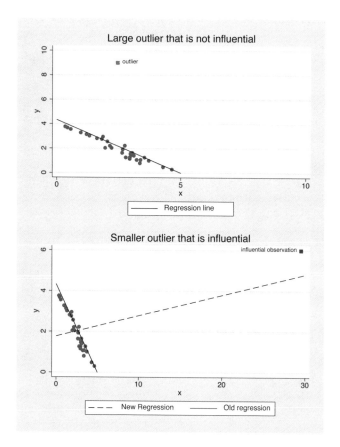

Figure 4.3: The distinction between an outlier and an influential observation.

Building on the analysis of residuals and influence in the linear regression model (see Fox 1991 and Weisberg 1980, Chapter 5 for details), Pregibon (1981) extended these ideas to the BRM.

4.4.1   Residuals

If we define the predicted probability for a given set of independent variables as

$$\pi_i = \Pr\left(y_i = 1 \mid \mathbf{x}_i\right)$$

then the deviations $y_i - \pi_i$ are heteroskedastic, with

$$\mathrm{Var}\left(y_i - \pi_i \mid \mathbf{x}_i\right) = \pi_i \left(1 - \pi_i\right)$$

This implies that the variance in a binary outcome is greatest when $\pi_i = .5$ and least as π_i approaches 0 or 1. For example, $.5\left(1 - .5\right) = .25$ and $.01\left(1 - .01\right) = .0099$. In other words, there is heteroskedasticity that depends on the probability of a positive outcome. This suggests the *Pearson residual*, which divides the residual $y - \widehat{\pi}$ by its standard deviation:

$$r_i = \frac{y_i - \widehat{\pi}_i}{\sqrt{\widehat{\pi}_i \left(1 - \widehat{\pi}_i\right)}}$$

Large values of r suggest a failure of the model to fit a given observation. Pregibon (1981) showed that the variance of r is not 1, as $\mathrm{Var}(y_i - \widehat{\pi}_i) \neq \widehat{\pi}_i \left(1 - \widehat{\pi}_i\right)$, and proposed the *standardized Pearson residual*

$$r_i^{\mathrm{Std}} = \frac{r_i}{\sqrt{1 - h_{ii}}}$$

where

$$h_{ii} = \widehat{\pi}_i \left(1 - \widehat{\pi}_i\right) \mathbf{x}_i \widehat{\mathrm{Var}}\left(\widehat{\beta}\right) \mathbf{x}_i' \tag{4.3}$$

While r^{Std} is preferred over r due to its constant variance, we find that the two residuals are often similar in practice. But, because r^{Std} is simple to compute in Stata, we recommend that you use this measure.

Example

An *index plot* is a useful way to examine residuals by simply plotting them against the observation number. The standardized residuals can be computed by specifying the `rs` option with `predict`. For example,

```
. logit lfp k5 k618 age wc hc lwg inc, nolog
  (output omitted )
. predict rstd, rs
. label var rstd "Standardized Residual"
. sort inc, stable
. generate index = _n
. label var index "Observation Number"
```

In this example, we first fit the logit model. Second, we use the `rs` option for `predict` to specify that we want standardized residuals, which are placed in a new variable that we have named `rstd`. Third, we sort the cases by `income`, so that observations are ordered from lowest to highest incomes. This results in a plot of residuals in which cases are ordered from low income to high income. The next line creates a new variable `index`, whose value for each observation is that observation's number (i.e., row) in the dataset. Note that `_n` on the right side of `generate` inserts the observation number. All that remains is to plot the residuals against the index using the commands[2]

```
. graph twoway scatter rstd index, xlabel(0(200)800) ylabel(-4(2)4)  ///
      xtitle("Observation Number") yline(0) msymbol(Oh)               ///
      ysize(2.7051) xsize(4.0413)
```

---

[2]The `///` is just a way of executing long lines in do-files. You should not type these characters if you are working from the Command window.

which produces the following index plot of standardized Pearson residuals:

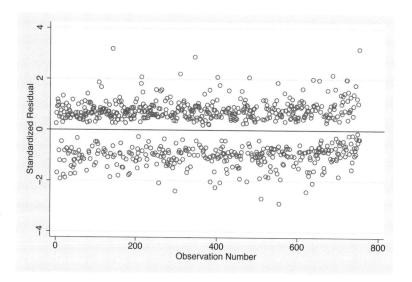

There is no hard-and-fast rule for what counts as a "large" residual. Indeed, in their detailed discussion of residuals and outliers in the binary regression model, Hosmer and Lemeshow (2000, 176) sagely caution that it is impossible to provide any absolute standard: "In practice, an assessment of 'large' is, of necessity, a judgment call based on experience and the particular set of data being analyzed".

One way to search for problematic residuals is to sort the residuals by the value of a variable that you think may be a problem for the model. In our example, we sorted the data by `income` before plotting. If this variable was primarily responsible for the lack of fit of some observations, the plot would show a disproportionate number of cases with large residuals among either the low income or the high income observations in our model. However, this does not appear to be the case for these data.

Still, in our plot, several residuals stand out as being large relative to the others. In such cases, it is important to identify the specific observations with large residuals for further inspection. We can do this by instructing `graph` to use the observation number to label each point in our plot. Recall that we just created a new variable called `index` whose value is equal to the observation number for each observation. We want the values of this index variable to be the marker symbols. We do this by labeling the marker with the index value and then placing the label over an invisible marker. In the command below, `msymbol(none)` makes the marker symbol invisible, `mlabel(index)` specifies that the variable `index` contains the labels, and `mlabposition(0)` causes the label to be positioned where the marker would have appeared. For example,

```
. graph twoway scatter rstd index, xlabel(0(200)800) ylabel(-4(2)4) ///
      xtitle("Observation Number") yline(0)                         ///
      msymbol(none) mlabel(index) mlabposition(0)                   ///
      ysize(2.7051) xsize(4.0413)
```

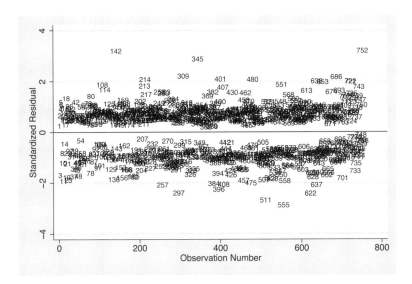

While labeling points with observations leads to chaos where there are many points, it effectively highlights and identifies the isolated cases. You can then easily list these cases. For example, observation 142 stands out and should be examined:

```
. list in 142, noobs
```

lfp	k5	k618	age	wc	hc	lwg	inc	rstd	index
inLF	1	2	36	NoCol	NoCol	-2.054124	11.2	3.191524	142

Alternatively, we can use **list** to list all observations with large residuals:

```
. list rstd index if rstd>2.5 | rstd<-2.5
```

	rstd	index
142.	3.191524	142
345.	2.873378	345
511.	-2.677243	511
555.	-2.871972	555
752.	3.192648	752

We can then check the listed cases to see if there are problems.

   Regardless of which method is used, further analyses of the highlighted cases might reveal either incorrectly coded data or some inadequacy in the specification of the model. Cases with large positive or negative residuals should *not* simply be discarded from the analysis, but rather should be examined to determine why they fit so poorly.

4.4.2   Influential cases

As shown in Figure 4.3, large residuals do not necessarily have a strong influence on the estimated parameters, and observations with relatively small residuals can have a large influence. *Influential* points are also sometimes called *high-leverage* points. These can be determined by examining the change in the estimated $\widehat{\beta}$ that occurs when the ith observation is deleted. While estimating a new logit for each case is usually impractical (although as the speed of computers increases, this may soon no longer be so), Pregibon (1981) derived an approximation that only requires fitting the model once. This measure summarizes the effect of removing the ith observation on the entire vector $\widehat{\beta}$, which is the counterpart to Cook's distance for the linear regression model. The measure is defined as

$$C_i = \frac{r_i^2 h_{ii}}{\left(1 - h_{ii}\right)^2}$$

where h_{ii} was defined in (4.3). In Stata, which refers to Cook's distance as `dbeta`, we can compute and plot Cook's distance as follows:

```
. predict cook, dbeta
. label var cook "Cook's Statistic"
. graph twoway scatter cook index, xlabel(0(200)800) ylabel(0(.1).3)  ///
       xtitle("Observation Number") yline(.1 .2)                       ///
       msymbol(none) mlabel(index) mlabposition(0)                     ///
       ysize(2.7051) xsize(4.0413)
```

These commands produce the following plot, which shows that cases 142, 309, and 752 merit further examination:

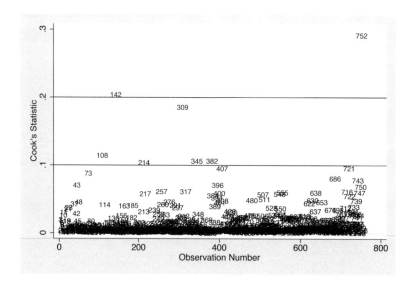

Methods for plotting residuals and outliers can be extended in many ways, including plots of different diagnostics against one another. Details of these plots are found in Cook and Weisberg (1999), Hosmer and Lemeshow (2000), and Landwehr et al. (1984).

4.5   Scalar measures of fit using fitstat

As discussed in Chapter 3, a scalar measure of fit can be useful in comparing competing models. Within a substantive area, measures of fit provide a *rough* index of whether a model is adequate. For example, if prior models of labor force participation routinely have values of .4 for a given measure of fit, you would expect that new analyses with a different sample or with revised measures of the variables would result in a similar value for that measure of fit. But, it is worth repeating that there is *no convincing evidence that selecting a model that maximizes the value of a given measure of fit results in a model that is optimal in any sense other than the model's having a larger value of that measure.* Details on these measures are presented in Chapter 3.

Example

To illustrate the use of scalar measures of fit, consider two models. M_1 contains our original specification of independent variables: k5, k618, age, wc, hc, lwg, and inc. M_2 drops the variables k618, hc, and lwg and adds agesq, which is the square of age. These models are fitted, and measures of fit are computed:

```
. quietly logit lfp k5 k618 age wc hc lwg inc, nolog
. estimates store model1
. quietly fitstat, save
. gen agesq = age*age
. quietly logit lfp k5 age agesq wc inc, nolog
. estimates store model2
```

We used quietly to suppress the output from logit and now use estimates table to combine the results from the two logits:

(Continued on next page)

```
. estimates table model1 model2, b(%9.3f) t
```

Variable	model1	model2
k5	-1.463	-1.380
	-7.43	-7.06
k618	-0.065	
	-0.95	
age	-0.063	0.057
	-4.92	0.50
wc	0.807	1.094
	3.51	5.50
hc	0.112	
	0.54	
lwg	0.605	
	4.01	
inc	-0.034	-0.032
	-4.20	-4.18
agesq		-0.001
		-1.00
_cons	3.182	0.979
	4.94	0.40

legend: b/t

The output from `fitstat` for M_1 was suppressed, but the results were saved to be listed by a second call to `fitstat` using the `dif` option:

```
. fitstat, dif
Measures of Fit for logit of lfp
```

	Current	Saved	Difference
Model:	logit	logit	
N:	753	753	0
Log-Lik Intercept Only:	-514.873	-514.873	0.000
Log-Lik Full Model:	-461.653	-452.633	-9.020
D:	923.306(747)	905.266(745)	18.040(2)
LR:	106.441(5)	124.480(7)	18.040(2)
Prob > LR:	0.000	0.000	0.000
McFadden's R2:	0.103	0.121	-0.018
McFadden's Adj R2:	0.092	0.105	-0.014
Maximum Likelihood R2:	0.132	0.152	-0.021
Cragg & Uhler's R2:	0.177	0.204	-0.028
McKelvey and Zavoina's R2:	0.182	0.217	-0.035
Efron's R2:	0.135	0.155	-0.020
Variance of y*:	4.023	4.203	-0.180
Variance of error:	3.290	3.290	0.000
Count R2:	0.677	0.693	-0.016
Adj Count R2:	0.252	0.289	-0.037
AIC:	1.242	1.223	0.019
AIC*n:	935.306	921.266	14.040
BIC:	-4024.871	-4029.663	4.791
BIC':	-73.321	-78.112	4.791

```
Difference of    4.791 in BIC' provides positive support for saved model.
Note: p-value for difference in LR is only valid if models are nested.
```

These results illustrate the limitations inherent in scalar measures of fit. M_2 deleted two variables that were not significant and one that was from M_1. It added a new variable that was not significant in the new model. Because the models are not nested, they cannot be compared using a difference of chi-squared test.[3] What do the fit statistics show? First, the values of the pseudo-R^2s are slightly larger for M_2, even though a significant variable was dropped and only a nonsignificant variable was added. *If you take the pseudo-R^2s as evidence for the "best" model, which we do not, there is some evidence preferring M_2.* Second, the BIC statistic is smaller for M_1, which provides support for that model. Following Raftery's (1996) guidelines, one would say that there is positive (neither weak nor strong) support for M_1.

4.6   Interpretation using predicted values

Because the BRM is nonlinear, no single approach to interpretation can fully describe the relationship between a variable and the outcome. We suggest that you try a variety of methods, with the goal of finding an elegant way to present the results that does justice to the complexities of the nonlinear model.

In general, the estimated parameters from the BRM do not provide directly useful information for understanding the relationship between the independent variables and the outcome. With the exception of the rarely used method of interpreting the latent variable (which we discuss in our treatment of ordinal models in Chapter 5), substantively meaningful interpretations are based on predicted probabilities and functions of those probabilities (e.g., ratios, differences). As shown in Figure 4.1, for a given set of values of the independent variables, the predicted probability in BRMs is defined as

$$\text{Logit: } \widehat{\Pr}\left(y = 1 \mid \mathbf{x}\right) = \Lambda\left(\mathbf{x}\widehat{\beta}\right) \qquad \text{Probit: } \widehat{\Pr}\left(y = 1 \mid \mathbf{x}\right) = \Phi\left(\mathbf{x}\widehat{\beta}\right)$$

where Λ is the cdf for the logistic distribution with variance $\pi^2/3$, and Φ is the cdf for the normal distribution with variance 1. For any set of values of the independent variables, the predicted probability can be computed. A variety of commands in Stata and our `pr*` commands make it very simple to work with these predicted probabilities.

4.6.1   Predicted probabilities with predict

After running `logit` or `probit`,

`predict` *newvarname* [`if` *exp*] [`in` *range*]

can be used to compute the predicted probability of a positive outcome for each observation, given the values on the independent variables for that observation. The predicted probabilities are stored in the new variable *newvarname*. The predictions are computed for all cases in memory that do not have missing values for the variables in the model,

---

[3]`fitstat, dif` computes the difference between all measures, even if the models are not nested. As with the Stata command `lrtest`, it is up to the user to determine if it makes sense to interpret the computed difference.

regardless of whether `if` and `in` had been used to restrict the estimation sample. For example, if you estimate `logit lfp k5 age if wc==1`, only 212 cases are used. But `predict` *newvarname* computes predictions for the entire dataset, 753 cases. If you only want predictions for the estimation sample, you can use the command `predict` *newvarname* `if e(sample)==1`.[4]

`predict` can be used to examine the range of predicted probabilities from your model. For example,

```
. predict prlogit
(option p assumed; Pr(lfp))
. summarize prlogit
```

Variable	Obs	Mean	Std. Dev.	Min	Max
prlogit	753	.5683931	.1944213	.0139875	.9621198

The message (`option p assumed; Pr(lfp)`) reflects that `predict` can compute many different quantities. Because we did not specify an option indicating which quantity to predict, option p for predicted probabilities was assumed, and the new variable `prlogit` was given the variable label `Pr(lfp)`. `summarize` computes summary statistics for the new variable and shows that the predicted probabilities in the sample range from .014 to .962, with a mean predicted probability of being in the labor force of .568.

We can use `dotplot` to plot the predicted probabilities for our sample:

```
. label var prlogit "Logit: Pr(lfp)"
. dotplot prlogit, ylabel(0(.2)1) ysize(2.7051) xsize(4.0413)
```

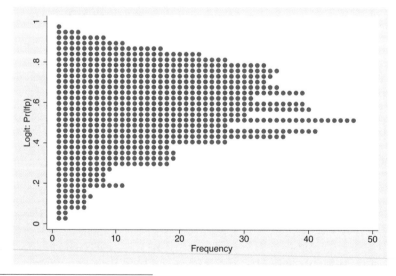

[4]Stata estimation commands create the variable `e(sample)`, indicating whether a case was used when fitting a model. Accordingly, the condition `if e(sample)==1` selects only cases used in the last estimation.

The plot clearly shows that the predicted probabilities for individual observations span almost the entire range from 0 to 1, but that roughly two-thirds of the observations have predicted probabilities between .40 and .80.

predict can also be used to demonstrate that the predictions from logit and probit models are essentially identical. Even though the two models make different assumptions about $\text{Var}(\varepsilon)$, these differences are absorbed in the relative magnitudes of the estimated coefficients. To see this, we first fit the two models and compute their predicted probabilities:

```
. use http://www.stata-press.com/data/lfr/binlfp2, clear
(Data from 1976 PSID-T Mroz)
. logit lfp k5 k618 age wc hc lwg inc, nolog
  (output omitted)
. predict prlogit
(option p assumed; Pr(lfp))
. label var prlogit "Logit: Pr(lfp)"
. probit lfp k5 k618 age wc hc lwg inc, nolog
  (output omitted)
. predict prprobit
(option p assumed; Pr(lfp))
. label var prprobit "Probit: Pr(lfp)"
```

Next, we check the correlation between the two sets of predicted values:

```
. pwcorr prlogit prprobit

              |  prlogit prprobit
    ----------+------------------
      prlogit |   1.0000
     prprobit |   0.9998   1.0000
```

The extremely high correlation is confirmed by plotting them against one another:

```
. graph twoway scatter prlogit prprobit,     ///
      xlabel(0(.25)1) ylabel(0(.25)1)        ///
      xline(.25(.25)1) yline(.25(.25)1)      ///
      plotregion(margin(zero)) msymbol(Oh)   ///
      ysize(4.0413) xsize(4.0413)
```

(Continued on next page)

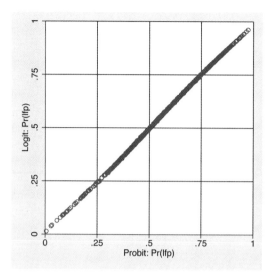

In terms of predictions, there is very little reason to prefer either logit or probit. If your substantive findings turn on whether you used logit or probit, *we would not place much confidence in either result*. In our own research, we tend to use logit, primarily because of the availability of interpretation in terms of odds and odds ratios (discussed below).

Overall, examining predicted probabilities for the cases in the sample provides an initial check of the model. To better understand and present the substantive findings, it is usually more effective to compute predictions at specific, substantively informative values. Our commands `prvalue`, `prtab`, and `prgen` are designed to make this very simple.

4.6.2   Individual predicted probabilities with prvalue

A table of probabilities for ideal types of people (or countries, cows, or whatever you are studying) can quickly summarize the effects of key variables. In our example of labor force participation, we could compute predicted probabilities of labor force participation for women in these three types of families:

- Young, low income and low education families with young children.

- Highly educated, middle aged couples with no children at home.

- An "average family" defined as having the mean on all variables.

This can be done with a series of calls to `prvalue` (see Chapter 3 for a discussion of options for this command):[5]

---

[5]`mean` is the default setting for the `rest()` option, so `rest(mean)` does not need to be specified. We include it in many of our examples anyway, because its use emphasizes that the results are contingent on specified values for *all* of the independent variables.

```
. * young, low income, low education families with young children.
. prvalue, x(age=35 k5=2 wc=0 hc=0 inc=15) rest(mean)

logit: Predictions for lfp
  Pr(y=inLF|x):        0.1318   95% ci: (0.0723,0.2282)
  Pr(y=NotInLF|x):     0.8682   95% ci: (0.7718,0.9277)
             k5       k618       age        wc        hc        lwg       inc
x=            2  1.3532537        35         0         0  1.0971148        15
```

We have set the values of the independent variables to those that define our first type of family, with other variables held at their mean. The output shows the predicted probability of working (.13), along with the chosen values for each variable. While the values of the independent variables can be suppressed with the **brief** option, it is safest to verify that they are correct. This process is repeated for the other ideal types:

```
. * highly education families with no children at home.
. prvalue, x(age=50 k5=0 k618=0 wc=1 hc=1) rest(mean)

logit: Predictions for lfp
  Pr(y=inLF|x):        0.7166   95% ci: (0.6266,0.7921)
  Pr(y=NotInLF|x):     0.2834   95% ci: (0.2079,0.3734)
             k5       k618       age        wc        hc        lwg       inc
x=            0          0        50         1         1  1.0971148  20.128965

.
. * an average person
. prvalue, rest(mean)

logit: Predictions for lfp
  Pr(y=inLF|x):        0.5778   95% ci: (0.5388,0.6159)
  Pr(y=NotInLF|x):     0.4222   95% ci: (0.3841,0.4612)
             k5       k618       age        wc        hc        lwg       inc
x=     .2377158  1.3532537  42.537849  .2815405  .39176627  1.0971148  20.128965
```

With predictions in hand, we can summarize the results and get a better general feel for the factors affecting a wife's labor force participation.

Ideal Type	Probability of LFP
Young, low income, and low education families with young children.	0.13
Highly educated, middle-aged couples with no children at home.	0.72
An "average" family	0.58

4.6.3   Tables of predicted probabilities with prtab

In some cases, the focus might be on two or three categorical independent variables. Predictions for all combinations of the categories of these variables could be presented in a table. For example,

Number of Young Children	Predicted Probability		
	Did Not Attend	Attended College	Difference
0	0.61	0.78	0.17
1	0.26	0.44	0.18
2	0.08	0.16	0.08
3	0.02	0.04	0.02

This table shows the strong effect on labor force participation of having young children and how the effect differs according to the wife's education. One way to construct such a table is by a series of calls to `prvalue` (we use the `brief` option to limit output):

```
. prvalue, x(k5=0 wc=0) rest(mean) brief
  Pr(y=inLF|x):        0.6069    95% ci: (0.5558,0.6558)
  Pr(y=NotInLF|x):     0.3931    95% ci: (0.3442,0.4442)
. prvalue, x(k5=1 wc=0) rest(mean) brief
  Pr(y=inLF|x):        0.2633    95% ci: (0.1994,0.3391)
  Pr(y=NotInLF|x):     0.7367    95% ci: (0.6609,0.8006)
. * and so on, ad nauseam...
```

Even for a simple table, this approach is tedious and error-prone. `prtab` automates the process by computing a table of predicted probabilities for all combinations of up to four categorical variables. For example,

```
. prtab k5 wc, rest(mean)
logit: Predicted probabilities of positive outcome for lfp
```

# kids < 6	Wife College: 1=yes 0=no	
	NoCol	College
0	0.6069	0.7758
1	0.2633	0.4449
2	0.0764	0.1565
3	0.0188	0.0412

```
             k5         k618        age          wc          hc         lwg          inc
x=    .2377158   1.3532537   42.537849   .2815405   .39176627   1.0971148   20.128965
```

4.6.4 Graphing predicted probabilities with prgen

When a variable of interest is continuous, you can either select values (e.g., quartiles) and construct a table or create a graph. For example, to examine the effects of income on labor force participation by age, we can use the estimated parameters to compute predicted probabilities as income changes for fixed values of age. This is shown in Figure 4.4. The command prgen creates data that can be graphed in this way. The first step is to generate the predicted probabilities for those aged 30:

```
. prgen inc, from(0) to(100) generate(p30) x(age=30) rest(mean) n(11)
logit: Predicted values as inc varies from 0 to 100.
          k5        k618      age       wc        hc        lwg       inc
x=    .2377158   1.3532537     30    .2815405  .39176627  1.0971148  20.128965
. label var p30p1 "Age 30"
```

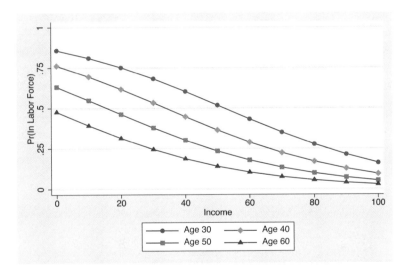

Figure 4.4: Graph of predicted probabilities created using prgen.

inc is the independent variable that we want to vary along the x-axis. The options that we use are

from(0) and to(100) specify the minimum and maximum values over which inc is to vary. The default is the variable's observed minimum and maximum values.

generate(p30) indicates the root name used in constructing new variables. prgen creates p30x that contains the values of inc that are used; p30p1 with the values of the probability of a 1 and p30p0 with values of the probability of a 0.

x(age=30) indicates that we want to hold the value of age at 30. By default, other
variables will be held at their mean unless rest() is used to specify some other
summary statistic.

n(11) indicates that 11 evenly spaced values of inc between 0 and 100 should be used.
You should choose the value that corresponds to the number of symbols you want
on your graph.

Additional calls of prgen are made holding age at different values:

```
. prgen inc, from(0) to(100) generate(p40) x(age=40) rest(mean) n(11)
  (output omitted)
. label var p40p1 "Age 40"
. prgen inc, from(0) to(100) generate(p50) x(age=50) rest(mean) n(11)
  (output omitted)
. label var p50p1 "Age 50"
. prgen inc, from(0) to(100) generate(p60) x(age=60) rest(mean) n(11)
  (output omitted)
. label var p60p1 "Age 60"
```

Listing the values for the first eleven observations in the dataset for some of the new
variables prgen has created may help you understand better what this command does:

```
. list p30p1 p40p1 p50p1 p60p1 p60x in 1/11
```

	p30p1	p40p1	p50p1	p60p1	p60x
1.	.8575829	.7625393	.6313345	.4773258	0
2.	.8101358	.6947005	.5482202	.3928797	10
3.	.7514627	.6172101	.462326	.3143872	20
4.	.6817801	.5332655	.3786113	.2452419	30
5.	.6028849	.4473941	.3015535	.187153	40
6.	.5182508	.36455	.2342664	.1402662	50
7.	.4325564	.289023	.1781635	.1036283	60
8.	.3507161	.2236366	.1331599	.0757174	70
9.	.2768067	.1695158	.0981662	.0548639	80
10.	.2133547	.1263607	.071609	.0395082	90
11.	.1612055	.0929622	.0518235	.0283215	100

The predicted probabilities of labor force participation for those average on all other
variables at ages 30, 40, 50, and 60 are in the first four columns. The clear negative
effect of age is shown by the increasingly small probabilities as we move across these
columns in any row. The last column indicates the value of income for a given row,
starting at 0 and ending at 100. We can see that the probabilities decrease as income
increases.

The following **graph** command generates the plot:

```
. graph twoway connected p30p1 p40p1 p50p1 p60p1 p60x, ///
    ytitle("Pr(In Labor Force)") ylabel(0(.25)1)     ///
    xtitle("Income")                                 ///
    ysize(2.7051) xsize(4.0413)
```

Because we have not used **graph** much yet, it is worth discussing some points that we find useful (also see the section on Graphics in Chapter 2).

1. Recall that /// is a way of entering long lines in do-files.

2. **graph twoway** is the command for plotting a dependent variable on the *y*-axis against an independent variable along the *x*-axis. **graph twoway connected** specifies that the symbols used to mark the individual points be connected.

3. The variables to plot are **p30p1 p40p1 p50p1 p60p1 p60x**, where **p60x**, the last variable in the list, is the variable for the horizontal axis. All variables before the last variable are plotted on the vertical axis.

4. The options **ytitle()** and **xtitle()** specify the axis titles.

5. The **ylabel()** specifies which points on the *y*-axis to label.

4.6.5 Changes in predicted probabilities

While graphs are very useful for showing how predicted probabilities are related to an independent variable, for even our simple example it is not practical to plot all possible combinations of the independent variables. And, in some cases, the plots show that a relationship is linear so that a graph is superfluous. In such circumstances, a useful summary measure is the change in the outcome as one variable changes, holding all other variables constant.

Marginal change

In economics, the *marginal effect* or *change* is commonly used:

$$\text{Marginal Change} = \frac{\partial \Pr(y = 1 \mid \mathbf{x})}{\partial x_k}$$

The marginal change is shown by the tangent to the probability curve in Figure 4.5. The value of the marginal effect depends on the level of all variables in the model. It is often computed with all variables held at their mean or by computing the marginal change for each observation in the sample and then averaging across all values.

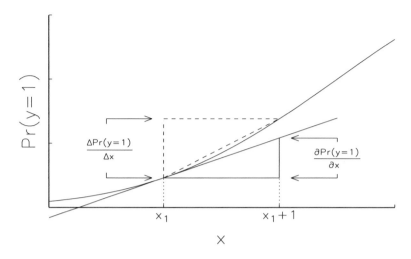

Figure 4.5: Marginal change compared with discrete change in the BRM.

Marginal change with prchange command

The command **prchange** computes the marginal at the values of the independent variables specified with **x()** or **rest()**. Running **prchange** without any options computes the marginal change (along with a lot of other things discussed below) with all variables at their mean. Or, we can compute the marginal at specific values of the independent variables, such as when **wc** = 1 and **age** = 40. Here, we request only the results for **age**:

```
. prchange age, x(wc=1 age=40) help
logit: Changes in Predicted Probabilities for lfp
        min->max       0->1      -+1/2    -+sd/2   MargEfct
age     -0.3940    -0.0017    -0.0121   -0.0971    -0.0121

          NotInLF      inLF
Pr(y|x)    0.2586    0.7414

            k5       k618       age        wc        hc       lwg       inc
   x=  .237716    1.35325        40         1   .391766   1.09711    20.129
sd(x)=  .523959    1.31987   8.07257   .450049   .488469   .587556   11.6348

 Pr(y|x): probability of observing each y for specified x values
 Avg|Chg|: average of absolute value of the change across categories
 Min->Max: change in predicted probability as x changes from its minimum to
           its maximum
    0->1: change in predicted probability as x changes from 0 to 1
   -+1/2: change in predicted probability as x changes from 1/2 unit below
          base value to 1/2 unit above
  -+sd/2: change in predicted probability as x changes from 1/2 standard
          dev below base to 1/2 standard dev above
 MargEfct: the partial derivative of the predicted probability/rate with
           respect to a given independent variable
```

In plots that we do not show (but that we encourage you to create using **prgen** and **graph**), we found that the relationship between age and the probability of being in the labor force was essentially linear for those who attend college. Accordingly, we can take the marginal computed by **prchange**, multiply it by 10 to get the amount of change over 10 years, and report that

> for women who attend college, a ten year increase in age decreases the probability of labor force participation by approximately .12, holding other variables at their mean.

When using the marginal, it is essential to keep two points in mind. First, the amount of change depends on the level of all variables. Second, as shown in Figure 4.5, the marginal is the instantaneous rate of change. In general, it does not equal the actual change for a given finite change in the independent variable unless you are in a region of the probability curve that is approximately linear. Such linearity justifies the interpretation given above.

Marginal change with mfx compute command

The marginal change can also be computed using **mfx compute**, where the **at()** option is used to set values of the independent variables. Below, we use **mfx compute** to estimate the marginal change for the same values that we used when calculating the marginal effect for **age** with **prchange** above:

```
. mfx compute, at(wc=1 age=40)
Marginal effects after logit
      y  = Pr(lfp) (predict)
         =  .74140317
```

variable	dy/dx	Std. Err.	z	P>|z|	[	95% C.I.	]	X
k5	-.2804763	.04221	-6.64	0.000	-.363212	-.197741		.237716
k618	-.0123798	.01305	-0.95	0.343	-.037959	.013199		1.35325
age	-.0120538	.00245	-4.92	0.000	-.016855	-.007252		40.0000
wc*	.1802113	.04742	3.80	0.000	.087269	.273154		1.00000
hc*	.0212952	.03988	0.53	0.593	-.056866	.099456		.391766
lwg	.1159345	.03229	3.59	0.000	.052643	.179226		1.09711
inc	-.0066042	.00163	-4.05	0.000	-.009802	-.003406		20.1290

```
(*) dy/dx is for discrete change of dummy variable from 0 to 1
```

mfx compute is particularly useful if you need estimates of the standard errors of the marginal effects; however, **mfx compute** computes the estimates using numerical methods, and for some models the command can take a long time.

Discrete change

Given the nonlinearity of the model, we prefer the *discrete change* in the predicted probabilities for a given change in an independent variable. To define discrete change, we need two quantities:

$\Pr(y = 1 \mid \mathbf{x}, x_k)$ is the probability of an event given $\mathbf{x}$, noting in particular the value of x_k.

$\Pr(y = 1 \mid \mathbf{x}, x_k + \delta)$ is the probability of the event with only x_k increased by some quantity δ.

Then, the *discrete change* for a change of δ in x_k equals

$$\frac{\Delta \Pr(y = 1 \mid \mathbf{x})}{\Delta x_k} = \Pr(y = 1 \mid \mathbf{x}, x_k + \delta) - \Pr(y = 1 \mid \mathbf{x}, x_k)$$

which can be interpreted that

> for a change in variable x_k from x_k to $x_k + \delta$, the predicted probability of an event changes by $\{\Delta \Pr(y = 1 \mid \mathbf{x})\}/\Delta x_k$, holding all other variables constant.

As shown in Figure 4.5, in general, the two measures of change are not equal. That is,

$$\frac{\partial \Pr(y = 1 \mid \mathbf{x})}{\partial x_k} \neq \frac{\Delta \Pr(y = 1 \mid \mathbf{x})}{\Delta x_k}$$

The measures differ because the marginal change is the instantaneous rate of change, while the discrete change is the amount of change in the probability for a given finite change in one independent variable. The two measures are similar, however, when the change occurs over a region of the probability curve that is roughly linear.

The value of the discrete change depends on

1. The start level of the variable that is being changed. For example, do you want to examine the effect of age beginning at 30? At 40? At 50?

2. The amount of change in that variable. Are you interested in the effect of a change of 1 year in age? Of 5 years? Of 10 years?

3. The level of all other variables in the model. Do you want to hold all variables at their mean? Or, do you want to examine the effect for women? Or, do you want to compute changes separately for men and women?

Accordingly, a decision must be made regarding each of these factors. See Chapter 3 for further discussion.

For our example, let's look at the discrete change with all variables held at their mean, which is computed by default by **prchange**, where the **help** option is used to get detailed descriptions of what the measures mean:

```
. prchange, help
logit: Changes in Predicted Probabilities for lfp
         min->max       0->1      -+1/2     -+sd/2   MargEfct
    k5    -0.6361    -0.3499    -0.3428    -0.1849   -0.3569
  k618    -0.1278    -0.0156    -0.0158    -0.0208   -0.0158
   age    -0.4372    -0.0030    -0.0153    -0.1232   -0.0153
    wc     0.1881     0.1881     0.1945     0.0884    0.1969
    hc     0.0272     0.0272     0.0273     0.0133    0.0273
   lwg     0.6624     0.1499     0.1465     0.0865    0.1475
   inc    -0.6415    -0.0068    -0.0084    -0.0975   -0.0084

           NotInLF      inLF
Pr(y|x)     0.4222    0.5778

              k5       k618        age         wc         hc        lwg        inc
    x=    .237716    1.35325    42.5378    .281541    .391766    1.09711    20.129
sd(x)=    .523959    1.31987    8.07257    .450049    .488469    .587556    11.6348

Pr(y|x): probability of observing each y for specified x values
Avg|Chg|: average of absolute value of the change across categories
Min->Max: change in predicted probability as x changes from its minimum to
          its maximum
   0->1: change in predicted probability as x changes from 0 to 1
  -+1/2: change in predicted probability as x changes from 1/2 unit below
          base value to 1/2 unit above
 -+sd/2: change in predicted probability as x changes from 1/2 standard
          dev below base to 1/2 standard dev above
MargEfct: the partial derivative of the predicted probability/rate with
          respect to a given independent variable
```

First, consider the results of changes from the minimum to the maximum. There is little to be learned by analyzing variables whose range of probabilities is small, such as hc, while age, k5, wc, lwg, and inc have *potentially* important effects. For these we can examine the value of the probabilities before and after the change by using the `fromto` option:

```
. prchange k5 age wc lwg inc, fromto
logit: Changes in Predicted Probabilities for lfp
           from:        to:       dif:       from:        to:       dif:       from:
          x=min      x=max    min->max        x=0        x=1       0->1      x-1/2
    k5    0.6596     0.0235    -0.6361     0.6596     0.3097    -0.3499     0.7398
   age    0.7506     0.3134    -0.4372     0.9520     0.9491    -0.0030     0.5854
    wc    0.5216     0.7097     0.1881     0.5216     0.7097     0.1881     0.4775
   lwg    0.1691     0.8316     0.6624     0.4135     0.5634     0.1499     0.5028
   inc    0.7326     0.0911    -0.6415     0.7325     0.7256    -0.0068     0.5820

             to:       dif:       from:        to:       dif:
          x+1/2      -+1/2    x-1/2sd    x+1/2sd     -+sd/2   MargEfct
    k5    0.3971    -0.3428     0.6675     0.4826    -0.1849    -0.3569
   age    0.5701    -0.0153     0.6382     0.5150    -0.1232    -0.0153
    wc    0.6720     0.1945     0.5330     0.6214     0.0884     0.1969
   lwg    0.6493     0.1465     0.5340     0.6204     0.0865     0.1475
   inc    0.5736    -0.0084     0.6258     0.5283    -0.0975    -0.0084

           NotInLF      inLF
Pr(y|x)     0.4222    0.5778

              k5       k618        age         wc         hc        lwg        inc
    x=    .237716    1.35325    42.5378    .281541    .391766    1.09711    20.129
sd(x)=    .523959    1.31987    8.07257    .450049    .488469    .587556    11.6348
```

We learn, for example, that varying `age` from its minimum of 30 to its maximum of 60 decreases the predicted probability from .75 to .31, a decrease of .44. Changing family income (`inc`) from its minimum to its maximum decreases the probability of a woman being in the labor force from .73 to .09. Interpreting other measures of change, the following interpretations can be made

> *Using the unit change labeled* `-+1/2`: For a woman who is average on all characteristics, an additional young child decreases the probability of employment by .34.

> *Using the standard deviation change labeled* `-+1/2sd`: A standard deviation change in age centered on the mean will decrease the probability of working by .12, holding other variables to their means.

> *Using a change from 0 to 1 labeled* `0->1`: If a woman attends college, her probability of being in the labor force is .18 greater than a woman who does not attend college, holding other variables at their mean.

What if you need to calculate discrete change for changes in the independent values that are not the default for `prchange` (e.g., a change of 10 years in age rather than 1 year)? This can be done in two ways:

Nonstandard discrete changes with prvalue command

The command `prvalue` can be used to calculate the change in the probability for a discrete change of any magnitude in an independent variable. Say that we want to calculate the effect of a ten-year increase in age for a 30-year-old woman who is average on all other characteristics:

```
. prvalue, x(age=30) save brief
  Pr(y=inLF|x):        0.7506   95% ci: (0.6771,0.8121)
  Pr(y=NotInLF|x):     0.2494   95% ci: (0.1879,0.3229)
. prvalue, x(age=40) dif brief
                     Current      Saved   Difference
  Pr(y=inLF|x):        0.6162     0.7506     -0.1345
  Pr(y=NotInLF|x):     0.3838     0.2494      0.1345
```

The `save` option preserves the results from the first call of `prvalue`. The second call adds the `dif` option to compute the differences between the two sets of predictions. We find that an increase in `age` from 30 to 40 years decreases a woman's probability of being in the labor force by .13.

Nonstandard discrete changes with prchange

Alternatively, we can use `prchange` with the `delta()` and `uncentered` options. `delta(#)` specifies that the discrete change is to be computed for a change of # units instead of a one-unit change. `uncentered` specifies that the change should be computed starting at the base value (i.e., values set by the `x()` and `rest()` options), rather than being centered on the base. In this case, we want an uncentered change of 10 units, starting at `age=30`:

```
. prchange age, x(age=30) uncentered delta(10) rest(mean) brief
        min->max      0->1    +delta       +sd  MargEfct
age     -0.4372    -0.0030   -0.1345   -0.1062   -0.0118
```

The result under the heading `+delta` is the same as what we just calculated using `prvalue`.

4.7   Interpretation using odds ratios with listcoef

Effects for the logit model, but *not* probit, can be interpreted in terms of changes in the odds. Recall that for binary outcomes, we typically consider the odds of observing a positive outcome versus a negative one:

$$\Omega = \frac{\Pr(y=1)}{\Pr(y=0)} = \frac{\Pr(y=1)}{1 - \Pr(y=1)}$$

Recall also that the log of the odds is called the *logit* and that the logit model is *linear in the logit*, meaning that the log odds are a linear combination of the xs and βs. For example, consider a logit model with three independent variables:

$$\ln\left\{\frac{\Pr(y=1\mid \mathbf{x})}{1 - \Pr(y=1\mid \mathbf{x})}\right\} = \ln\Omega(\mathbf{x}) = \beta_0 + \beta_1 x_1 + \beta_2 x_2 + \beta_3 x_3$$

We can interpret the coefficients as indicating that

> for a unit change in x_k, we expect the logit to change by β_k, holding all other variables constant.

This interpretation does *not* depend on the level of the other variables in the model. The problem is that a change of β_k in the log odds has little substantive meaning for most people (including the authors of this book). Alternatively, by taking the exponential of both sides of this equation, we can create a model that is multiplicative instead of linear, but in which the outcome is the more intuitive measure, the odds:

$$\Omega\left(\mathbf{x}, x_2\right) = e^{\beta_0} e^{\beta_1 x_1} e^{\beta_2 x_2} e^{\beta_3 x_3}$$

where we take particular note of the value of x_2. If we let x_2 change by 1,

$$\Omega\left(\mathbf{x}, x_2+1\right) = e^{\beta_0} e^{\beta_1 x_1} e^{\beta_2 (x_2+1)} e^{\beta_3 x_3}$$

$$= e^{\beta_0} e^{\beta_0} e^{\beta_1 x_1} e^{\beta_2 x_2} e^{\beta_2} e^{\beta_3 x_3}$$

which leads to the *odds ratio:*

$$\frac{\Omega\left(\mathbf{x}, x_2 + 1\right)}{\Omega\left(\mathbf{x}, x_2\right)} = \frac{e^{\beta_0} e^{\beta_1 x_1} e^{\beta_2 x_2} e^{\beta_2} e^{\beta_3 x_3}}{e^{\beta_0} e^{\beta_1 x_1} e^{\beta_2 x_2} e^{\beta_3 x_3}} = e^{\beta_2}$$

Accordingly, we can interpret the exponential of the coefficient as follows:

> For a unit change in x_k, the *odds* are expected to change by a factor of $\exp(\beta_k)$, holding all other variables constant.

For $\exp(\beta_k) > 1$, you could say that the odds are "$\exp(\beta_k)$ times larger"; for $\exp(\beta_k) < 1$, you could say that the odds are "$\exp(\beta_k)$ times smaller". We can evaluate the effect of a standard deviation change in x_k instead of a unit change:

> For a standard deviation change in x_k, the odds are expected to change by a factor of $\exp(\beta_k \times s_k)$, holding all other variables constant.

The odds ratios for both a unit and a standard deviation change of the independent variables can be obtained with `listcoef`:

```
. listcoef, help
logit (N=753): Factor Change in Odds
  Odds of: inLF vs NotInLF
```

lfp	b	z	P>|z|	e^b	e^bStdX	SDofX
k5	-1.46291	-7.426	0.000	0.2316	0.4646	0.5240
k618	-0.06457	-0.950	0.342	0.9375	0.9183	1.3199
age	-0.06287	-4.918	0.000	0.9391	0.6020	8.0726
wc	0.80727	3.510	0.000	2.2418	1.4381	0.4500
hc	0.11173	0.542	0.588	1.1182	1.0561	0.4885
lwg	0.60469	4.009	0.000	1.8307	1.4266	0.5876
inc	-0.03445	-4.196	0.000	0.9661	0.6698	11.6348

```
       b = raw coefficient
       z = z-score for test of b=0
    P>|z| = p-value for z-test
     e^b = exp(b) = factor change in odds for unit increase in X
 e^bStdX = exp(b*SD of X) = change in odds for SD increase in X
   SDofX = standard deviation of X
```

Some examples of interpretations are as follows:

> For each additional young child, the odds of being employed decrease by a factor of .23, holding all other variables constant.

> For a standard deviation increase in the log of the wife's expected wages, the odds of being employed are 1.43 times greater, holding all other variables constant.

> Being ten years older decreases the odds by a factor of .53 $(= e^{[-.063] \times 10})$, holding all other variables constant.

Other ways of computing odds ratios Odds ratios can also be computed with the
or option for `logit`. This approach does not, however, report the odds ratios for
a standard deviation change in the independent variables.

Multiplicative coefficients

When interpreting the odds ratios, remember that they are multiplicative. This means
that positive effects are greater than one and negative effects are between zero and
one. *Magnitudes of positive and negative effects should be compared by taking the
inverse of the negative effect (or vice versa).* For example, a positive factor change of
2 has the same magnitude as a negative factor change of .5 = 1/2. Thus, a coefficient
of .1 = 1/10 indicates a stronger effect than a coefficient of 2. Another consequence
of the multiplicative scale is that to determine the effect on the odds of the event not
occurring, you simply take the inverse of the effect on the odds of the event occurring.
`listcoef` will automatically calculate this for you if you specify the **reverse** option:

```
. listcoef, reverse
logit (N=753): Factor Change in Odds
  Odds of: NotInLF vs inLF
```

lfp	b	z	P>|z|	e^b	e^bStdX	SDofX
k5	-1.46291	-7.426	0.000	4.3185	2.1522	0.5240
k618	-0.06457	-0.950	0.342	1.0667	1.0890	1.3199
age	-0.06287	-4.918	0.000	1.0649	1.6612	8.0726
wc	0.80727	3.510	0.000	0.4461	0.6954	0.4500
hc	0.11173	0.542	0.588	0.8943	0.9469	0.4885
lwg	0.60469	4.009	0.000	0.5462	0.7010	0.5876
inc	-0.03445	-4.196	0.000	1.0350	1.4930	11.6348

Note that the header indicates that these are now the factor changes in the odds of
`NotInLF` versus `inLF`, whereas before we computed the factor change in the odds of
`inLF` versus `NotInLF`. We can interpret the result for `k5` as follows:

> For each additional child, the odds of not being employed are increased by
> a factor of 4.3 (= 1/.23), holding other variables constant.

Effect of the base probability

The interpretation of the odds ratio assumes that the other variables have been held
constant, but it does not require that they be held at any specific values. While the odds
ratio seems to resolve the problem of nonlinearity, it is essential to keep the following in
mind: *A constant factor change in the odds does not correspond to a constant change
or constant factor change in the probability.* For example, if the odds are 1/100, the

corresponding probability is .01.[6] If the odds double to 2/100, the probability increases only by approximately .01. Depending on your substantive purposes, this small change may be trivial or quite important (such as when you identify a risk factor that makes it twice as likely that a subject will contract a fatal disease). Meanwhile, if the odds are 1/1 and double to 2/1, the probability increases by .167. Accordingly, the meaning of a given factor change in the odds depends on the predicted probability, which in turn depends on the levels of all variables in the model.

Percent change in the odds

Instead of a multiplicative or factor change in the outcome, some people prefer the percent change,

$$100 \left\{ \exp\left(\beta_k \times \delta\right) - 1 \right\}$$

which is listed by `listcoef` with the `percent` option.

```
. listcoef, percent
logit (N=753): Percentage Change in Odds
    Odds of: inLF vs NotInLF
```

lfp	b	z	P>|z|	%	%StdX	SDofX
k5	-1.46291	-7.426	0.000	-76.8	-53.5	0.5240
k618	-0.06457	-0.950	0.342	-6.3	-8.2	1.3199
age	-0.06287	-4.918	0.000	-6.1	-39.8	8.0726
wc	0.80727	3.510	0.000	124.2	43.8	0.4500
hc	0.11173	0.542	0.588	11.8	5.6	0.4885
lwg	0.60469	4.009	0.000	83.1	42.7	0.5876
inc	-0.03445	-4.196	0.000	-3.4	-33.0	11.6348

With this option, the interpretations would be the following:

> For each additional young child, the odds of being employed decrease by 77%, holding all other variables constant.

> A standard deviation increase in the log of the wife's expected wages increases the odds of being employed by 43%, holding all other variables constant.

Percentage and factor change provide the same information; which you use for the binary model is a matter of preference. While we both tend to prefer percentage change, methods for the graphical interpretation of the multinomial logit model (Chapter 6) only work with factor change coefficients.

---

[6]The formula for computing probabilities from odds is $p = \frac{\Omega}{1+\Omega}$.

Additional note If you report the odds ratios instead of the untransformed coefficients, the 95% confidence interval of the odds ratio is typically reported instead
of the standard error. The reason is that the odds ratio is a nonlinear transformation of the logit coefficient, so the confidence interval is asymmetric. For example,
if the logit coefficient is .75 with a standard error of .25, the 95% interval around
the logit coefficient is appoximately [.26, 1.24], but the confidence interval around
the odds ratio exp(.75)=2.11 is [exp(.26)=1.30, exp(1.24)=3.45]. Using the or
option with the `logit` command reports as odds ratios and includes confidence
intervals.

4.8   Other commands for binary outcomes

Logit and probit models are the most commonly used models for binary outcomes and
are the only ones that we consider in this book, but other models exist that can be
fitted in Stata. Among them, `cloglog` assumes a complementary log-log distribution
for the errors instead of a logistic or normal distribution. `scobit` fits a logit model that
relaxes the assumption that the marginal change in the probability is greatest when
$\Pr(y = 1) = .5$. `hetprob` allows the assumed variance of the errors in the probit model
to vary as a function of the independent variables. `blogit` and `bprobit` fit logit and
probit models on grouped ("blocked") data. Further details on all of these models can
be found in the appropriate entries in the Stata manuals.

5 Models for Ordinal Outcomes

The categories of an ordinal variable can be ranked, but the distances between the categories are unknown. For example, in survey research, opinions are often ranked as strongly agree, agree, disagree, and strongly disagree, without an assumption that the distance from strongly agreeing and agreeing is the same as the distance from agree to disagree. Educational attainments can be ordered as elementary education, high school diploma, college diploma, and graduate or professional degree. Ordinal variables also commonly result from limitations of data availability that require a coarse categorization of a variable that could, in principle, have been measured on an interval scale. For example, we might have a measure of income that is simply low, medium, or high.

Ordinal variables are often coded as consecutive integers from 1 to the number of categories. Perhaps as a consequence of this coding, it is tempting to analyze ordinal outcomes with the linear regression model. However, an ordinal dependent variable violates the assumptions of the LRM, which can lead to incorrect conclusions, as demonstrated strikingly by McKelvey and Zavoina (1975, 117) and Winship and Mare (1984, 521–523). Accordingly, with ordinal outcomes it is much better to use models that avoid the assumption that the distances between categories are equal. While many different models have been designed for ordinal outcomes, in this chapter we focus on the logit and probit versions of the *ordinal regression model* (ORM), introduced by McKelvey and Zavoina (1975) in terms of an underlying latent variable and in biostatistics by McCullagh (1980) who referred to the logit version as the *proportional odds model*.

As with the binary regression model, the ORM is nonlinear, and the magnitude of the change in the outcome probability for a given change in one of the independent variables depends on the levels of all of the independent variables. As with the BRM, the challenge is to summarize the effects of the independent variables in a way that fully reflects key substantive processes without overwhelming and distracting detail. For ordinal outcomes, as well as for the models for nominal outcomes in Chapter 6, the difficulty of this task is increased by having more than two outcomes to explain.

Before proceeding, we caution that researchers should think carefully before concluding that their outcome is indeed ordinal. Simply because the values of a variable *can* be ordered, do not assume that the variable *should* be analyzed as ordinal. A variable that can be ordered when considered for one purpose could be unordered or ordered differently when used for another purpose. Miller and Volker (1985) show how different assumptions about the ordering of occupations resulted in different conclusions. A variable might also reflect ordering on more than one dimension, such as attitude scales that reflect both the intensity of opinion and the direction of opinion. Moreover, surveys commonly include the category "don't know", which probably does not correspond to the middle category in a scale, even though analysts might be tempted to treat it this way. Overall, when the proper ordering is ambiguous, the models for nominal outcomes discussed in Chapter 6 should be considered.

We begin by reviewing the statistical model, followed by an examination of testing, fit, and methods of interpretation. These discussions are intended as a review for those who are familiar with the models. For a complete discussion, see Long (1997). We end the chapter by considering several less-common models for ordinal outcomes, which can be fitted using ado-files that others have developed. As always, you can obtain sample do-files and data files by downloading the `spostst4` package (see Chapter 1 for details).

5.1   The statistical model

The ORM can be developed in different ways, each of which leads to the same form of the model. These approaches to the model parallel those for the BRM. Indeed, the BRM can be viewed as a special case of the ordinal model in which the ordinal outcome has only two categories.

5.1.1   A latent variable model

The ordinal regression model is commonly presented as a latent variable model. Defining y^* as a latent variable ranging from $-\infty$ to ∞, the *structural model* is

$$y_i^* = \mathbf{x}_i\beta + \varepsilon_i$$

Or, for the case of a single independent variable,

$$y_i^* = \alpha + \beta x_i + \varepsilon_i$$

where i is the observation and ε is a random error, as discussed further below.

The *measurement model* for binary outcomes is expanded to divide y^* into J ordinal categories,

$$y_i = m \quad \text{if } \tau_{m-1} \le y_i^* < \tau_m \quad \text{for } m = 1 \text{ to } J$$

where the *cut-points* τ_1 through τ_{J-1} are estimated. (Some authors refer to these as thresholds.) We assume $\tau_0 = -\infty$ and $\tau_J = \infty$ for reasons that will be clear shortly.

To illustrate the measurement model, consider the example that is used in this chapter. People are asked to respond to the following statement:

> A working mother can establish just as warm and secure of a relationship with her child as a mother who does not work.

Possible responses are: $1 =$ Strongly Disagree (SD), $2 =$ Disagree (D), $3 =$ Agree (A), and $4 =$ Strongly Agree (SA). The continuous latent variable can be thought of as the *propensity* to agree that working mothers can be warm and secure mothers. The observed response categories are tied to the latent variable by the measurement model:

$$y_i = \begin{cases} 1 \Rightarrow \text{SD} & \text{if } \tau_0 = -\infty \leq y_i^* < \tau_1 \\ 2 \Rightarrow \text{D} & \text{if } \tau_1 \leq y_i^* < \tau_2 \\ 3 \Rightarrow \text{A} & \text{if } \tau_2 \leq y_i^* < \tau_3 \\ 4 \Rightarrow \text{SA} & \text{if } \tau_3 \leq y_i^* < \tau_4 = \infty \end{cases}$$

Thus, when the latent y^* crosses a cut-point, the observed category changes.

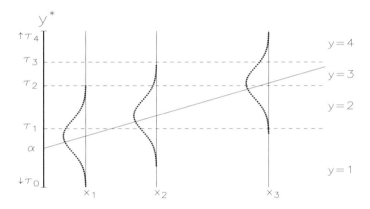

Figure 5.1: Relationship between observed y and latent y^* in ordinal regression model with a single independent variable.

For a single independent variable, the structural model is $y^* = \alpha + \beta x + \varepsilon$, which is plotted in Figure 5.1 along with the cut-points for the measurement model. This figure is similar to that for the binary regression model, except that there are now three horizontal lines representing the cut-points τ_1, τ_2, and τ_3. The three cut-points lead to four levels of y that are labeled on the right-hand side of the graph.

The probability of an observed outcome for a given value of x is the area under the curve between a pair of cut-points. For example, the probability of observing $y = m$ for given values of the xs corresponds to the region of the distribution where y^* falls between τ_{m-1} and τ_m:

$$\Pr(y = m \mid \mathbf{x}) = \Pr(\tau_{m-1} \leq y^* < \tau_m \mid \mathbf{x})$$

Substituting $\mathbf{x}\beta + \varepsilon$ for y^* and using some algebra leads to the standard formula for the predicted probability in the ORM,

$$\Pr(y = m \mid \mathbf{x}) = F(\tau_m - \mathbf{x}\beta) - F(\tau_{m-1} - \mathbf{x}\beta) \tag{5.1}$$

where F is the cdf for ε. In ordinal probit, F is normal with $\mathrm{Var}(\varepsilon) = 1$; in ordinal logit, F is logistic with $\mathrm{Var}(\varepsilon) = \pi^2/3$. Note that for $y = 1$, the second term on the right drops out because $F(-\infty - \mathbf{x}\beta) = 0$, and for $y = J$, the first term equals $F(\infty - \mathbf{x}\beta) = 1$.

Comparing these equations with those for the BRM shows that the ORM is identical to the binary regression model, with one exception. To show this, we fit Chapter 4's binary model for labor-force participation using both `logit` and `ologit` (the command for ordinal logit):

```
. use http://www.stata-press.com/data/lfr/binlfp2, clear
(Data from 1976 PSID-T Mroz)
. logit lfp k5 k618 age wc hc lwg inc, nolog
(output omitted)
. estimates store logit
. ologit lfp k5 k618 age wc hc lwg inc, nolog
(output omitted)
. estimates store ologit
```

To compare the coefficients, we combine them using `estimates table`; this leads to the following table:

```
. estimates table logit ologit, b(%9.3f) t label varwidth(30)
```

Variable	logit	ologit
# kids < 6	-1.463	-1.463
	-7.43	-7.43
# kids 6-18	-0.065	-0.065
	-0.95	-0.95
Wife's age in years	-0.063	-0.063
	-4.92	-4.92
Wife College: 1=yes 0=no	0.807	0.807
	3.51	3.51
Husband College: 1=yes 0=no	0.112	0.112
	0.54	0.54
Log of wife's estimated wages	0.605	0.605
	4.01	4.01
Family income excluding wife's	-0.034	-0.034
	-4.20	-4.20
_cut1		-3.182
		-4.94
Constant	3.182	
	4.94	

                                                                    legend: b/t

The slope coefficients and their standard errors are identical, but for `logit` an intercept is reported (i.e., the coefficient associated with `_cons`), while for `ologit` the constant is replaced by the cut-point labeled `_cut1`, which is equal but of opposite sign.

This difference is due to how the two models are identified. As the ORM has been presented, there are "too many" free parameters; that is, you cannot estimate $J - 1$ thresholds and the constant, too. For a unique set of ML estimates to exist, an identifying assumption needs to be made about either the intercept or one of the cut-points. In

Stata, the ORM is identified by assuming that the intercept is 0, and the values of all cut-points are estimated. Some statistics packages for the ORM instead fix one of the cut-points to 0 and estimate the intercept. In presenting the BRM, we immediately assumed that the value that divided y^* into observed 0s and 1s was 0. In effect, we identified the model by assuming a threshold of 0. While different parameterizations can be confusing, keep in mind that the slope coefficients and predicted probabilities are the same under either parameterization (see Long 1997, 122–23 for further details).

5.1.2 A nonlinear probability model

The ordinal regression model can also be developed as a nonlinear probability model without appealing to the idea of a latent variable. Here, we show how this can be done for the ordinal logit model. First, we define the odds that an outcome is less than or equal to m versus greater than m given $\mathbf{x}$:

$$\Omega_{\leq m | > m}(\mathbf{x}) \equiv \frac{\Pr(y \leq m \mid \mathbf{x})}{\Pr(y > m \mid \mathbf{x})} \quad \text{for } m = 1, J - 1$$

For example, we could compute the odds of disagreeing or strongly disagreeing (i.e., $m \leq 2$) versus agreeing or strongly agreeing ($m > 2$). The log of the odds is assumed to equal

$$\ln \Omega_{\leq m | > m}(\mathbf{x}) = \tau_m - \mathbf{x}\beta \tag{5.2}$$

For a single independent variable and three categories (where we are fixing the intercept to equal 0),

$$\ln \frac{\Pr(y \leq 1 \mid \mathbf{x})}{\Pr(y > 1 \mid \mathbf{x})} = \tau_1 - \beta_1 x_1$$

$$\ln \frac{\Pr(y \leq 2 \mid \mathbf{x})}{\Pr(y > 2 \mid \mathbf{x})} = \tau_2 - \beta_1 x_1$$

While it may seem confusing that the model subtracts βx rather than adding it, this is a consequence of computing the logit of $y \leq m$ versus $y > m$. While we agree that it would be simpler to stick with $\tau_m + \beta x$, this is not the way the model is normally presented.

5.2 Estimation using ologit and oprobit

The ordered logit and probit models can be fitted with the following commands:

<u>ologit</u> *depvar* $\big[$*indepvars*$\big]$ $\big[$*weight*$\big]$ $\big[$**if** *exp*$\big]$ $\big[$**in** *range*$\big]$ $\big[$, <u>table</u> <u>r</u>obust
    <u>cl</u>uster(*varname*) <u>s</u>core(*newvarlist* | *stub*∗) <u>level</u>(#) <u>nolog</u> $\big]$

<u>oprobit</u> *depvar* $\big[$*indepvars*$\big]$ $\big[$*weight*$\big]$ $\big[$**if** *exp*$\big]$ $\big[$**in** *range*$\big]$ $\big[$, <u>table</u> <u>r</u>obust
    <u>cl</u>uster(*varname*) <u>s</u>core(*newvarlist* | *stub*∗) <u>level</u>(#) <u>nolog</u> $\big]$

In our experience, these models take more steps to converge than the models for either binary or nominal outcomes.

Variable lists

depvar is the dependent variable. The specific values assigned to the outcome categories are irrelevant, except that larger values are assumed to correspond to "higher" outcomes. For example, if you had three outcomes, you could use the values 1, 2, and 3, or -1.23, 2.3, and 999. Up to 50 outcomes are allowed in Intercooled Stata and Stata/SE; 20 outcomes are allowed in Small Stata.

indepvars is a list of independent variables. If *indepvars* is not included, Stata fits a model with only cut-points.

Specifying the estimation sample

if and in qualifiers can be used to restrict the estimation sample. For example, if you want to fit an ordered logit model for only those in the 1989 sample, you could specify `ologit warm age ed prst male white if yr89==1`.

Listwise deletion Stata excludes cases in which there are missing values for any of the variables in the model. Accordingly, if two models are fitted using the same dataset but have different sets of independent variables, it is possible to have different samples. We recommend that you use `mark` and `markout` (discussed in Chapter 3) to explicitly remove cases with missing data.

Weights

Both `ologit` and `oprobit` can be used with `fweights`, `pweights`, and `iweights`. See Chapter 3 for further details.

Options

`table` lists the equations for predicted probabilities and reports the *observed* percent of cases for each category in the estimation sample. For example,

warm	Probability	Observed
SD	Pr(xb+u<_cut1)	0.1295
D	Pr(_cut1<xb+u<_cut2)	0.3153
A	Pr(_cut2<xb+u<_cut3)	0.3733
SA	Pr(_cut3<xb+u)	0.1819

`robust` indicates that robust variance estimates are to be used. When `cluster()` is specified, robust standard errors are automatically used. See Chapter 3 for further details.

cluster(*varname)* specifies that the observations are independent across the groups specified by unique values of *varname* but not necessarily within the groups. See Chapter 3 for further details.

score(*newvarlist | stub**) creates k new variables, where k is the number of possible outcomes. Each new variable contains the contributions to the score for each equation in the model; see [U] **23.15 Obtaining scores**.

level(*#)* specifies the level of the confidence interval for estimated parameters. By default, Stata uses a 95% interval. You can also change the default level, say, to a 90% interval, with the command set level 90.

nolog suppresses the iteration history.

5.2.1 Example of attitudes toward working mothers

Our example is based on a question from the 1977 and 1989 General Social Survey. As we have already described, respondents were asked to evaluate the following statement: "A working mother can establish just as warm and secure of a relationship with her child as a mother who does not work". Responses were coded as: 1 = Strongly Disagree (SD), 2 = Disagree (D), 3 = Agree (A), and 4 = Strongly Agree (SA). A complete description of the data can be obtained by using describe, summarize, and tabulate:

```
. use http://www.stata-press.com/data/lfr/ordwarm2
(77 & 89 General Social Survey)
. describe warm yr89 male white age ed prst
              storage  display   value
variable name  type    format    label     variable label

warm           byte    %10.0g    SD2SA     Mom can have warm relations
                                             with child
yr89           byte    %10.0g    yrlbl     Survey year: 1=1989 0=1977
male           byte    %10.0g    sexlbl    Gender: 1=male 0=female
white          byte    %10.0g    racelbl   Race: 1=white 0=not white
age            byte    %10.0g              Age in years
ed             byte    %10.0g              Years of education
prst           byte    %10.0g              Occupational prestige
. sum warm yr89 male white age ed prst
    Variable |      Obs        Mean    Std. Dev.       Min        Max

        warm |     2293    2.607501    .9282156          1          4
        yr89 |     2293    .3986044    .4897178          0          1
        male |     2293    .4648932    .4988748          0          1
       white |     2293    .8765809    .3289894          0          1
         age |     2293    44.93546    16.77903         18         89

          ed |     2293    12.21805    3.160827          0         20
        prst |     2293    39.58526    14.49226         12         82
```

```
. tab warm
```

Mom can have warm relations with child	Freq.	Percent	Cum.
SD	297	12.95	12.95
D	723	31.53	44.48
A	856	37.33	81.81
SA	417	18.19	100.00
Total	2,293	100.00	

Using these data, we fitted the model

$$\Pr(\text{warm} = m \mid \mathbf{x}_i) = F(\tau_m - \mathbf{x}\beta) - F(\tau_{m-1} - \mathbf{x}\beta)$$

where

$$\mathbf{x}\beta = \beta_{\text{yr89}}\text{yr89} + \beta_{\text{male}}\text{male} + \beta_{\text{white}}\text{white} + \beta_{\text{age}}\text{age} + \beta_{\text{prst}}\text{prst}$$

Here is the output from `ologit` and `oprobit`, which we combine using `outreg`:

```
. ologit warm yr89 male white age ed prst, nolog
```

Ordered logit estimates				Number of obs	=	2293
				LR chi2(6)	=	301.72
				Prob > chi2	=	0.0000
Log likelihood = -2844.9123				Pseudo R2	=	0.0504

warm	Coef.	Std. Err.	z	P>|z|	[95% Conf. Interval]	
yr89	.5239025	.0798988	6.56	0.000	.3673037	.6805013
male	-.7332997	.0784827	-9.34	0.000	-.8871229	-.5794766
white	-.3911595	.1183808	-3.30	0.001	-.6231815	-.1591374
age	-.0216655	.0024683	-8.78	0.000	-.0265032	-.0168278
ed	.0671728	.015975	4.20	0.000	.0358624	.0984831
prst	.0060727	.0032929	1.84	0.065	-.0003813	.0125267
_cut1	-2.465362	.2389126	(Ancillary parameters)			
_cut2	-.630904	.2333155				
_cut3	1.261854	.2340179				

```
. estimates store ologit
```

```
. oprobit warm yr89 male white age ed prst, nolog
Ordered probit estimates                      Number of obs   =       2293
                                              LR chi2(6)      =     294.32
                                              Prob > chi2     =     0.0000
Log likelihood =  -2848.611                   Pseudo R2       =     0.0491
```

warm	Coef.	Std. Err.	z	P>|z|	[95% Conf. Interval]	
yr89	.3188147	.0468519	6.80	0.000	.2269867	.4106427
male	-.4170287	.0455459	-9.16	0.000	-.5062971	-.3277603
white	-.2265002	.0694773	-3.26	0.001	-.3626733	-.0903272
age	-.0122213	.0014427	-8.47	0.000	-.0150489	-.0093937
ed	.0387234	.0093241	4.15	0.000	.0204485	.0569982
prst	.003283	.001925	1.71	0.088	-.0004899	.0070559
_cut1	-1.428578	.1387742	(Ancillary parameters)			
_cut2	-.3605589	.1369219				
_cut3	.7681637	.1370564				

```
. estimates store oprobit
```

The information in the header and the table of coefficients is in the same form as discussed in Chapter 3.

The estimated coefficients have been combined using `estimates table`. The first call of the program saves the coefficients from `ologit` to the file `05lgtpbt.out`, while the second call using `append` adds the coefficients from `oprobit`. After making a few edits to the file, we get

```
. estimates table ologit oprobit, b(%9.3f) t label varwidth(30)
```

Variable	ologit	oprobit
Survey year: 1=1989 0=1977	0.524	0.319
	6.56	6.80
Gender: 1=male 0=female	-0.733	-0.417
	-9.34	-9.16
Race: 1=white 0=not white	-0.391	-0.227
	-3.30	-3.26
Age in years	-0.022	-0.012
	-8.78	-8.47
Years of education	0.067	0.039
	4.20	4.15
Occupational prestige	0.006	0.003
	1.84	1.71
_cut1	-2.465	-1.429
	-10.32	-10.29
_cut2	-0.631	-0.361
	-2.70	-2.63
_cut3	1.262	0.768
	5.39	5.60

```
                                            legend: b/t
```

As with the BRM, the estimated coefficients differ from logit to probit by a factor of
about 1.7, reflecting the differing scaling of the ordered logit and ordered probit models.
Values of the z-tests are very similar because they are not affected by the scaling,
but they are not identical because of slight differences in the shape of the assumed
distribution of the errors.

5.2.2 Predicting perfectly

If the dependent variable does not vary within one of the categories of an independent
variable, there will be a problem with estimation. To see what happens, let's transform
the prestige variable prst into a dummy variable:

```
. gen dumprst = (prst<20 & warm==1)
. tab dumprst warm, miss
```

	Mom can have warm relations with child				
dumprst	SD	D	A	SA	Total
0	257	723	856	417	2,253
1	40	0	0	0	40
Total	297	723	856	417	2,293

In all cases where dumprst is 1, respondents have values of SD for warm. That is, if
you know dumprst is 1 you can predict perfectly that warm is 1 (i.e., SD). While we
purposely constructed dumprst so this would happen, perfect prediction can also occur
in real data. If we estimate the ORM using dumprst rather than prst,

```
. ologit warm yr89 male white age ed dumprst, nolog
```

Ordered logit estimates

Number of obs	= 2293
LR chi2(6)	= 447.02
Prob > chi2	= 0.0000

Log likelihood = -2772.2621

Pseudo R2	= 0.0746

warm	Coef.	Std. Err.	z	P>|z|	[95% Conf.	Interval]
yr89	.5268578	.0805997	6.54	0.000	.3688853	.6848303
male	-.7251825	.0792896	-9.15	0.000	-.8805872	-.5697778
white	-.4240687	.1197416	-3.54	0.000	-.658758	-.1893795
age	-.0210592	.0024462	-8.61	0.000	-.0258536	-.0162648
ed	.072143	.0133133	5.42	0.000	.0460494	.0982366
dumprst	-34.58373	1934739	-0.00	1.000	-3792053	3791983
_cut1	-2.776233	.243582	(Ancillary parameters)			
_cut2	-.8422903	.2363736				
_cut3	1.06148	.236561				

note: 40 observations completely determined. Standard errors questionable.

The note at the bottom of the output above indicates the problem. In practice, the
next step would be to delete the 40 cases in which dumprst equals 1 (you could use
the command drop if dumprst==1 to do this) and refit the model without dumprst.
This corresponds to what is done automatically for binary models fitted by logit and
probit.

5.3 Hypothesis testing with test and lrtest

Hypothesis tests of regression coefficients can be evaluated with the z-statistics in the estimation output, with `test` for Wald tests of simple and complex hypotheses, and with `lrtest` for the corresponding likelihood-ratio tests. We will briefly review each.

5.3.1 Testing individual coefficients

If the assumptions of the model hold, the ML estimators from `ologit` and `oprobit` are distributed asymptotically normally. The hypothesis $H_0: \beta_k = \beta^*$ can be tested with $z = \left(\widehat{\beta}_k - \beta^*\right) / \widehat{\sigma}_{\widehat{\beta}_k}$. Under the assumptions justifying ML, if H_0 is true, then z is distributed approximately normally with a mean of 0 and a variance of 1 for large samples. For example, consider the results for the variable `male` from the `ologit` output above:

```
. ologit warm male yr89 white age ed prst, nolog
  (output omitted)
```

warm	Coef.	Std. Err.	z	P>|z|	[95% Conf. Interval]
male	-.7332997	.0784827	-9.34	0.000	-.8871229 -.5794766

```
  (output omitted)
```

We conclude that

> gender significantly affects attitudes toward working mothers ($z = -9.34, p < 0.01$ for a two-tailed test).

Either a one-tailed or a two-tailed test can be used as discussed in Chapter 4.

The z-test in the output of estimation commands is a Wald test, which can also be computed using `test`. For example, to test $H_0: \beta_{\texttt{male}} = 0$,

```
. test male
 ( 1)  male = 0
          chi2( 1) =    87.30
        Prob > chi2 =    0.0000
```

We conclude that

> gender significantly affects attitudes toward working mothers ($X^2 = 87.30, df = 1, p < 0.01$).

The value of a chi-squared test with 1 degree of freedom is identical to the square of the corresponding z-test, which can be demonstrated with the `display` command:

```
. display "z*z=" -9.343*-9.343
z*z=87.291649
```

An LR test is computed by comparing the log likelihood from a full model with that of a restricted model. To test a single coefficient, we begin by fitting the full model:

```
. ologit warm yr89 male white age ed prst, nolog
Ordered logit estimates                     Number of obs   =        2293
                                            LR chi2(6)      =      301.72
                                            Prob > chi2     =      0.0000
Log likelihood = -2844.9123                 Pseudo R2       =      0.0504
   (output omitted)
. estimates store fmodel
```

Then we fit the model, excluding `male`:

```
. ologit warm yr89 white age ed prst, nolog
Ordered logit estimates                     Number of obs   =        2293
                                            LR chi2(5)      =      212.98
                                            Prob > chi2     =      0.0000
Log likelihood =  -2889.278                 Pseudo R2       =      0.0355
   (output omitted)
. estimates store nmodel
. lrtest fmodel nmodel
likelihood-ratio test                       LR chi2(1)   =       88.73
(Assumption: nmodel nested in fmodel)       Prob > chi2 =      0.0000
```

The resulting LR test can be interpreted that

the effect of being male is significant at the .01 level $(LRX^2 = 88.73, df = 1, p < .01)$.

5.3.2 Testing multiple coefficients

We can also test a complex hypothesis that involves more than one coefficient. For example, our model has three demographic variables: age, `white`, and `male`. To test that all of the demographic factors are simultaneously equal to zero, H_0: $\beta_{age} = \beta_{white} = \beta_{male} = 0$, we can use either a Wald or an LR test. For the Wald test, we fit the full model as before and then type

```
. test age white male
 ( 1)   age = 0
 ( 2)   white = 0
 ( 3)   male = 0
         chi2(  3) =   166.62
       Prob > chi2 =     0.0000
```

We conclude that

the hypothesis that the demographic effects of age, race, and gender are simultaneously equal to zero can be rejected at the .01 level $(X^2 = 166.62, df = 3, p < .01)$.

`test` can also be used to test the equality of effects as shown in Chapter 4.

To compute an LR test of multiple coefficients, we first fit the full model and save the results with `lrtest, saving(0)`. Then, to test H_0: $\beta_{\text{age}} = \beta_{\text{white}} = \beta_{\text{male}} = 0$, we fit the model that excludes these three variables and run `lrtest`:

```
. ologit warm yr89 male white age ed prst, nolog
  (output omitted )

. estimates store fmodel

. ologit warm yr89 ed prst, nolog
  (output omitted )

. estimates store nmodel

. lrtest fmodel nmodel
likelihood-ratio test                            LR chi2(3)  =    171.58
(Assumption: nmodel nested in fmodel)            Prob > chi2 =    0.0000
```

We conclude that

> the hypothesis that the demographic effects of age, race, and gender are simultaneously equal to zero can be rejected at the .01 level $(X^2 = 171.58, df = 3, p < .01)$.

In our experience, the Wald and LR tests usually lead to the same decisions. When there are differences, they generally occur when the tests are near the cutoff for statistical significance. Given that the LR test is invariant to reparameterization, we prefer the LR test.

5.4 Scalar measures of fit using fitstat

As we discuss at greater length in Chapter 3, scalar measures of fit can be useful in comparing competing models (see also Long 1997, 85–113). Several different measures can be computed after either `ologit` or `oprobit` with the SPost command `fitstat`:

```
. ologit warm yr89 male white age ed prst, nolog
  (output omitted )

. fitstat
Measures of Fit for ologit of warm
Log-Lik Intercept Only:    -2995.770    Log-Lik Full Model:       -2844.912
D(2284):                    5689.825    LR(6):                       301.716
                                        Prob > LR:                     0.000
McFadden's R2:                 0.050    McFadden's Adj R2:             0.047
Maximum Likelihood R2:         0.123    Cragg & Uhler's R2:            0.133
McKelvey and Zavoina's R2:     0.127
Variance of y*:                3.768    Variance of error:             3.290
Count R2:                      0.432    Adj Count R2:                  0.093
AIC:                           2.489    AIC*n:                      5707.825
BIC:                      -11982.891    BIC':                       -255.291
```

Using simulations, both Hagle and Mitchell (1992) and Windmeijer (1995) find that, for ordinal outcomes, McKelvey and Zavoina's R^2 most closely approximates the R^2 obtained by fitting the linear regression model on the underlying latent variable.

5.5   Converting to a different parameterization*

Earlier, we noted that different software packages use different parameterizations to identify the model. Stata sets $\beta_0 = 0$ and estimates τ_1, while some programs fix $\tau_1 = 0$ and estimate β_0. While all quantities of interest for the purpose of interpretation (e.g., predicted probabilities) are the same under both parameterizations, it is useful to see how Stata can fit the model under either parameterization. The key to understanding how this is done is the equation

$$\Pr\left(y = m \mid \mathbf{x}\right) = F\left(\left[\tau_m - \delta\right] - \left[\beta_0 - \delta\right] - \mathbf{x}\beta\right) - F\left(\left[\tau_{m-1} - \delta\right] - \left[\beta_0 - \delta\right] - \mathbf{x}\beta\right)$$

Without further constraints, it is only possible to estimate the differences $\tau_m - \delta$ and $\beta_0 - \delta$. Stata assumes $\delta = \beta_0$, which forces the estimate of β_0 to be 0, while some other programs assume $\delta = \tau_1$, which forces the estimate of τ_1 to be 0. For example,

Model Parameter	Stata's Estimate	Alternative Parameterization
β_0	$\beta_0 - \beta_0 = 0$	$\beta_0 - \tau_1$
τ_1	$\tau_1 - \beta_0$	$\tau_1 - \tau_1 = 0$
τ_2	$\tau_2 - \beta_0$	$\tau_2 - \tau_1$
τ_3	$\tau_3 - \beta_0$	$\tau_3 - \tau_1$

While you would only need to compute the alternative parameterization if you wanted to compare your results with those produced by another statistics package, seeing how this is done illustrates why the intercept and thresholds are arbitrary. To estimate the alternative parameterization, we use lincom to estimate the difference between Stata's estimates (see page 158) and the estimated value of the first cut-point:

```
. ologit warm yr89 male white age ed prst, nolog
  (output omitted)
. * intercept
. lincom 0 - _b[_cut1]
  ( 1)  - _cut1 = 0
```

warm	Coef.	Std. Err.	z	P>|z|	[95% Conf. Interval]
(1)	2.465362	.2389126	10.32	0.000	1.997102 2.933622

Here, we are computing the alternative parameterization of the intercept. ologit assumes that $\beta_0 = 0$, so we simply estimate $0 - \tau_1$; that is, 0-_b[_cut1]. The trick is that the cut-points are contained in the vector _b[ ], with the index for these scalars specified as _cut1, _cut2, and _cut3. For the thresholds, we are estimating $\tau_2 - \tau_1$ and $\tau_3 - \tau_1$, which correspond to _b[_cut2]-_b[_cut1] and _b[_cut3]-_b[_cut1]:

```
. * cutpoint 2
. lincom _b[_cut2] - _b[_cut1]
 ( 1)  - _cut1 + _cut2 = 0
```

warm	Coef.	Std. Err.	z	P>|z|	[95% Conf. Interval]
(1)	1.834458	.0630432	29.10	0.000	1.710895 1.95802

```
. * cutpoint 3
. lincom _b[_cut3] - _b[_cut1]
 ( 1)  - _cut1 + _cut3 = 0
```

warm	Coef.	Std. Err.	z	P>|z|	[95% Conf. Interval]
(1)	3.727216	.0826215	45.11	0.000	3.565281 3.889151

The estimate of $\tau_1 - \tau_1$ is, of course, 0.

5.6 The parallel regression assumption

Before discussing interpretation, it is important to understand an assumption that is implicit in the ORM, known both as the *parallel regression assumption* and, for the ordinal logit model, as the *proportional odds assumption*. Using (5.1), the ORM can be written as

$$\Pr(y = 1 \mid \mathbf{x}) = F(\tau_m - \mathbf{x}\beta)$$
$$\Pr(y = m \mid \mathbf{x}) = F(\tau_m - \mathbf{x}\beta) - F(\tau_{m-1} - \mathbf{x}\beta) \text{ for } m = 2 \text{ to } J - 1$$
$$\Pr(y = J \mid \mathbf{x}) = 1 - F(\tau_{m-1} - \mathbf{x}\beta)$$

These equations can be used to compute the cumulative probabilities, which have the simple form

$$\Pr(y \le m \mid \mathbf{x}) = F(\tau_m - \mathbf{x}\beta) \quad \text{for } m = 1 \text{ to } J - 1$$

This equation shows that the ORM is equivalent to $J - 1$ binary regressions with the critical assumption that the slope coefficients are identical across each regression.

For example, with four outcomes and a single independent variable, the equations are

$$\Pr(y \le 1 \mid \mathbf{x}) = F(\tau_1 - \beta x)$$
$$\Pr(y \le 2 \mid \mathbf{x}) = F(\tau_2 - \beta x)$$
$$\Pr(y \le 3 \mid \mathbf{x}) = F(\tau_3 - \beta x)$$

The intercept α is not in the equation since it has been assumed to equal 0 to identify the model. These equations lead to the following figure:[1]

---

[1]This plot illustrates how **graph** can be used to construct graphs that are not based on real data. The commands for this graph are contained in **st8ch5.do**, which is part of the package **spostst8**. See Chapter 1 for details.

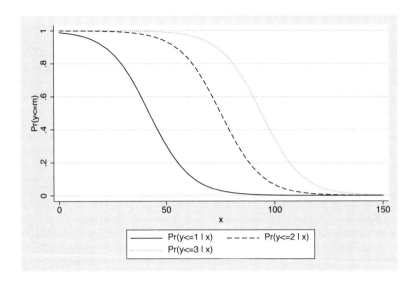

Each probability curve differs *only* in being shifted to the left or right. That is, they are parallel as a consequence of the assumption that the βs are equal for each equation.

This figure suggests that the parallel regression assumption can be tested by comparing the estimate from the $J-1$ binary regressions,

$$\Pr(y \le m \mid \mathbf{x}) = F(\tau_m - \mathbf{x}\beta_m) \quad \text{for } m = 1, J-1$$

where the βs are allowed to differ across the equations. The parallel regression assumption implies that $\beta_1 = \beta_2 = \cdots = \beta_{J-1}$. To the degree that the parallel regression assumption holds, the coefficients $\widehat{\beta}_1, \widehat{\beta}_2, \ldots, \widehat{\beta}_{J-1}$ should be "close". There are two commands in Stata that perform this test.

An approximate LR test

The command `omodel` (Wolfe and Gould 1998) is not part of official Stata but can be obtained by typing `net search omodel` and following the prompts. `omodel` computes an approximate LR test. Essentially, this method compares the log likelihood from `ologit` (or `oprobit`) with that obtained from pooling $J-1$ binary models fitted with `logit` (or `probit`), making an adjustment for the correlation between the binary outcomes defined by $y \le m$. The syntax is

`omodel` [logit | probit] *depvar* [*varlist*] [*weight*] [if *exp*] [in *range*]

where the subcommand `logit` or `probit` indicates whether ordered logit or ordered probit is to be used. For example,

```
. omodel logit warm yr89 male white age ed prst
 (same output as for ologit warm yr89 male white age ed prst)
Approximate likelihood-ratio test of proportionality of odds
across response categories:
        chi2(12) =     48.91
      Prob > chi2 =    0.0000
```

In this case, the parallel regression assumption can be rejected at the .01 level.

A Wald test

The LR test is an omnibus test that the coefficients for all variables are simultaneously equal. Accordingly, you cannot determine whether the coefficients for some variables are identical across the binary equations while coefficients for other variables differ. To this end, a Wald test by Brant (1990) is useful since it tests the parallel regression assumption for each variable individually. The messy details of computing this test are found in Brant (1990) or Long (1997, 143–144). In Stata the test is computed quickly with **brant**, which is part of **SPost**. After running **ologit** (**brant** does not work with **oprobit**), you run **brant** with the syntax:

brant $\left[\, , \ \texttt{detail} \,\right]$

The **detail** option provides a table of coefficients from each of the binary models. For example,

```
. brant, detail
Estimated coefficients from j-1 binary regressions
              y>1          y>2          y>3
  yr89     .9647422     .56540626     .31907316
  male   -.30536425   -.69054232    -1.0837888
 white   -.55265759   -.31427081    -.39299842
   age    -.0164704   -.02533448    -.01859051
    ed    .10479624     .05285265     .05755466
  prst   -.00141118     .00953216     .00553043
 _cons   1.8584045     .73032873    -1.0245168
```

Brant Test of Parallel Regression Assumption

Variable	chi2	p>chi2	df
All	49.18	0.000	12
yr89	13.01	0.001	2
male	22.24	0.000	2
white	1.27	0.531	2
age	7.38	0.025	2
ed	4.31	0.116	2
prst	4.33	0.115	2

```
A significant test statistic provides evidence that the parallel
regression assumption has been violated.
```

The chi-squared of 49.18 for the Brant test is very close to the value of 48.91 from the LR test. However, the Brant test shows that the largest violations are for `yr89` and `male`, which suggests that there may be problems related to these variables.

Caveat regarding the parallel regression assumption

In our experience, the parallel regression assumption is frequently violated. When the assumption of parallel regressions is rejected, alternative models should be considered that do not impose the constraint of parallel regressions. Violation of the parallel regression assumption is not a rationale for using ordinary least squares regression since the assumptions implied by the application of the LRM to ordinal data are even stronger. Alternative models that can be considered include models for nominal outcomes discussed in Chapter 6 or other models for ordinal outcomes discussed in Section 5.9.

5.7   Residuals and outliers using predict

While no methods for detecting influential observations and outliers have been developed specifically for the ORM, Hosmer and Lemeshow (2000, 305) suggest applying the methods for binary models to the $J - 1$ cumulative probabilities that were discussed in the last section. As noted by Hosmer and Lemeshow, the disadvantage of this approach is that you are only evaluating an approximation to the model you have fitted, because the coefficients of the binary models differ from those fitted in the ordinal model. But, if the parallel regression assumption is *not* rejected, you can be more confident in the results of your residual analysis.

To illustrate this approach, we start by generating three binary variables corresponding to `warm < 2`, `warm < 3`, and `warm < 4`:

```
. gen warmlt2 = (warm<2) if warm <.
. gen warmlt3 = (warm<3) if warm <.
. gen warmlt4 = (warm<4) if warm <.
```

For example, `warmlt3` is 1 if `warm` equals 1 or 2, else 0. Next, we estimate binary logits for `warmlt2`, `warmlt3`, and `warmlt4` using the same independent variables as in our original `ologit` model. After estimating each logit, we generate standardized residuals using `predict` (for a detailed discussion of generating and inspecting these residuals, see Chapter 4):

```
* warm < 2
. logit warmlt2 yr89 male white age ed prst
  (output omitted)
. predict rstd_lt2, rs
* warm < 3
. logit warmlt3 yr89 male white age ed prst
  (output omitted)
```

```
. predict rstd_lt3, rs
* warm < 4
. logit warmlt4 yr89 male white age ed prst
  (output omitted)
. predict rstd_lt4, rs
```

Next, we create an index plot for each of the three binary equations. For example, using the results from the logit of `warmlt3`,

```
. sort prst
. gen index = _n
. graph twoway scatter rstd_lt3 index, yline(0) ylabel(-4(2)4)
    xtitle("Observation Number")  xlabel(0(500)2293)
    ysize(2.6558) xsize(4.0413) msymbol(Oh)
```

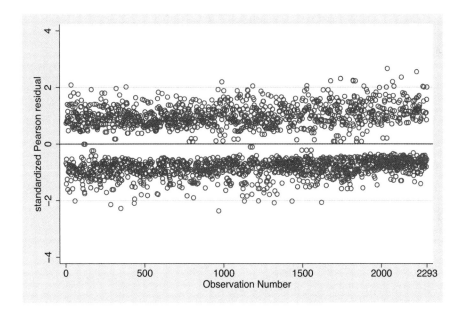

Given the size of the dataset, no residual stands out as being especially large.

5.8 Interpretation

If the idea of a latent variable makes substantive sense, simple interpretations are possible by rescaling y^* to compute standardized coefficients that can be used just like coefficients for the linear regression model. If the focus is on the categories of the ordinal variable (e.g., what affects the likelihood of strongly agreeing), the methods illustrated for the BRM can be extended to multiple outcomes. Because the ORM is nonlinear in the outcome probabilities, no single approach can fully describe the relationship between a variable and the outcome probabilities. Consequently, you should consider each of these methods before deciding which approach is most effective in your application. For purposes of illustration, we continue to use the example of attitudes toward working mothers. Keep in mind, however, that the test of the parallel regression assumption suggests that this model is not appropriate for these data.

5.8.1   Marginal change in y*

In the ORM, $y^* = \mathbf{x}\beta + \varepsilon$, and the marginal change in y^* with respect to x_k is

$$\frac{\partial y^*}{\partial x_k} = \beta_k$$

Because y^* is latent (and hence its metric is unknown), the marginal change cannot be interpreted without standardizing by the estimated standard deviation of y^*,

$$\widehat{\sigma}_{y^*}^2 = \widehat{\beta}'\widehat{\mathrm{Var}}\left(\mathbf{x}\right)\widehat{\beta} + \mathrm{Var}\left(\varepsilon\right)$$

where $\widehat{\mathrm{Var}}\left(\mathbf{x}\right)$ is the covariance matrix for the observed xs, $\widehat{\beta}$ contains ML estimates, and $\mathrm{Var}(\varepsilon) = 1$ for ordered probit and $\pi^2/3$ for ordered logit. Then, the y^*-*standardized* coefficient for x_k is

$$\beta_k^{Sy^*} = \frac{\beta_k}{\sigma_{y^*}}$$

which can be interpreted that

> for a unit increase in x_k, y^* is expected to increase by $\beta_k^{Sy^*}$ standard deviations, holding all other variables constant.

The *fully standardized coefficient* is

$$\beta_k^S = \frac{\sigma_k \beta_k}{\sigma_{y^*}} = \sigma_k \beta_k^{Sy^*}$$

which can be interpreted that

> for a standard deviation increase in x_k, y^* is expected to increase by β_k^S standard deviations, holding all other variables constant.

These coefficients can be computed with `listcoef` using the `std` option. For example, after fitting the ordered logit model,

```
. listcoef, std help
ologit (N=2293): Unstandardized and Standardized Estimates
  Observed SD: .9282156
    Latent SD: 1.9410634
```

warm	b	z	P>|z|	bStdX	bStdY	bStdXY	SDofX
yr89	0.52390	6.557	0.000	0.2566	0.2699	0.1322	0.4897
male	-0.73330	-9.343	0.000	-0.3658	-0.3778	-0.1885	0.4989
white	-0.39116	-3.304	0.001	-0.1287	-0.2015	-0.0663	0.3290
age	-0.02167	-8.778	0.000	-0.3635	-0.0112	-0.1873	16.7790
ed	0.06717	4.205	0.000	0.2123	0.0346	0.1094	3.1608
prst	0.00607	1.844	0.065	0.0880	0.0031	0.0453	14.4923

```
       b = raw coefficient
       z = z-score for test of b=0
   P>|z| = p-value for z-test
   bStdX = x-standardized coefficient
   bStdY = y-standardized coefficient
  bStdXY = fully standardized coefficient
   SDofX = standard deviation of X
```

If we think of the dependent variable as measuring "support" for mothers in the workplace, then the effect of the year of the interview can be interpreted as indicating that

> in 1989 support was .27 standard deviations higher than in 1977, holding all other variables constant.

To consider the effect of education,

> Each standard deviation increase in education increases support by .11 standard deviations, holding all other variables constant.

5.8.2 Predicted probabilities

For the most part, we prefer interpretations based in one way or another on predicted probabilities. These probabilities can be estimated with the formula

$$\widehat{\Pr}\left(y = m \mid \mathbf{x}\right) = F\left(\widehat{\tau}_m - \mathbf{x}\widehat{\beta}\right) - F\left(\widehat{\tau}_{m-1} - \mathbf{x}\widehat{\beta}\right)$$

with cumulative probabilities computed as

$$\widehat{\Pr}\left(y \le m \mid \mathbf{x}\right) = F\left(\tau_m - \mathbf{x}\widehat{\beta}\right)$$

The values of $\mathbf{x}$ can be based on observations in the sample or can be hypothetical values of interest. The most basic command for computing probabilities is predict, but our SPost commands can be used to compute predicted probabilities in particularly useful ways.

5.8.3   Predicted probabilities with predict

After fitting a model with `ologit` or `oprobit`, a useful first step is to compute the in sample predictions with the command

predict *newvar1* [*newvar2* [*newvar3*...]] [if *exp*] [in *range*]

where you indicate one new variable name for each category of the dependent variable. For instance, in the following example `predict` specifies that the variables `SDwarm`, `Dwarm`, `Awarm`, and `SAwarm` should be created with predicted values for the four outcome categories:

```
. ologit warm yr89 male white age ed prst, nolog
(output omitted)
. predict SDlogit Dlogit Alogit SAlogit
(option p assumed; predicted probabilities)
```

The message (`option p assumed; predicted probabilities`) reflects that `predict` can compute many different quantities. Because we did not specify an option indicating which quantity to predict, option `p` for predicted probabilities was assumed.

An easy way to see the range of the predictions is with `dotplot`, one of our favorite commands for quickly checking data:

```
. label var SDwarm "Pr(SD)"
. label var Dwarm "Pr(D)"
. label var Awarm "Pr(A)"
. label var SAwarm "Pr(SA)"
. dotplot SDwarm Dwarm Awarm SAwarm, ylabel(0(.25).75) ysize(2.0124) xsize(3.039)
```

which leads to the following plot:

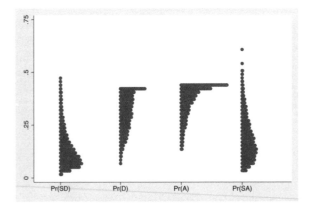

The predicted probabilities for the extreme categories tend to be less than .25, with the majority of predictions for the middle categories falling between .25 and .5. In only a few cases is the probability of any outcome greater than .5.

Examining predicted probabilities within the sample provides a first, quick check of the model. To understand and present the substantive findings, however, it is usually more effective to compute predictions at specific, substantively informative values. Our commands `prvalue`, `prtab`, `prgen`, and `prchange` are designed to make this simple.

5.8.4   Individual predicted probabilities with prvalue

Predicted probabilities for individuals with a particular set of characteristics can be computed with `prvalue`. For example, we might want to examine the predicted probabilities for individuals with the following characteristics:

- Working class men in 1977 who are near retirement.

- Young, highly educated women with prestigious jobs.

- An "average" individual in 1977.

- An "average" individual in 1989.

Each of these can be easily computed with `prvalue` (see Chapter 3 for a discussion of options for this command). The predicted probabilities for older, working class men are

```
. prvalue, x(yr89=0 male=1 prst=20 age=64 ed=16) rest(mean)
ologit: Predictions for warm
  Pr(y=SD|x):         0.2317
  Pr(y=D|x):          0.4221
  Pr(y=A|x):          0.2723
  Pr(y=SA|x):         0.0739
          yr89      male     white       age        ed      prst
  x=          0         1  .8765809        64        16        20
```

For young, highly educated women with prestigious jobs, they are

```
. prvalue, x(yr89=1 male=0 prst=80 age=30 ed=24) rest(mean) brief
  Pr(y=SD|x):         0.0164
  Pr(y=D|x):          0.0781
  Pr(y=A|x):          0.3147
  Pr(y=SA|x):         0.5908
```

and so on, for other sets of values.

There are several points about using `prvalue` that are worth emphasizing. First, we have set the values of the independent variables that define our hypothetical person using the `x()` and `rest()` options. The output from the first call of `prvalue` lists the values that have been set for all independent variables. This allows you to verify that `x()` and `rest()` did what you intended. For the second call, we added the `brief` option. This suppresses the output showing the levels of the independent variables. If you use this option, be certain that you have correctly specified the levels of all variables. Second, the output of `prvalue` labels the categories according to the value labels assigned to the dependent variable. For example, `Pr(y=SD | x): 0.2317`. As it is very easy to be confused about the outcome categories when using these models, it is prudent to assign clear value labels to your dependent variable (we describe how to assign value labels in Chapter 2).

We can summarize the results in a table that lists the ideal types and provides a clear indication of which variables are important:

| | **Outcome Category** | | | |
Ideal Type	*SD*	*D*	*A*	*SA*
Working class men in 1977 who are near retirement.	0.23	0.42	0.27	0.07
Young, highly educated women in 1989 with				
prestigious jobs.	0.02	0.08	0.32	0.59
An "average individual" in 1977.	0.13	0.36	0.37	0.14
An "average individual" in 1989.	0.08	0.28	0.43	0.21

5.8.5   Tables of predicted probabilities with prtab

In other cases, it can be useful to compute predicted probabilities for all combinations of a set of categorical independent variables. For example, the ideal types illustrate the importance of gender and the year when the question was asked. Using `prtab`, we can easily show the degree to which these variables affect opinions for those average on other characteristics.

```
. prtab yr89 male, novarlbl
ologit: Predicted probabilities for warm
Predicted probability of outcome 1 (SD)

             male
   yr89   Women     Men

   1977   0.0989   0.1859
   1989   0.0610   0.1191

Predicted probability of outcome 2 (D)

             male
   yr89   Women     Men

   1977   0.3083   0.4026
   1989   0.2282   0.3394
```

(*tables for other outcomes omitted*)

Tip Sometimes the output of `prtab` is clearer without the variable labels. These can be suppressed with the `novarlbl` option.

The output from `prtab` can be rearranged into a table that clearly shows that men are more likely than women to strongly disagree or disagree with the proposition that working mothers can have as warm of relationships with their children as mothers who do not work. The table also shows that between 1977 and 1989 there was a movement for both men and women toward more positive attitudes about working mothers.

1977	SD	D	A	SA
Men	0.19	0.40	0.32	0.10
Women	0.10	0.31	0.41	0.18
Difference	0.09	0.09	-0.09	-0.08

1989	SD	D	A	SA
Men	0.12	0.34	0.39	0.15
Women	0.06	0.23	0.44	0.27
Difference	0.06	0.11	-0.05	-0.12

Change from 1977 to 1989				
	SD	D	A	SA
Men	-0.07	-0.06	0.07	0.05
Women	-0.04	-0.08	0.03	0.09

5.8.6 Graphing predicted probabilities with prgen

Graphing predicted probabilities for each outcome can also be useful for the ORM. In this example, we consider women in 1989 and show how predicted probabilities are affected by age. Of course, the plot could also be constructed for other sets of characteristics. The predicted probabilities as `age` ranges from 20 to 80 are generated by `prgen`:

```
. prgen age, from(20) to(80) generate(w89) x(male=0 yr89=1) ncases(13)
ologit: Predicted values as age varies from 20 to 80.
           yr89       male      white        age         ed       prst
    x=        1          0    .8765809  44.935456  12.218055  39.585259
```

You should be familiar with how `x()` operates, but it is useful to review the other options:

`from(20)` and `to(80)` specify the minimum and maximum values over which `inc` is to vary. The default is the variable's minimum and maximum values.

generate(w89) is the root name for the new variables.

ncases(13) indicates that 13 evenly spaced values of age between 20 and 80 should
   be generated.

In our example, w89x contains values of age ranging from 20 to 80. The p# variables
contain the predicted probability for outcome # (e.g., w89p2 is the predicted probability
of outcome 2). With ordinal outcomes, prgen also computes cumulative probabilities
(i.e., summed) that are indicated by s (e.g., w89s2 is the sum of the predicted probability
of outcomes 1 and 2). A list of the variables that are created should make this clear:

```
. desc w89*

                  storage  display    value
   variable name  type     format     label      variable label

   w89x           float    %9.0g                  Changing value of age
   w89p1          float    %9.0g                  pr(SD) [1]
   w89s1          float    %9.0g                  pr(y<=1)
   w89p2          float    %9.0g                  pr(D) [2]
   w89s2          float    %9.0g                  pr(y<=2)
   w89p3          float    %9.0g                  pr(A) [3]
   w89s3          float    %9.0g                  pr(y<=3)
   w89p4          float    %9.0g                  pr(SA) [4]
   w89s4          float    %9.0g                  pr(y<=4)
```

Although prgen assigns variable labels to the variables it creates, we can change these
to improve the look of the plot that we are creating. Specifically,

```
. label var w89p1 "SD"
. label var w89p2 "D"
. label var w89p3 "A"
. label var w89p4 "SA"
. label var w89s1 "SD"
. label var w89s2 "SD or D"
. label var w89s3 "SD, D or A"
```

First, we plot the probabilities of individual outcomes using graph. Because the graph
command is long, we use /// to allow the commands to be longer than one line in our
do-file.

```
. // step 1: graph predicted probabilities
. graph twoway connected w89p1 w89p2 w89p3 w89p4 w89x,        ///
         title("Panel A: Predicted Probabilities")            ///
         xtitle("Age") xlabel(20(10)80) ylabel(0(.25).50)     ///
         yscale(noline) ylabel("") xline(44.93)               ///
         ytitle("") name(graph1, replace)
```

This graph command plots the four predicted probabilities against generated values
for age contained in w89x. Standard options for graph are used to specify the axes
and labels. The vertical line specified by xline(44.93) marks the average age in the
sample. This line is used to illustrate the marginal effect discussed in Section 5.8.7.
Option name(graph1, replace) saves the graph in memory under the name graph1 so
that we can combine it with the next graph, which plots the cumulative probabilities:

```
. // step 2: graph cumulative probabilities
. graph twoway connected w89s1 w89s2 w89s3 w89x,              ///
        title("Panel B: Cumulative Probabilities")            ///
        xtitle("Age") xlabel(20(10)80) ylabel(0(.25)1)        ///
        yscale(noline) ylabel("") name(graph2, replace)       ///
        ytitle("")
```

Next, we combine these two graphs (see Chapter 2 for details on combining graphs):

```
. // step 3: combine graphs
. graph combine graph1 graph2, col(1) iscale(*.9) imargin(small)     ///
      ysize(4.31) xsize(3.287)
```

This leads to Figure 5.2. Panel A plots the predicted probabilities and shows that with age the probability of SA decreases rapidly, while the probability of D (and to a lesser degree SD) increases. Panel B plots the cumulative probabilities. both panels present the same information, which one you use is largely a matter of personal preference.

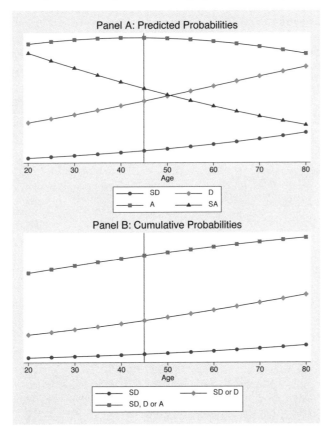

Figure 5.2: Plot of predicted probabilities for the ordered logit model.

5.8.7   Changes in predicted probabilities

When there are many variables in the model, it is impractical to plot them all. In such cases, measures of change in the outcome probabilities are useful for summarizing the effects of each variable. Before proceeding, however, we hasten to note that values of both discrete and marginal change depend on the levels of all variables in the model. We return to this point shortly.

Marginal change with prchange

The marginal change in the probability is computed as

$$\frac{\partial \Pr(y = m \mid \mathbf{x})}{\partial x_k} = \frac{\partial F(\tau_m - \mathbf{x}\beta)}{\partial x_k} - \frac{\partial F(\tau_{m-1} - \mathbf{x}\beta)}{\partial x_k}$$

which is the slope of the curve relating x_k to $\Pr(y=m|\mathbf{x})$, holding all other variables constant. In our example, we consider the marginal effect of age $(\partial \Pr(y = m \mid \mathbf{x}) /\partial \texttt{age})$ for women in 1989 who are average on all other variables. This corresponds to the slope of the curves in Panel A of Figure 5.2 evaluated at the vertical line (recall that this line is drawn at the average age in the sample). The marginal is computed with prchange, where we specify that only the coefficients for age should be computed:

```
. prchange age, x(male=0 yr89=1) rest(mean)

ologit: Changes in Predicted Probabilities for warm

age
              Avg|Chg|           SD           D            A           SA
 Min->Max    .16441458    .10941909    .21941006  -.05462247   -.27420671
    -+1/2    .00222661    .00124099    .00321223   -.0001803   -.00427291
   -+sd/2     .0373125     .0208976    .05372739  -.00300205   -.07162295
 MargEfct    .00222662    .00124098    .00321226  -.00018032   -.00427292

                    SD           D            A           SA
 Pr(y|x)    .06099996    .22815652    .44057754    .27026597

                  yr89      male     white       age        ed      prst
       x=           1         0    .876581   44.9355   12.2181   39.5853
    sd(x)=    .489718   .498875   .328989    16.779   3.16083   14.4923
```

The first thing to notice is the row labeled `Pr(y|x)`, which is the predicted probabilities at the values set by `x()` and `rest()`. In Panel A, these probabilities correspond to the intersection of the vertical line and the probability curves. The row `MargEfct` lists the slopes of the probability curves at the point of intersection with the vertical line in the figure. For example, the slope for SD (shown with circles) is .00124, while the slope for A (shown with squares) is negative and small. As with the BRM, the size of the slope indicates the instantaneous rate of change, but does not correspond exactly to the amount of change in the probability for a change of one unit in the independent variable. However, when the probability curve is approximately linear, the marginal effect can be used to summarize the effect of a unit change in the variable on the probability of an outcome.

Marginal change with mfx compute

Marginal change can also be computed using `mfx compute`, where `at()` is used to set values of the independent variables. Unlike `prchange`, `mfx` does not allow you to compute effects for a subset of the independent variables. And it only estimates the marginal effects for one outcome category at a time, where the category is specified with the option `predict(outcome(#))`. Using the same values for the independent variables as in the example above, we obtain the following results:

```
. mfx compute, at(male=0 yr89=1) predict(outcome(1))

Marginal effects after ologit
      y  = Pr(warm==1) (predict, outcome(1))
         =  .06099996
```

variable	dy/dx	Std. Err.	z	P>|z|	[95% C.I.]	X
yr89*	-.0378526	.00601	-6.30	0.000	-.049633 -.026072	1.00000
male*	.0581355	.00731	7.95	0.000	.043803 .072468	0.00000
white*	.0197511	.0055	3.59	0.000	.008972 .03053	.876581
age	.001241	.00016	7.69	0.000	.000925 .001557	44.9355
ed	-.0038476	.00097	-3.96	0.000	-.005754 -.001941	12.2181
prst	-.0003478	.00019	-1.83	0.068	-.000721 .000025	39.5853

(*) dy/dx is for discrete change of dummy variable from 0 to 1

The marginal for `age` is .001241, which matches the result obtained from `prchange`. The advantage of `mfx` is that it computes standard errors and confidence intervals.

Discrete change with prchange

As the marginal change can be misleading when the probability curve is changing rapidly or when an independent variable is a dummy variable, we prefer using discrete change (`mfx` computes discrete change for independent variables that are binary but not for other independent variables). The discrete change is the change in the predicted probability for a change in x_k from the start value x_S to the end value x_E (e.g., a change from $x_k = 0$ to $x_k = 1$). Formally,

$$\frac{\Delta \Pr(y = m \mid \mathbf{x})}{\Delta x_k} = \Pr(y = m \mid \mathbf{x}, x_k = x_E) - \Pr(y = m \mid \mathbf{x}, x_k = x_S)$$

where $\Pr(y = m \mid \mathbf{x}, x_k)$ is the probability that $y = m$ given $\mathbf{x}$, noting a specific value for x_k. The change is interpreted as indicating that

> when x_k changes from x_S to x_E, the predicted probability of outcome m changes by $\{\Delta \Pr(y = m \mid \mathbf{x})\}/\Delta x_k$, holding all other variables at $\mathbf{x}$.

The value of the discrete change depends on (1) the value at which x_k starts, (2) the amount of change in x_k, and (3) the values of all other variables. Most frequently, each continuous variable except x_k is held at its mean. For dummy independent variables, the change could be computed for both values of the variable. For example, we could compute the discrete change for `age` separately for men and women.

In our example, the discrete change coefficients for `male`, `age`, and `prst` for women in 1989, with other variables at their mean, are computed as follows:

```
. prchange male age prst, x(male=0 yr89=1) rest(mean)
ologit: Changes in Predicted Probabilities for warm
male
            Avg|Chg|         SD          D          A          SA
    0->1    .08469636    .05813552    .11125721  -.05015317  -.11923955
age
            Avg|Chg|         SD          D          A          SA
Min->Max    .16441458    .10941909    .21941006  -.05462247  -.27420671
  -+1/2     .00222661    .00124099    .00321223   -.0001803  -.00427291
  -+sd/2    .0373125     .0208976     .05372739  -.00300205  -.07162295
MargEfct    .00222662    .00124098    .00321226  -.00018032  -.00427292
prst
            Avg|Chg|         SD          D          A          SA
Min->Max    .04278038   -.02352008   -.06204067   .00013945   .08542132
  -+1/2     .00062411   -.00034784   -.00090037   .00005054   .00119767
  -+sd/2    .00904405   -.00504204   -.01304607   .00073212   .01735598
MargEfct    .0006241    -.00034784   -.00090038   .00005054   .00119767
                  SD          D          A          SA
Pr(y|x)     .06099996   .22815652   .44057754   .27026597

             yr89       male      white       age        ed       prst
    x=          1          0    .876581   44.9355   12.2181   39.5853
 sd(x)=    .489718    .498875   .328989    16.779   3.16083   14.4923
```

For variables that are not binary, the discrete change can be interpreted for a unit change centered on the mean, for a standard deviation change centered on the mean, or as the variable changes from its minimum to its maximum value. The following are two examples:

> For a standard deviation increase in age, the probability of disagreeing increases by .05, holding other variables constant at their means.

> Moving from the minimum prestige to the maximum prestige changes the predicted probability of strongly agreeing by .08, holding all other variables constant at their means.

The J discrete change coefficients for a variable can be summarized by computing the average of the *absolute values* of the changes across all of the outcome categories:

$$\overline{\Delta} = \frac{1}{J} \sum_{j=1}^{J} \left| \frac{\Delta \Pr(y = j \mid \overline{\mathbf{x}})}{\Delta x_k} \right|$$

The absolute value must be used because the sum of the changes without taking the absolute value is necessarily zero. These are labeled as `Avg | Chg |`. For example, the effect of being a male is on average 0.08, which is larger than the average effect of a standard deviation change in either age or occupational prestige.

Computing discrete change for a 10-year increase in age

In the example above, age was measured in years. Not surprisingly, the change in the predicted probability for a one-year increase in age is trivially small. But, to characterize the effect of age, we could report the effect of a ten-year change in age.

Warning It is tempting to compute the discrete change for a ten-year change in age by simply multiplying the one-year discrete change by 10. This will give you approximately the right answer *if* the probability curve is nearly linear over the range of change. But, when the curve is not linear, simply multiplying can give very misleading results and even the wrong sign. To be safe, don't do it!

The delta(#) option for prchange computes the discrete change as an independent value changes from #/2 units below the base value to #/2 above. In our example, we use delta(10) and set the base value of age to its mean:

```
. prchange age, x(male=0 yr89=1) rest(mean) delta(10)
ologit: Changes in Predicted Probabilities for warm
(Note: d = 10)
age
              Avg|Chg|          SD            D          A           SA
Min->Max      .16441458    .10941909    .21941006   -.05462247   -.27420671
   -+d/2      .02225603    .01242571    .03208634   -.00179818   -.04271388
  -+sd/2       .0373125     .0208976    .05372739   -.00300205   -.07162295
MargEfct      .00222662    .00124098    .00321226   -.00018032   -.00427292
                    SD           D           A          SA
Pr(y|x)     .06099996   .22815652   .44057754   .27026597
               yr89       male      white        age         ed       prst
    x=            1          0    .876581    44.9355    12.2181    39.5853
 sd(x)=     .489718    .498875    .328989     16.779    3.16083    14.4923
```

For females interviewed in 1989, the results in the -+d/2 row show the changes in the predicted probabilities associated with a ten-year increase in age centered on the mean.

5.8.8   Odds ratios using listcoef

For ologit, but not oprobit, we can interpret the results using odds ratios. Earlier, (5.2) defined the ordered logit model as

$$\Omega_{\leq m|>m}\left(\mathbf{x}\right) = \exp\left(\tau_m - \mathbf{x}\beta\right)$$

For example, with four outcomes we would simultaneously estimate three equations:

$$\Omega_{\leq 1|>1}\left(\mathbf{x}\right) = \exp\left(\tau_1 - \mathbf{x}\beta\right)$$
$$\Omega_{\leq 2|>2}\left(\mathbf{x}\right) = \exp\left(\tau_2 - \mathbf{x}\beta\right)$$
$$\Omega_{\leq 3|>3}\left(\mathbf{x}\right) = \exp\left(\tau_3 - \mathbf{x}\beta\right)$$

Using the same approach as shown for binary logit, the effect of a change in x_k of 1 equals

$$\frac{\Omega_{\leq m | > m}(\mathbf{x}, x_k + 1)}{\Omega_{\leq m | > m}(\mathbf{x}, x_k)} = e^{-\beta_k} = \frac{1}{e^{\beta_k}}$$

which can be interpreted as indicating that

> for a unit increase in x_k, the odds of an outcome being less than or equal to m is changed by the factor $\exp(-\beta_k)$, holding all other variables constant.

The value of the odds ratio does *not* depend on the value of m, which is why the parallel regression assumption is also known as the proportional odds assumption. We could interpret the odds ratio as follows:

> For a unit increase in x_k, the odds of a lower outcome compared with a higher outcome are changed by the factor $\exp(-\beta_k)$, holding all other variables constant.

or, for a change in x_k of δ,

$$\frac{\Omega_{\leq m | > m}(\mathbf{x}, x_k + \delta)}{\Omega_{\leq m | > m}(\mathbf{x}, x_k)} = \exp(-\delta \times \beta_k) = \frac{1}{\exp(\delta \times \beta_k)}$$

so that

> for an increase of δ in x_k, the odds of lower outcome compared with a higher outcome change by the factor $\exp(-\delta \times \beta_k)$, holding all other variables constant.

In these results, we are discussing factor changes in the odds of lower outcomes compared with higher outcomes. This is done because the model is traditionally written as $\ln \Omega_{\leq m | > m}(\mathbf{x}) = \tau_m - \mathbf{x}\beta$, which leads to the factor change coefficient of $\exp(-\beta_k)$. For purposes of interpretation, we could just as well consider the factor change in the odds of higher versus lower values; that is, changes in the odds $\Omega_{> m | \leq m}(\mathbf{x})$. This would equal $\exp(\beta_k)$.

The odds ratios for both a unit and a standard deviation change of the independent variables can be computed with `listcoef`, which lists the factor changes in the odds of higher versus lower outcomes. Here, we request coefficients for only `male` and `age`:

```
. ologit warm yr89 male white age ed prst, nolog
  (output omitted)
. listcoef male age, help
ologit (N=2293): Factor Change in Odds

  Odds of: >m vs <=m
```

warm	b	z	P>|z|	e^b	e^bStdX	SDofX
male	-0.73330	-9.343	0.000	0.4803	0.6936	0.4989
age	-0.02167	-8.778	0.000	0.9786	0.6952	16.7790

```
          b = raw coefficient
          z = z-score for test of b=0
      P>|z| = p-value for z-test
        e^b = exp(b) = factor change in odds for unit increase in X
   e^bStdX = exp(b*SD of X) = change in odds for SD increase in X
      SDofX = standard deviation of X
```

or to compute percent changes in the odds,

```
. listcoef male age, help percent
ologit (N=2293): Percentage Change in Odds

  Odds of: >m vs <=m
```

warm	b	z	P>|z|	%	%StdX	SDofX
male	-0.73330	-9.343	0.000	-52.0	-30.6	0.4989
age	-0.02167	-8.778	0.000	-2.1	-30.5	16.7790

```
          b = raw coefficient
          z = z-score for test of b=0
      P>|z| = p-value for z-test
          % = percent change in odds for unit increase in X
      %StdX = percent change in odds for SD increase in X
      SDofX = standard deviation of X
```

These results can be interpreted as follows:

> The odds of having more positive attitudes towards working mothers are
> .48 times smaller for men than women, holding all other variables constant.
> Equivalently, the odds of having more positive values are 52 percent smaller
> for men than women, holding other variables constant.

> For a standard deviation increase in age, the odds of having more positive
> attitudes decrease by a factor of .69, holding all other variables constant.

When presenting odds ratios, our experience is that people find it easier to understand the results if you talk about *increases* in the odds rather than *decreases*. That is, it is clearer to say, "The odds increased by a factor of 2" than to say, "The odds decreased by a factor of .5". If you agree, then you can reverse the order when presenting odds. For example, we could say that

The odds of having more *negative* attitudes towards working mothers are 2.08 times larger for men than women, holding all other variables constant.

This new factor change, 2.08, is just the inverse of the old value .48 (that is, $1/.48$). listcoef computes the odds of a lower category versus a higher category if you specify the reverse option:

```
. listcoef male, reverse
ologit (N=2293): Factor Change in Odds
  Odds of: <=m vs >m
```

warm	b	z	P>|z|	e^b	e^bStdX	SDofX
male	-0.73330	-9.343	0.000	2.0819	1.4417	0.4989

Notice that the output now says Odds of: <=m vs >m instead of Odds of: >m vs <=m, as it did earlier.

When interpreting the odds ratios, it is important to keep in mind two points that are discussed in detail in Chapter 4. First, because odds ratios are multiplicative coefficients, *positive and negative effects should be compared by taking the inverse of the negative effect* (or vice versa). For example, a negative factor change of .5 has the same magnitude as a positive factor change of $2 = 1/.5$. Second, the interpretation only assumes that the other variables have been held constant, not held at any specific values (as was required for discrete change). But, *a constant factor change in the odds does not correspond to a constant change or constant factor change in the probability.*

5.9   Less-common models for ordinal outcomes

Stata can also be used to fit several less-commonly used models for ordinal outcomes. In concluding this chapter, we describe these models briefly and note their commands for estimation. Our SPost commands do not work with these models. For gologit and ocratio, this is mainly because these commands do not fully incorporate the new methods of returning information that were introduced with Stata 6.

5.9.1   Generalized ordered logit model

The parallel regression assumption results from assuming the same coefficient vector β for all comparisons in the $J - 1$ equations

$$\ln \Omega_{\leq m|>m}(\mathbf{x}) = \tau_m - \mathbf{x}\beta$$

where $\Omega_{\leq m|>m}(\mathbf{x}) = \{\Pr(y \leq m \mid \mathbf{x})\} / \{\Pr(y > m \mid \mathbf{x})\}$. The generalized ordered logit model (GOLM) allows β to differ for each of the $J - 1$ comparisons. That is,

$$\ln \Omega_{\leq m|>m}(\mathbf{x}) = \tau_m - \mathbf{x}\beta_m \quad \text{for } j = 1 \text{ to } J - 1$$

where predicted probabilities are computed as

$$\Pr\left(y=1\mid\mathbf{x}\right) = \frac{\exp\left(\tau_1 - \mathbf{x}\beta_1\right)}{1 + \exp\left(\tau_1 - \mathbf{x}\beta_1\right)}$$

$$\Pr\left(y=j\mid\mathbf{x}\right) = \frac{\exp\left(\tau_j - \mathbf{x}\beta_j\right)}{1 + \exp\left(\tau_j - \mathbf{x}\beta_j\right)} - \frac{\exp\left(\tau_{j-1} - \mathbf{x}\beta_{j-1}\right)}{1 + \exp\left(\tau_{j-1} - \mathbf{x}\beta_{j-1}\right)} \quad \text{for } j = 2 \text{ to } J - 1$$

$$\Pr\left(y=J\mid\mathbf{x}\right) = 1 - \frac{\exp\left(\tau_{J-1} - \mathbf{x}\beta_{J-1}\right)}{1 + \exp\left(\tau_{J-1} - \mathbf{x}\beta_{J-1}\right)}$$

To ensure that the $\Pr\left(y=j\mid\mathbf{x}\right)$ is between 0 and 1, the condition

$$\left(\tau_j - \mathbf{x}\beta_j\right) \geq \left(\tau_{j-1} - \mathbf{x}\beta_{j-1}\right)$$

must hold. Once predicted probabilities are computed, all of the approaches used to interpret the ORM results can be readily applied. This model has been discussed by Clogg and Shihadeh (1994, 146–147), Fahrmeir and Tutz (1994, 91), and McCullagh and Nelder (1989, 155). It can be fitted in Stata with the add-on command `gologit` (Fu 1998). To obtain this command, enter `net search gologit` and follow the prompts to download it.

5.9.2 The stereotype model

The *stereotype ordered regression model* (SORM) was proposed by Anderson (1984) in response to the restrictive assumption of parallel regressions in the ordered regression model. The model, which can be fitted with `mclest`—written by Hendrickx (2000) (type `net search mclest` to download)—is a compromise between allowing the coefficients for each independent variable to vary by outcome category and restricting them to be identical across all outcomes. The SORM is defined as[2]

$$\ln \frac{\Pr\left(y=q\mid\mathbf{x}\right)}{\Pr\left(y=r\mid\mathbf{x}\right)} = \left(\alpha_q - \alpha_r\right)\beta_0 + \left(\phi_q - \phi_r\right)\left(\mathbf{x}\beta\right) \tag{5.3}$$

where β_0 is the intercept, and β is a vector of coefficients associated with the independent variables; as β_0 is included in the equation, it is not included in β. The αs and ϕs are scale factors associated with the outcome categories. The model allows the coefficients associated with each independent variable to differ by a scalar factor that depends on the pair of outcomes on the left-hand side of the equation. Similarly, the αs allow different intercepts for each pair of outcomes. As the model stands, there are too many unconstrained αs and ϕs for the parameters to be uniquely determined. The model can be identified in a variety of ways. For example, we can assume that $\phi_1 = 1$, $\phi_J = 0$, $\alpha_1 = 1$, and $\alpha_J = 0$. Or, using the approach from loglinear models for ordinal outcomes, the model is identified by the constraints $\sum_{j=1}^{J} \phi_j = 0$ and $\sum_{j=1}^{J} \phi_j^2 = 1$.

---

[2]The stereotype model can be set up in several different ways. For example, in some presentations, it is assumed that $\beta_0 = 0$ and fewer constraints are imposed on the αs. Here we parameterize the model to highlight its links to other models that we consider.

See DiPrete (1990) for further discussion. To ensure ordinality of the outcomes, $\phi_1 = 1 > \phi_2 > \cdots > \phi_{J-1} > \phi_J = 0$ must hold. Note that `mclest` does *not* impose this inequality constraint during estimation.

Equation (5.3) can be used to compute the predicted probabilities:

$$\Pr\left(y = m \mid \mathbf{x}\right) = \frac{\exp\left(\alpha_m \beta_0 + \phi_m \mathbf{x}\beta\right)}{\sum_{j=1}^{J} \exp\left(\alpha_j \beta_0 + \phi_j \mathbf{x}\beta\right)}$$

This formula can be used for interpreting the model using methods discussed above. The model can also be interpreted in terms of the effect of a change in x_k on the odds of outcome q versus r. After rewriting (5.3) in terms of odds,

$$\Omega_{q|r}\left(\mathbf{x},x_k\right) = \frac{\Pr\left(y = q \mid \mathbf{x},x_k\right)}{\Pr\left(y = r \mid \mathbf{x},x_k\right)} = \exp\left\{\left(\alpha_q - \alpha_r\right)\beta_0 + \left(\phi_q - \phi_r\right)\left(\mathbf{x}\beta\right)\right\}$$

It is easy to show that

$$\frac{\Omega_{q|r}\left(\mathbf{x},x_k + 1\right)}{\Omega_{q|r}\left(\mathbf{x},x_k\right)} = e^{(\phi_q - \phi_r)\beta_k} = \left(\frac{e^{\phi_q}}{e^{\phi_r}}\right)^{\beta_k}$$

Thus, the effect of x_k on the odds of q versus r differs across outcome comparisons according to the scaling coefficients ϕ.

5.9.3   The continuation ratio model

The *continuation ratio model* was proposed by Fienberg (1980, 110) and was designed for ordinal outcomes in which the categories represent the progression of events or stages in some process though which an individual can advance. For example, the outcome could be faculty rank, where the stages are assistant professor, associate professor, and full professor. A key characteristic of the process is that an individual must pass through each stage. For example, to become an associate professor you must be an assistant professor; to be a full professor, an associate professor. While there are versions of this model based on other binary models (e.g., probit), here we consider the logit version.

If $\Pr\left(y = m \mid \mathbf{x}\right)$ is the probability of being in stage m given $\mathbf{x}$ and $\Pr\left(y > m \mid \mathbf{x}\right)$ is the probability of being in a stage later than m, the continuation ratio model for the log odds is

$$\ln\left\{\frac{\Pr\left(y = m \mid \mathbf{x}\right)}{\Pr\left(y > m \mid \mathbf{x}\right)}\right\} = \tau_m - \mathbf{x}\beta \quad \text{for } m = 1 \text{ to } J - 1$$

where the βs are constrained to be equal across outcome categories, while the constant term τ_m differs by stage. As with other logit models, we can also express the model in terms of the odds:

$$\frac{\Pr\left(y = m \mid \mathbf{x}\right)}{\Pr\left(y > m \mid \mathbf{x}\right)} = \exp\left(\tau_m - \mathbf{x}\beta\right)$$

Accordingly, $\exp\left(-\beta_k\right)$ can be interpreted as the effect of a unit increase in x_k on the odds of being in m compared with being in a higher category given that an individual is in category m or higher, holding all other variables constant. From this equation, the predicted probabilities can be computed as

$$\Pr\left(y = m \mid \mathbf{x}\right) = \frac{\exp\left(\tau_m - \mathbf{x}\beta\right)}{\prod_{j=1}^{m}\left\{1 + \exp\left(\tau_j - \mathbf{x}\beta\right)\right\}} \quad \text{for } m = 1 \text{ to } J - 1$$

$$\Pr\left(y = J \mid \mathbf{x}\right) = 1 - \sum_{j=1}^{J-1} \Pr\left(y = j \mid \mathbf{x}\right)$$

These predicted probabilities can be used for interpreting the model. In Stata, this model can be fitted using `ocratio` by Wolfe (1998); type `net search ocratio` and follow the prompts to download.

6   Models for Nominal Outcomes

An outcome is nominal when the categories are assumed to be unordered. For example, marital status can be grouped nominally into the categories of divorced, never married, married, or widowed. Occupations might be organized as professional, white collar, blue collar, craft, and menial, which is the example we use in this chapter. Other examples include reasons for leaving the parents' home, the organizational context of scientific work (e.g., industry, government, and academia), and the choice of language in a multilingual society. Further, in some cases a researcher might prefer to treat an outcome as nominal, even though it is ordered or partially ordered. For example, if the response categories are strongly agree, agree, disagree, strongly disagree, and don't know, the category "don't know" invalidates models for ordinal outcomes. Or, you might decide to use a nominal regression model when the assumption of parallel regressions is rejected. In general, if you have concerns about the ordinality of the dependent variable, the potential loss of efficiency in using models for nominal outcomes is outweighed by avoiding potential bias.

This chapter focuses on two closely related models for nominal outcomes. The *multinomial logit model* (MNLM) is the most frequently used nominal regression model. In this model, the effects of the independent variables are allowed to differ for each outcome and are similar to the generalized ordered logit model discussed in the last chapter. In the *conditional logit model* (CLM), characteristics of the outcomes are used to predict which choice is made. While probit versions of these models are theoretically possible, issues of computation and identification limit their use (Keane 1992).

The biggest challenge in using the MNLM is that the model includes a lot of parameters, and it is easy to be overwhelmed by the complexity of the results. This complexity is compounded by the nonlinearity of the model, which leads to the same difficulties of interpretation found for models in prior chapters. While fitting the model is straightforward, interpretation involves many challenges that are the focus of this chapter. We begin by reviewing the statistical model, followed by a discussion of testing, fit, and finally methods of interpretation. These discussions are intended as a review for those who are familiar with the models. For a complete discussion, see Long (1997). As always, you can obtain sample do-files and data files by downloading the `spostst8` package (see Chapter 1 for details).

6.1   The multinomial logit model

The MNLM can be thought of as simultaneously estimating binary logits for all comparisons among the dependent categories. For example, let occ3 be a nominal outcome with the categories M for manual jobs, W for white collar jobs, and P for professional jobs. Assume that there is a single independent variable ed measuring years of education. We can examine the effect of ed on occ3 by estimating three binary logits,

$$\ln\left\{\frac{\Pr(P\mid\mathbf{x})}{\Pr(M\mid\mathbf{x})}\right\} = \beta_{0,P\mid M} + \beta_{1,P\mid M}\,\mathrm{ed}$$

$$\ln\left\{\frac{\Pr(W\mid\mathbf{x})}{\Pr(M\mid\mathbf{x})}\right\} = \beta_{0,W\mid M} + \beta_{1,W\mid M}\,\mathrm{ed}$$

$$\ln\left\{\frac{\Pr(P\mid\mathbf{x})}{\Pr(W\mid\mathbf{x})}\right\} = \beta_{0,P\mid W} + \beta_{1,P\mid W}\,\mathrm{ed}$$

where the subscripts to the βs indicate which comparison is being made (e.g., $\beta_{1,P\mid M}$ is the coefficient for the first independent variable for the comparison of P and M).

The three binary logits include redundant information. Because $\ln\frac{a}{b} = \ln a - \ln b$, the following equality must hold:

$$\ln\left\{\frac{\Pr(P\mid\mathbf{x})}{\Pr(M\mid\mathbf{x})}\right\} - \ln\left\{\frac{\Pr(W\mid\mathbf{x})}{\Pr(M\mid\mathbf{x})}\right\} = \ln\left\{\frac{\Pr(P\mid\mathbf{x})}{\Pr(W\mid\mathbf{x})}\right\}$$

This implies that

$$\beta_{0,P\mid M} - \beta_{0,W\mid M} = \beta_{0,P\mid W} \tag{6.1}$$
$$\beta_{1,P\mid M} - \beta_{1,W\mid M} = \beta_{1,P\mid W}$$

In general, with J outcomes, only $J - 1$ binary logits need to be estimated. Estimates for the remaining coefficients can be computed using equalities of the sort shown in (6.1).

The problem with estimating the MNLM by estimating a series of binary logits is that each binary logit is based on a different sample. For example, in the logit comparing P with M, those in W are dropped. To see this, we can look at the output from a series of binary logits. First, we estimate a binary logit comparing manual and professional workers:

```
. use http://www.stata-press.com/data/lfr/nomintro2, clear
(1982 General Social Survey)

. tab prof_man, miss
```

prof_man	Freq.	Percent	Cum.
Manual	184	54.60	54.60
Prof	112	33.23	87.83
.	41	12.17	100.00
Total	337	100.00	

```
. logit prof_man ed, nolog
```

Logit estimates				Number of obs	=	296
				LR chi2(1)	=	139.78
				Prob > chi2	=	0.0000
Log likelihood = -126.43879				Pseudo R2	=	0.3560

prof_man	Coef.	Std. Err.	z	P>|z|	[95% Conf. Interval]	
ed	.7184599	.0858735	8.37	0.000	.550151	.8867688
_cons	-10.19854	1.177457	-8.66	0.000	-12.50632	-7.89077

Notice that 41 cases are missing for **prof_man** and have been deleted. These correspond to respondents who have white collar occupations. In the same way, the next two binary logits also exclude cases corresponding to the excluded category:

```
. tab wc_man, miss
```

wc_man	Freq.	Percent	Cum.
Manual	184	54.60	54.60
WhiteCol	41	12.17	66.77
.	112	33.23	100.00
Total	337	100.00	

```
. logit wc_man ed, nolog
```

Logit estimates				Number of obs	=	225
				LR chi2(1)	=	16.00
				Prob > chi2	=	0.0001
Log likelihood = -98.818194				Pseudo R2	=	0.0749

wc_man	Coef.	Std. Err.	z	P>|z|	[95% Conf. Interval]	
ed	.3418255	.0934517	3.66	0.000	.1586636	.5249875
_cons	-5.758148	1.216291	-4.73	0.000	-8.142035	-3.374262

```
. tab prof_wc, miss
```

prof_wc	Freq.	Percent	Cum.
WhiteCol	41	12.17	12.17
Prof	112	33.23	45.40
.	184	54.60	100.00
Total	337	100.00	

```
. logit prof_wc ed, nolog
```

Logit estimates				Number of obs	=	153
				LR chi2(1)	=	23.34
				Prob > chi2	=	0.0000
Log likelihood = -77.257045				Pseudo R2	=	0.1312

prof_wc	Coef.	Std. Err.	z	P>|z|	[95% Conf. Interval]	
ed	.3735466	.0874469	4.27	0.000	.2021538	.5449395
_cons	-4.332833	1.227293	-3.53	0.000	-6.738283	-1.927382

The results from the binary logits can be compared with the output from `mlogit`, the command that estimates the MNLM:

```
. tab occ3, miss
      occ3 |      Freq.     Percent        Cum.
-----------+-----------------------------------
    Manual |        184       54.60       54.60
  WhiteCol |         41       12.17       66.77
      Prof |        112       33.23      100.00
-----------+-----------------------------------
     Total |        337      100.00
. mlogit occ3 ed, nolog
Multinomial logistic regression                  Number of obs   =        337
                                                 LR chi2(2)      =     145.89
                                                 Prob > chi2     =     0.0000
Log likelihood = -248.14786                      Pseudo R2       =     0.2272

------------------------------------------------------------------------------
      occ3 |      Coef.   Std. Err.      z    P>|z|     [95% Conf. Interval]
-----------+------------------------------------------------------------------
WhiteCol   |
        ed |   .3000735   .0841358     3.57   0.000     .1351703    .4649767
     _cons |  -5.232602   1.096086    -4.77   0.000    -7.380892   -3.084312
-----------+------------------------------------------------------------------
Prof       |
        ed |   .7195673   .0805117     8.94   0.000     .5617671    .8773674
     _cons |  -10.21121   1.106913    -9.22   0.000    -12.38072   -8.041698
------------------------------------------------------------------------------
(Outcome occ3==Manual is the comparison group)
```

The output from `mlogit` is divided into two panels. The top panel is labeled `WhiteCol`, which is the value label for the second category of the dependent variable; the second panel is labeled `Prof`, which corresponds to the third outcome category. The key to understanding the two panels is the last line of output: `Outcome occ3==Manual is the comparison group`. This means that the panel `WhiteCol` presents coefficients from the comparison of W to M. The second panel labeled `Prof` holds the comparison of P to M. Accordingly, the top panel should be compared with the coefficients from the binary logit for W and M (outcome variable `wc_man`) listed above. For example, the coefficient for the comparison of W to M from `mlogit` is $\widehat{\beta}_{1,W|M} = .3000735$ with $z = 3.567$, while the `logit` estimate is $\widehat{\beta}_{1,W|M} = .3418255$ with $z = 3.658$. Overall, the estimates from the binary model are close to those from the MNLM but not exactly the same.

Next, notice that while theoretically $\beta_{1,P|M} - \beta_{1,W|M} = \beta_{1,P|W}$, the estimates from the *binary* logits are $\widehat{\beta}_{1,P|M} - \widehat{\beta}_{1,W|M} = .7184599 - .3418255 = .3766344$, which does not equal the binary logit estimate $\widehat{\beta}_{1,P|W} = .3735466$. The general point is that a series of binary logits using `logit` does *not* impose the constraints among coefficients that are implicit in the definition of the model. When fitting the model with `mlogit`, the constraints are imposed. Indeed, the output from `mlogit` only presents two of the three comparisons from our example, namely, W versus M and P versus M. The remaining comparison, W versus P, is the difference between the two sets of estimated coefficients. Details on using `listcoef` to automatically compute the remaining comparisons are given below.

6.1.1   Formal statement of the model

Formally, the MNLM can be written as

$$\ln \Omega_{m|b}\left(\mathbf{x}\right) = \ln \frac{\Pr\left(y = m \mid \mathbf{x}\right)}{\Pr\left(y = b \mid \mathbf{x}\right)} = \mathbf{x}\beta_{m|b} \quad \text{for } m = 1 \text{ to } J$$

where b is the base category, which is also referred to as the comparison group. As $\ln \Omega_{b|b}\left(\mathbf{x}\right) = \ln 1 = 0$, it must hold that $\beta_{b|b} = \mathbf{0}$. That is, the log odds of an outcome compared with itself are always 0, and thus the effects of any independent variables must also be 0. These J equations can be solved to compute the predicted probabilities:

$$\Pr\left(y = m \mid \mathbf{x}\right) = \frac{\exp\left(\mathbf{x}\beta_{m|b}\right)}{\sum_{j=1}^{J} \exp\left(\mathbf{x}\beta_{j|b}\right)}$$

While the predicted probability will be the same regardless of the base category b, changing the base category can be confusing since the resulting output from `mlogit` appears to be quite different. For example, suppose you have three outcomes and fit the model with outcome 1 as the base category. Your probability equations would be

$$\Pr\left(y = m \mid \mathbf{x}\right) = \frac{\exp\left(\mathbf{x}\beta_{m|1}\right)}{\sum_{j=1}^{J} \exp\left(\mathbf{x}\beta_{j|1}\right)}$$

and you would obtain estimates $\widehat{\beta}_{2|1}$ and $\widehat{\beta}_{3|1}$, where $\beta_{1|1} = \mathbf{0}$. If someone else set up the model with base category 2, their equations would be

$$\Pr\left(y = m \mid \mathbf{x}\right) = \frac{\exp\left(\mathbf{x}\beta_{m|2}\right)}{\sum_{j=1}^{J} \exp\left(\mathbf{x}\beta_{j|2}\right)}$$

and they would obtain $\widehat{\beta}_{1|2}$ and $\widehat{\beta}_{3|2}$, where $\beta_{2|2} = \mathbf{0}$. While the estimated parameters are different, they are only different *parameterizations* that provide the same predicted probabilities. The confusion arises only if you are not clear about which parameterization you are using. Unfortunately, some software packages—but *not* Stata—make it very difficult to tell which set of parameters is being estimated. We return to this issue when we discuss how Stata's `mlogit` parameterizes the model in the next section.

6.2   Estimation using mlogit

The multinomial logit model is fitted with the following command:

<u>mlogit</u> *depvar* [*indepvars*] [*weight*] [if *exp*] [in *range*] [, <u>b</u>asecategory(#)
   <u>c</u>onstraints(*clist*) <u>r</u>obust <u>cl</u>uster(*varname*) <u>sc</u>ore(*newvarlist* | *stub**)
   <u>l</u>evel(#) <u>rrr</u> <u>noc</u>onstant <u>nolog</u>]

In our experience, the model converges very quickly, even when there are many outcome categories and independent variables.

Variable lists

depvar is the dependent variable. The actual values taken on by the dependent variable are irrelevant. For example, if you had three outcomes, you could use the values 1, 2, and 3 or -1, 0, and 999. Up to 50 outcomes are allowed in Intercooled Stata, and 20 outcomes are allowed in Small Stata.

indepvars is a list of independent variables. If *indepvars* is not included, Stata fits a model with only constants.

Specifying the estimation sample

if and in qualifiers can be used to restrict the estimation sample. For example, if you want to fit the model with only white respondents, use the command `mlogit occ ed exper if white==1`.

Listwise deletion Stata excludes cases in which there are missing values for any of the variables. Accordingly, if two models are fitted using the same dataset but have different sets of independent variables, it is possible to have different samples. We recommend that you use `mark` and `markout` (discussed in Chapter 3) to explicitly remove cases with missing data.

Weights

`mlogit` can be used with `fweights`, `pweights`, and `iweights`. In Chapter 3, we provide a brief discussion of the different types of weights and how weights are specified in Stata's syntax.

Options

`basecategory(#)` specifies the value of *depvar* that is the base category (i.e., reference group) for the coefficients that are listed. This determines how the model is parameterized. If the `basecategory` option is not specified, the most frequent category in the estimation sample is chosen as the base. The base category is always reported immediately below the estimates; for example, `Outcome occ3==Manual is the comparison group`.

`constraints(clist)` specifies the linear constraints to be applied during estimation. The default is to perform unconstrained estimation. Constraints are defined with the `constraint` command. This option is illustrated in Section 6.3.3 when we discuss an LR test for combining outcome categories.

robust indicates that robust variance estimates are to be used. When cluster() is specified, robust standard errors are automatically used. See Chapter 3 for further details.

cluster(*varname*) specifies that the observations be independent across the groups specified by unique values of *varname* but not necessarily independent within the groups. See Chapter 3 for further details.

score(*newvarlist* | *stub**) creates $k-1$ new variables, where k is the number of observed outcomes. Each new variable contains the contributions to the score for an equation in the model; see [U] **23.15 Obtaining scores.**

level(#) specifies the level of the confidence interval for estimated parameters. By default, Stata uses 95% intervals. You can also change the default level, say, to a 90% interval, with the command set level 90.

rrr reports the estimated coefficients transformed to relative risk ratios, defined as $\exp(b)$ rather than b, along with standard errors and confidence intervals for these ratios.

noconstant excludes the constant terms from the model.

nolog suppresses the iteration history.

6.2.1 Example of occupational attainment

The 1982 General Social Survey asked respondents their occupation, which we recoded into five broad categories: menial jobs (M), blue collar jobs (B), craft jobs (C), white collar jobs (W), and professional jobs (P). Three independent variables are considered: white indicating the race of the respondent, ed measuring years of education, and exper measuring years of work experience.

```
. sum white ed exper
```

Variable	Obs	Mean	Std. Dev.	Min	Max
white	337	.9169139	.2764227	0	1
ed	337	13.09496	2.946427	3	20
exper	337	20.50148	13.95936	2	66

The distribution among outcome categories is

```
. tab occ, missing
```

Occupation	Freq.	Percent	Cum.
Menial	31	9.20	9.20
BlueCol	69	20.47	29.67
Craft	84	24.93	54.60
WhiteCol	41	12.17	66.77
Prof	112	33.23	100.00
Total	337	100.00	

Using these variables, the following MNLM was fitted:

$$\ln \Omega_{M|P}(\mathbf{x}_i) = \beta_{0,M|P} + \beta_{1,M|P}\texttt{white} + \beta_{2,M|P}\texttt{ed} + \beta_{3,M|P}\texttt{exper}$$
$$\ln \Omega_{B|P}(\mathbf{x}_i) = \beta_{0,B|P} + \beta_{1,B|P}\texttt{white} + \beta_{2,B|P}\texttt{ed} + \beta_{3,B|P}\texttt{exper}$$
$$\ln \Omega_{C|P}(\mathbf{x}_i) = \beta_{0,C|P} + \beta_{1,C|P}\texttt{white} + \beta_{2,C|P}\texttt{ed} + \beta_{3,C|P}\texttt{exper}$$
$$\ln \Omega_{W|P}(\mathbf{x}_i) = \beta_{0,W|P} + \beta_{1,W|P}\texttt{white} + \beta_{2,W|P}\texttt{ed} + \beta_{3,W|P}\texttt{exper}$$

where we specify the fifth category P as the base category:

```
. mlogit occ white ed exper, basecategory(5) nolog
Multinomial logistic regression              Number of obs   =        337
                                             LR chi2(12)     =     166.09
                                             Prob > chi2     =     0.0000
    Log likelihood = -426.80048             Pseudo R2       =     0.1629
```

occ	Coef.	Std. Err.	z	P>|z|	[95% Conf. Interval]
Menial						
white	-1.774306	.7550543	-2.35	0.019	-3.254186	-.2944273
ed	-.7788519	.1146293	-6.79	0.000	-1.003521	-.5541826
exper	-.0356509	.018037	-1.98	0.048	-.0710028	-.000299
_cons	11.51833	1.849356	6.23	0.000	7.893659	15.143
BlueCol						
white	-.5378027	.7996033	-0.67	0.501	-2.104996	1.029391
ed	-.8782767	.1005446	-8.74	0.000	-1.07534	-.6812128
exper	-.0309296	.0144086	-2.15	0.032	-.05917	-.0026893
_cons	12.25956	1.668144	7.35	0.000	8.990061	15.52907
Craft						
white	-1.301963	.647416	-2.01	0.044	-2.570875	-.0330509
ed	-.6850365	.0892996	-7.67	0.000	-.8600605	-.5100126
exper	-.0079671	.0127055	-0.63	0.531	-.0328693	.0169351
_cons	10.42698	1.517943	6.87	0.000	7.451864	13.40209
WhiteCol						
white	-.2029212	.8693072	-0.23	0.815	-1.906732	1.50089
ed	-.4256943	.0922192	-4.62	0.000	-.6064407	-.2449479
exper	-.001055	.0143582	-0.07	0.941	-.0291967	.0270866
_cons	5.279722	1.684006	3.14	0.002	1.979132	8.580313

```
(Outcome occ==Prof is the comparison group)
```

Methods of testing coefficients and interpretation of the estimates will be considered after we discuss the effects of using different base categories.

6.2.2   Using different base categories

By default, `mlogit` sets the base category to the outcome with the most observations. Alternatively, as illustrated in the last example, you can select the base category with `basecategory()`. `mlogit` then reports coefficients for the effect of each independent variable on each category relative to the base category. However, you should also examine the effects on other pairs of outcome categories. For example, you might be interested in how race affects the allocation of workers between `Craft` and `BlueCol` (e.g., $\beta_{1,B|C}$), which was not estimated in the output listed above. While this coefficient can be estimated by rerunning `mlogit` with a different base category (e.g., `mlogit occ white ed exper, basecategory(3)`), it is easier to use `listcoef`, which presents estimates for *all* combinations of outcome categories. Since `listcoef` can generate a lot of output, we illustrate two options that limit which coefficients are listed. First, you can include a list of variables, and only coefficients for those variables will be listed. For example,

```
. listcoef white, help
mlogit (N=337): Factor Change in the Odds of occ

Variable: white (sd=      .28)
```

Odds comparing Group 1 vs Group 2	b	z	P>|z|	e^b	e^bStdX
Menial -BlueCol	-1.23650	-1.707	0.088	0.2904	0.7105
Menial -Craft	-0.47234	-0.782	0.434	0.6235	0.8776
Menial -WhiteCol	-1.57139	-1.741	0.082	0.2078	0.6477
Menial -Prof	-1.77431	-2.350	0.019	0.1696	0.6123
BlueCol -Menial	1.23650	1.707	0.088	3.4436	1.4075
BlueCol -Craft	0.76416	1.208	0.227	2.1472	1.2352
BlueCol -WhiteCol	-0.33488	-0.359	0.720	0.7154	0.9116
BlueCol -Prof	-0.53780	-0.673	0.501	0.5840	0.8619
Craft -Menial	0.47234	0.782	0.434	1.6037	1.1395
Craft -BlueCol	-0.76416	-1.208	0.227	0.4657	0.8096
Craft -WhiteCol	-1.09904	-1.343	0.179	0.3332	0.7380
Craft -Prof	-1.30196	-2.011	0.044	0.2720	0.6978
WhiteCol-Menial	1.57139	1.741	0.082	4.8133	1.5440
WhiteCol-BlueCol	0.33488	0.359	0.720	1.3978	1.0970
WhiteCol-Craft	1.09904	1.343	0.179	3.0013	1.3550
WhiteCol-Prof	-0.20292	-0.233	0.815	0.8163	0.9455
Prof -Menial	1.77431	2.350	0.019	5.8962	1.6331
Prof -BlueCol	0.53780	0.673	0.501	1.7122	1.1603
Prof -Craft	1.30196	2.011	0.044	3.6765	1.4332
Prof -WhiteCol	0.20292	0.233	0.815	1.2250	1.0577

```
        b = raw coefficient
        z = z-score for test of b=0
    P>|z| = p-value for z-test
      e^b = exp(b) = factor change in odds for unit increase in X
 e^bStdX = exp(b*SD of X) = change in odds for SD increase in X
```

Or, you can limit the output to those coefficients that are significant at a given level using the `pvalue(#)` option, which specifies that only coefficients significant at the # significance level or smaller will be printed. For example,

(Continued on next page)

```
. listcoef, pvalue(.05)
mlogit (N=337): Factor Change in the Odds of occ when P>|z| < 0.05
Variable: white (sd=     .28)
```

Odds comparing Group 1 vs Group 2	b	z	P>|z|	e^b	e^bStdX
Menial -Prof	-1.77431	-2.350	0.019	0.1696	0.6123
Craft -Prof	-1.30196	-2.011	0.044	0.2720	0.6978
Prof -Menial	1.77431	2.350	0.019	5.8962	1.6331
Prof -Craft	1.30196	2.011	0.044	3.6765	1.4332

```
Variable: ed (sd=     2.9)
```

Odds comparing Group 1 vs Group 2	b	z	P>|z|	e^b	e^bStdX
Menial -WhiteCol	-0.35316	-3.011	0.003	0.7025	0.3533
Menial -Prof	-0.77885	-6.795	0.000	0.4589	0.1008
BlueCol -Craft	-0.19324	-2.494	0.013	0.8243	0.5659
BlueCol -WhiteCol	-0.45258	-4.425	0.000	0.6360	0.2636
BlueCol -Prof	-0.87828	-8.735	0.000	0.4155	0.0752
Craft -BlueCol	0.19324	2.494	0.013	1.2132	1.7671
Craft -WhiteCol	-0.25934	-2.773	0.006	0.7716	0.4657
Craft -Prof	-0.68504	-7.671	0.000	0.5041	0.1329
WhiteCol-Menial	0.35316	3.011	0.003	1.4236	2.8308
WhiteCol-BlueCol	0.45258	4.425	0.000	1.5724	3.7943
WhiteCol-Craft	0.25934	2.773	0.006	1.2961	2.1471
WhiteCol-Prof	-0.42569	-4.616	0.000	0.6533	0.2853
Prof -Menial	0.77885	6.795	0.000	2.1790	9.9228
Prof -BlueCol	0.87828	8.735	0.000	2.4067	13.3002
Prof -Craft	0.68504	7.671	0.000	1.9838	7.5264
Prof -WhiteCol	0.42569	4.616	0.000	1.5307	3.5053

```
Variable: exper (sd=     14)
```

Odds comparing Group 1 vs Group 2	b	z	P>|z|	e^b	e^bStdX
Menial -Prof	-0.03565	-1.977	0.048	0.9650	0.6079
BlueCol -Prof	-0.03093	-2.147	0.032	0.9695	0.6494
Prof -Menial	0.03565	1.977	0.048	1.0363	1.6449
Prof -BlueCol	0.03093	2.147	0.032	1.0314	1.5400

6.2.3   Predicting perfectly

mlogit handles perfect prediction somewhat differently than the estimations commands for binary and ordinal models that we have discussed. logit and probit automatically remove the observations that imply perfect prediction and compute estimates accordingly. ologit and oprobit keep these observations in the model, fit the z for the problem variable as 0, and provide an incorrect LR chi-squared but also warn that a given number of observations are completely determined. You should delete these observations and refit the model. mlogit is just like ologit and oprobit, except that *you do not receive a warning message*. You will see, however, that all coefficients associated with the variable causing the problem have $z = 0$ (and $p > |z| = 1$). You should refit the model, excluding the problem variable and deleting the observations that imply the perfect predictions. Using the `tabulate` command to generate a cross-tabulation of the problem variable and the dependent variable should reveal the combination that results in perfect prediction.

6.3 Hypothesis testing of coefficients

In the MNLM, you can test individual coefficients with the reported z-statistics, with a Wald test using `test`, or with an LR test using `lrtest`. As the methods of testing a single coefficient that were discussed in Chapters 4 and 5 still apply fully, they are not considered further here. However, in the MNLM there are new reasons for testing groups of coefficients. First, testing that a variable has no effect requires a test that $J - 1$ coefficients are simultaneously equal to zero. Second, testing whether the independent variables as a group differentiate between two outcomes requires a test of K coefficients. This section focuses on these two kinds of tests.

Caution Regarding Specification Searches Given the difficulties of interpretation that are associated with the MNLM, it is tempting to search for a more parsimonious model by excluding variables or combining outcome categories based on a sequence of tests. Such a search requires great care. First, these tests involve multiple coefficients. While the overall test might indicate that *as a group* the coefficients are not significantly different from zero, an *individual* coefficient can still be substantively and statistically significant. Accordingly, you should examine the individual coefficients involved in each test before deciding to revise your model. Second, as with all searches that use repeated, sequential tests, there is a danger of overfitting the data. When models are constructed based on prior testing using the same data, significance levels should only be used as rough guidelines.

6.3.1 mlogtest for tests of the MNLM

While the tests in this section can be computed using `test` or `lrtest`, in practice this is tedious. The `mlogtest` command (Freese and Long 2000) makes the computation of these tests very simple. The syntax is

mlogtest $\big[$, <u>a</u>ll lr <u>w</u>ald <u>c</u>ombine lrcomb <u>set</u>(*varlist*$\big[\backslash$ *varlist*$\big[\backslash\ldots\big]\big]$) <u>ii</u>a

    <u>h</u>ausman <u>sm</u>hsiao <u>det</u>ail <u>b</u>ase $\big]$

Options

    lr requests LR tests for each independent variable.

wald requests Wald tests for each independent variable.

combine requests Wald tests of whether dependent categories can be combined.

lrcomb requests LR tests of whether dependent categories can be combined. These tests
use constrained estimation and overwrite constraint #999 if it is already defined.

set(*varlist* [\ *varlist* [\ ...]]) specifies that a set of variables is to be considered together
for the LR test or Wald test. \ is used to specify multiple sets of variables. For
example, **mlogtest, lr set(age age2 \ iscat1 iscat2)** computes one LR test for
the hypothesis that the effects of **age** and **age2** are jointly 0 and a second LR test
that the effects of **iscat1** and **iscat2** are jointly 0.

Other options for **mlogtest** are discussed later in the chapter.

6.3.2   Testing the effects of the independent variables

With J dependent categories, there are $J - 1$ nonredundant coefficients associated with
each independent variable x_k. For example, in our logit on occupation, there are four
coefficients associated with **ed**: $\beta_{2,M|P}$, $\beta_{2,B|P}$, $\beta_{2,C|P}$, and $\beta_{2,W|P}$. The hypothesis that
x_k does not affect the dependent variable can be written as

$$H_0\text{: } \beta_{k,1|b} = \cdots = \beta_{k,J|b} = 0$$

where b is the base category. Because $\beta_{k,b|b}$ is necessarily 0, the hypothesis imposes
constraints on $J - 1$ parameters. This hypothesis can be tested with either a Wald or
an LR test.

A likelihood-ratio test

The LR test involves 1) fitting the full model, including all of the variables, resulting in
the likelihood-ratio statistic LR^2_F; 2) fitting the restricted model that excludes variable
x_k, resulting in LR^2_R; and 3) computing the difference $LR^2_{RvsF} = LR^2_F - LR^2_R$, which is
distributed as chi-squared with $J-1$ degrees of freedom if the null hypothesis is true.
This can be done using **lrtest**:

```
. use http://www.stata-press.com/data/lfr/nomocc2, clear
(1982 General Social Survey)

. mlogit occ white ed exper, basecategory(5) nolog
  (output omitted)

. estimates store fmodel

. mlogit occ ed exper, basecategory(5) nolog
  (output omitted)

. estimates store nmodel_white

. lrtest fmodel nmodel_white

likelihood-ratio test                              LR chi2(4)  =     8.10
(Assumption: nmodel_white nested in fmodel)        Prob > chi2 =   0.0881

. mlogit occ white exper, basecategory(5) nolog
  (and so on)
```

While using `lrtest` is straightforward, the command `mlogtest, lr` is even simpler because it automatically computes the tests for all variables by making repeated calls to `lrtest`:

```
. mlogit occ white ed exper, basecategory(5) nolog
  (output omitted )
. mlogtest, lr
**** Likelihood-ratio tests for independent variables
  Ho: All coefficients associated with given variable(s) are 0.
      occ |       chi2   df    P>chi2
 ---------+-----------------------------
    white |      8.095    4    0.088
       ed |    156.937    4    0.000
    exper |      8.561    4    0.073
 ---------+-----------------------------
```

The results of the LR test, regardless of how they are computed, can be interpreted as follows:

> The effect of race on occupation is significant at the .10 level but not at the .05 level ($X^2 = 8.10, df = 4, p = .09$). The effect of education is significant at the .01 level ($X^2 = 156.94, df = 4, p < .01$).

Or, it can be stated more formally:

> The hypothesis that all of the coefficients associated with education are simultaneously equal to 0 can be rejected at the .01 level ($X^2 = 156.94, df = 4, p < .01$).

A Wald test

Although the LR test is generally considered superior, if the model is complex or the sample is very large, the computational costs of the LR test can be prohibitive. Alternatively, K Wald tests can be computed using `test` without fitting additional models. For example,

```
. mlogit occ white ed exper, basecategory(5) nolog
  (output omitted )
. test white
 ( 1)  [Menial]white = 0
 ( 2)  [BlueCol]white = 0
 ( 3)  [Craft]white = 0
 ( 4)  [WhiteCol]white = 0
            chi2(  4) =      8.15
          Prob > chi2 =    0.0863
```

```
. test ed

 ( 1)   [Menial]ed = 0
 ( 2)   [BlueCol]ed = 0
 ( 3)   [Craft]ed = 0
 ( 4)   [WhiteCol]ed = 0

           chi2(  4) =    84.97
         Prob > chi2 =    0.0000

. test exper

 ( 1)   [Menial]exper = 0
 ( 2)   [BlueCol]exper = 0
 ( 3)   [Craft]exper = 0
 ( 4)   [WhiteCol]exper = 0

           chi2(  4) =     7.99
         Prob > chi2 =    0.0918
```

The output from **test** makes explicit which coefficients are being tested. Here, we see the way in which Stata labels parameters in models with multiple equations. For example, [Menial]white is the coefficient for the effect of white in the equation comparing the outcome Menial with the base category Prof; [BlueCol]white is the coefficient for the effect of white in the equation comparing the outcome BlueCol with the base category Prof.

As with the LR test, **mlogtest, wald** automates this process:

```
. mlogtest, wald

**** Wald tests for independent variables

 Ho: All coefficients associated with given variable(s) are 0.
```

occ	chi2	df	P>chi2
white	8.149	4	0.086
ed	84.968	4	0.000
exper	7.995	4	0.092

These tests can be interpreted in the same way as illustrated for the LR test above.

Testing multiple independent variables

The logic of the Wald or LR tests can be extended to test that the effects of two or more independent variables are simultaneously zero. For example, the hypothesis to test that x_k and x_ℓ have no effect is

$$H_0: \beta_{k,1|b} = \cdots = \beta_{k,J|b} = \beta_{\ell,1|b} = \cdots = \beta_{\ell,J|b} = 0$$

The set(*varlist* $[\backslash$ *varlist* $[\backslash \ldots]]$) option in **mlogtest** specifies which variables are to be simultaneously tested. For example, to test the hypothesis that the effects of ed and exper are simultaneously equal to 0, we could use **lrtest** as follows:

```
. mlogit occ white ed exper, basecategory(5) nolog
  (output omitted)
```

```
. estimates store fmodel
. mlogit occ white, basecategory(5) nolog
  (output omitted)
. estimates store nmodel
. lrtest fmodel nmodel
likelihood-ratio test                              LR chi2(8)  =      160.77
(Assumption: nmodel nested in fmodel)              Prob > chi2 =      0.0000
```

or, using `mlogtest`,

```
. mlogit occ white ed exper, basecategory(5) nolog
  (output omitted)
. mlogtest, lr set(ed exper)
**** Likelihood-ratio tests for independent variables
 Ho: All coefficients associated with given variable(s) are 0.
```

occ	chi2	df	P>chi2
white	8.095	4	0.088
ed	156.937	4	0.000
exper	8.561	4	0.073
set_1:	160.773	8	0.000
ed			
exper			

6.3.3   Tests for combining dependent categories

If none of the independent variables significantly affect the odds of outcome m versus outcome n, we say that m and n are *indistinguishable* with respect to the variables in the model (Anderson 1984). Outcomes m and n being indistinguishable corresponds to the hypothesis that

$$H_0: \beta_{1,m|n} = \cdots \beta_{K,m|n} = 0$$

which can be tested with either a Wald or an LR test. In our experience, the two tests provide very similar results. If two outcomes are indistinguishable with respect to the variables in the model, then you can obtain more efficient estimates by combining them. To test whether categories are indistinguishable, you can use `mlogtest`.

A Wald test for combining outcomes

The command `mlogtest, combine` computes Wald tests of the null hypothesis that two categories can be combined for all combinations of outcome categories. For example,

```
. mlogit occ white ed exper, basecategory(5) nolog
  (output omitted)
```

```
. mlogtest, combine

**** Wald tests for combining outcome categories

  Ho: All coefficients except intercepts associated with given pair
      of outcomes are 0 (i.e., categories can be collapsed).

  Categories tested |    chi2   df   P>chi2
 -------------------+-------------------------
    Menial- BlueCol |   3.994    3   0.262
    Menial-   Craft |   3.203    3   0.361
    Menial-WhiteCol |  11.951    3   0.008
    Menial-    Prof |  48.190    3   0.000
    BlueCol-   Craft |   8.441   3   0.038
    BlueCol-WhiteCol |  20.055   3   0.000
    BlueCol-    Prof |  76.393   3   0.000
      Craft-WhiteCol |   8.892   3   0.031
      Craft-    Prof |  60.583   3   0.000
   WhiteCol-    Prof |  22.203   3   0.000
```

For example, we can reject the hypothesis that outcomes Menial and Prof are indistinguishable, while we cannot reject that Menial and BlueCol are indistinguishable.

Using test [category]*

The mlogtest command computes the tests for combining categories with the test command. For example, to test that Menial is indistinguishable from the base category Prof, type

```
. test [Menial]
 ( 1)  [Menial]white = 0
 ( 2)  [Menial]ed = 0
 ( 3)  [Menial]exper = 0
           chi2(  3) =    48.19
         Prob > chi2 =    0.0000
```

which matches the results from mlogtest in row Menial-Prof. [*category*] in test is used to indicate which equation is being referenced in multiple equation commands. mlogit is a multiple equation command because it is in effect estimating $J - 1$ binary logit equations.

The test is more complicated when neither category is the base category. For example, to test that m and n are indistinguishable when the base category b is neither m nor n, the hypothesis you want to test is

$$H_0\colon \left(\beta_{1,m|b} - \beta_{1,n|b}\right) = \cdots = \left(\beta_{K,m|b} - \beta_{K,n|b}\right) = 0$$

That is, you want to test the difference between two sets of coefficients. This can be done with test [*category1*=*category2*]. For example, to test if Menial and Craft can be combined, type

```
. test [Menial=Craft]
 ( 1)  [Menial]white - [Craft]white = 0
 ( 2)  [Menial]ed - [Craft]ed = 0
 ( 3)  [Menial]exper - [Craft]exper = 0
           chi2(  3) =      3.20
         Prob > chi2 =    0.3614
```

Again, the results are identical to those from `mlogtest`.

An LR test for combining outcomes

An LR test of combining m and n can be computed by first fitting the full model with no constraints, with the resulting LR statistic LR_F^2. Then, we fit a restricted model M_R in which category m is used as the base category and all the coefficients except the constant in the equation for category n are constrained to 0, with the resulting test statistic LR_R^2. The test statistic is the difference $LR_{RvsF}^2 = LR_F^2 - LR_R^2$, which is distributed as chi-squared with K degrees of freedom. The command `mlogtest, lrcomb` computes $J \times (J-1)$ tests for all pairs of outcome categories. For example,

```
. mlogit occ white ed exper, basecategory(5) nolog
  (output omitted)
. mlogtest, lrcomb

**** LR tests for combining outcome categories

Ho: All coefficients except intercepts associated with given pair
    of outcomes are 0 (i.e., categories can be collapsed).

Categories tested |    chi2   df   P>chi2
------------------+-----------------------
  Menial- BlueCol |   4.095    3    0.251
  Menial-   Craft |   3.376    3    0.337
  Menial-WhiteCol |  13.223    3    0.004
  Menial-    Prof |  64.607    3    0.000
  BlueCol-   Craft |   9.176    3    0.027
 BlueCol-WhiteCol |  22.803    3    0.000
  BlueCol-    Prof | 125.699    3    0.000
   Craft-WhiteCol |   9.992    3    0.019
    Craft-    Prof |  95.889    3    0.000
 WhiteCol-    Prof |  26.736    3    0.000
```

Using constraint with lrtest*

The command `mlogtest, lrcomb` computes the test by using the powerful `constraint` command. To illustrate this, we use the test comparing `Menial` and `BlueCol` reported by `mlogtest, lrcomb` above. First, we fit the full model and save the results of `lrtest`:

```
. mlogit occ white ed exper, nolog
  (output omitted)
. estimates store fmodel
```

Second, we define a constraint using the command

```
. constraint define 999 [Menial]
```

This defines constraint 999, where the number is arbitrary. The expression [Menial] indicates that all the coefficients except the constant from the Menial equation should be constrained to 0. Third, we refit the model with this constraint. The base category must be BlueCol, so that the coefficients indicated by [Menial] are comparisons of BlueCol and Menial:

```
. mlogit occ exper ed white, base(2) constraint(999) nolog
```

Multinomial logistic regression					Number of obs	=	337
					LR chi2(9)	=	161.99
					Prob > chi2	=	0.0000
Log likelihood = -428.84791					Pseudo R2	=	0.1589

```
 ( 1)   [Menial]exper = 0
 ( 2)   [Menial]ed = 0
 ( 3)   [Menial]white = 0
```

occ	Coef.	Std. Err.	z	P>|z|	[95% Conf. Interval]	
Menial						
exper	(dropped)					
ed	(dropped)					
white	(dropped)					
_cons	-.8001193	.2162194	-3.70	0.000	-1.223901	-.3763371
Craft						
exper	.0242824	.0113959	2.13	0.033	.0019469	.0466179
ed	.1599345	.0693853	2.31	0.021	.0239418	.2959273
white	-.2381783	.4978563	-0.48	0.632	-1.213959	.7376021
_cons	-1.969087	1.054935	-1.87	0.062	-4.036721	.098547
WhiteCol						
exper	.0312007	.0143598	2.17	0.030	.0030561	.0593454
ed	.4195709	.0958978	4.38	0.000	.2316147	.607527
white	.8829927	.843371	1.05	0.295	-.7699841	2.535969
_cons	-7.140306	1.623401	-4.40	0.000	-10.32211	-3.958498
Prof						
exper	.032303	.0133779	2.41	0.016	.0060827	.0585233
ed	.8445092	.093709	9.01	0.000	.6608429	1.028176
white	1.097459	.6877939	1.60	0.111	-.2505923	2.44551
_cons	-12.42143	1.569897	-7.91	0.000	-15.49837	-9.344489

```
(Outcome occ==BlueCol is the comparison group)
```

mlogit requires the option constraint(999) to indicate that estimation should impose this constraint. The output clearly indicates which constraints have been imposed. Finally, we use lrtest to compute the test:

```
. estimates store nmodel
```

```
. lrtest fmodel nmodel
```

| likelihood-ratio test | LR chi2(3) | = | 4.09 |
| (Assumption: nmodel nested in fmodel) | Prob > chi2 | = | 0.2514 |

6.4   Independence of irrelevant alternatives

Both the MNLM and the conditional logit model (discussed below) make the assumption known as the *independence of irrelevant alternatives* (IIA). Here, we describe the assumption in terms of the MNLM. In this model,

$$\frac{\Pr(y = m \mid \mathbf{x})}{\Pr(y = n \mid \mathbf{x})} = \exp\left(\mathbf{x}\left[\beta_{m|b} - \beta_{n|b}\right]\right)$$

where the odds do not depend on other outcomes that are available. In this sense, these alternative outcomes are "irrelevant". What this means is that adding or deleting outcomes does not affect the odds among the remaining outcomes. This point is often made with the red bus/blue bus example. Suppose that you have the choice of a red bus or a car to get to work and that the odds of taking a red bus compared with a car are 1:1. IIA implies that the odds will remain 1:1 between these two alternatives, even if a new *blue* bus company comes to town that is identical to the red bus company, except for the color of the bus. Thus, the probability of driving a car can be made arbitrarily small by adding enough different colors of buses! More reasonably, we might expect that the odds of a red bus compared with a car would be reduced to 1:2 since half of those riding the red bus would be expected to ride the blue bus.

There are two tests of the IIA assumption. Hausman and McFadden (1984) proposed a Hausman-type test and McFadden, Tye, and Train (1976) proposed an approximate likelihood-ratio test that was improved by Small and Hsiao (1985). For both the Hausman and the Small–Hsiao tests, multiple tests of IIA are possible. Assuming that the MNLM is estimated with base category b, $J - 1$ tests can be computed by excluding each of the remaining categories to form the restricted model. By changing the base category, a test can also be computed that excludes b. The results of the test differ, depending on which base category was used to fit the model. See Zhang and Hoffman (1993) or Long (1997, Chapter 6) for further information.

Hausman test of IIA

The Hausman test of IIA involves the following steps:

1. Fit the full model with all J outcomes included, with estimates in $\widehat{\beta}_F$.

2. Fit a restricted model by eliminating one or more outcome categories, with estimates in $\widehat{\beta}_R$.

3. Let $\widehat{\beta}_F^*$ be a subset of $\widehat{\beta}_F$ after eliminating coefficients not fitted in the restricted model. The test statistic is

$$H = \left(\widehat{\beta}_R - \widehat{\beta}_F^*\right)' \left\{\widehat{\mathrm{Var}}\left(\widehat{\beta}_R\right) - \widehat{\mathrm{Var}}\left(\widehat{\beta}_F^*\right)\right\}^{-1} \left(\widehat{\beta}_R - \widehat{\beta}_F^*\right)$$

where H is asymptotically distributed as chi-squared with degrees of freedom equal to the rows in $\widehat{\beta}_R$ if IIA is true. Significant values of H indicate that the IIA assumption has been violated.

The Hausman test of IIA can be computed with `mlogtest`. In our example, the results are

```
. mlogit occ white ed exper, basecategory(5) nolog
  (output omitted)
. mlogtest, hausman base
**** Hausman tests of IIA assumption
Ho: Odds(Outcome-J vs Outcome-K) are independent of other alternatives.
 Omitted  |    chi2    df   P>chi2   evidence
----------+------------------------------------
   Menial |    7.324   12   0.835    for Ho
  BlueCol |    0.320   12   1.000    for Ho
    Craft |  -14.436   12   1.000    for Ho
 WhiteCol |   -5.541   11   1.000    for Ho
     Prof |   -0.119   12   1.000    for Ho
----------+
```

Five tests of IIA are reported. The first four correspond to excluding one of the four nonbase categories. The fifth test, in row `Prof`, is computed by refitting the model using the largest remaining category as the base category.[1] While none of the tests reject the H_0 that IIA holds, the results differ considerably, depending on the category considered. Further, three of the test statistics are negative, which we find to be very common. Hausman and McFadden (1984, 1226) note this possibility and conclude that a negative result is evidence that IIA has *not* been violated. A further sense of the variability of the results can be seen by rerunning `mlogit` with a different base category and then running `mlogtest, hausman base`.

Small and Hsiao test of IIA

To compute Small and Hsiao's test, the sample is divided randomly into two subsamples of about equal size. The unrestricted MNLM is estimated on both subsamples, where $\widehat{\beta}_u^{S_1}$ contains estimates from the unrestricted model on the first subsample, and $\widehat{\beta}_u^{S_2}$ is its counterpart for the second subsample. A weighted average of the coefficients is computed as

$$\widehat{\beta}_u^{S_1 S_2} = \left(\frac{1}{\sqrt{2}}\right)\widehat{\beta}_u^{S_1} + \left\{1 - \left(\frac{1}{\sqrt{2}}\right)\right\}\widehat{\beta}_u^{S_2}$$

Next, a restricted sample is created from the second subsample by eliminating all cases with a chosen value of the dependent variable. The MNLM is estimated using the restricted sample, yielding the estimates $\widehat{\beta}_r^{S_2}$ and the likelihood $L(\widehat{\beta}_r^{S_2})$. The Small–Hsiao statistic is

$$SH = -2\left\{L(\widehat{\beta}_u^{S_1 S_2}) - L(\widehat{\beta}_r^{S_2})\right\}$$

which is asymptotically distributed as a chi-squared with the degrees of freedom equal to $K + 1$, where K is the number of independent variables.

---

[1]Even though `mlogtest` fits other models to compute various tests, when the command ends it restores the estimates from your original model. Accordingly, other commands that require results from your original `mlogit`, such as `predict` and `prvalue`, will still work correctly.

To compute the Small–Hsiao test, you use the command `mlogtest, smhsiao` (our program uses code from `smhsiao` by Nick Winter, available at the SSC-IDEAS archive). For example,

```
. mlogtest, smhsiao
**** Small-Hsiao tests of IIA assumption
 Ho: Odds(Outcome-J vs Outcome-K) are independent of other alternatives.
   Omitted |   lnL(full)  lnL(omit)    chi2   df   P>chi2   evidence

    Menial |   -182.140   -169.907   24.466    4    0.000   against Ho
   BlueCol |   -148.711   -140.054   17.315    4    0.002   against Ho
     Craft |   -131.801   -119.286   25.030    4    0.000   against Ho
  WhiteCol |   -161.436   -148.550   25.772    4    0.000   against Ho
```

The results vary considerably from those of the Hausman tests. In this case, each test indicates that IIA has been violated.

Because the Small–Hsiao test requires randomly dividing the data into subsamples, the results will differ with successive calls of the command, as the sample will be divided differently. To obtain test results that can be replicated, you must explicitly set the seed used by the random number generator. For example,

```
. set seed 8675309
. mlogtest, smhsiao
**** Small-Hsiao tests of IIA assumption
 Ho: Odds(Outcome-J vs Outcome-K) are independent of other alternatives.
   Omitted |   lnL(full)  lnL(omit)    chi2   df   P>chi2   evidence

    Menial |   -169.785   -161.523   16.523    4    0.002   against Ho
   BlueCol |   -131.900   -125.871   12.058    4    0.017   against Ho
     Craft |   -136.934   -129.905   14.058    4    0.007   against Ho
  WhiteCol |   -155.364   -150.239   10.250    4    0.036   against Ho
```

Advanced: setting the random seed The random numbers that divide the sample for the Small–Hsiao test are based on Stata's `uniform()` function, which uses a pseudo-random number generator. This generator creates a sequence of numbers based on a seed number. While these numbers appear to be random, exactly the same sequence will be generated each time you start with the same seed number. In this sense (and some others), these numbers are pseudo-random rather than random. If you specify the seed with `set seed #`, you can ensure that you will be able to replicate your results later. See the *User's Guide* for further details.

Conclusions regarding tests of IIA

Our experience with these tests is that they often give inconsistent results and provide little guidance to violations of the IIA assumption. Unfortunately, there do not appear to be simulation studies that examine their small sample properties. Perhaps as a result of the practical limitations of these tests, McFadden (1973) suggested that IIA implies that the multinomial and conditional logit models should only be used in cases where the outcome categories "can plausibly be assumed to be distinct and weighed independently in the eyes of each decision maker". Similarly, Amemiya (1981, 1517) suggests that the MNLM works well when the alternatives are dissimilar. Care in specifying the model to involve distinct outcomes that are not substitutes for one another seems to be reasonable, albeit unfortunately ambiguous, advice.

6.5 Measures of fit

As with the binary and ordinal models, scalar measures of fit for the MNLM model can be computed with the SPost command `fitstat`. The same caveats against overstating the importance of these scalar measures apply here as to the other models we consider (see also Chapter 3). To examine the fit of individual observations, you can estimate the series of binary logits implied by the multinomial logit model and use the established methods of examining the fit of observations to binary logit estimates. This is the same approach that was recommended in Chapter 5 for ordinal models.

6.6 Interpretation

While the MNLM is a mathematically simple extension of the binary model, interpretation is made difficult by the large number of possible comparisons. Even in our simple example with five outcomes, we have many possible comparisons: $M|P$, $B|P$, $C|P$, $W|P$, $M|W$, $B|W$, $C|W$, $M|C$, $B|C$, and $M|B$. It is tedious to write all of the comparisons, let alone to interpret each of them for each of the independent variables. Thus, the key to interpretation is to avoid being overwhelmed by the many comparisons. Most of the methods we propose are very similar to those for ordinal outcomes, and accordingly, these are treated very briefly. However, methods of plotting discrete changes and factor changes are new, so these are considered in greater detail.

6.6.1 Predicted probabilities

Predicted probabilities can be computed with the formula

$$\widehat{\Pr}\left(y = m \mid \mathbf{x}\right) = \frac{\exp\left(\mathbf{x}\widehat{\beta}_{m|J}\right)}{\sum_{j=1}^{J} \exp\left(\mathbf{x}\widehat{\beta}_{j|J}\right)}$$

where **x** can contain values from individuals in the sample or hypothetical values. The most basic command for computing probabilities is `predict`, but we also illustrate a series of SPost commands that compute predicted probabilities in useful ways.

6.6.2 Predicted probabilities with predict

After fitting the model with `mlogit`, the predicted probabilities within the sample can be calculated with the command

predict *newvar1* $\big[$ *newvar2* ... $\big[$ *newvarJ* $\big]$ $\big]$ $\big[$if *exp*$\big]$ $\big[$in *range*$\big]$

where you must provide one new variable name for each of the J categories of the dependent variable, ordered from the lowest to highest numerical values. For example,

```
. mlogit occ white ed exper, basecategory(5) nolog
  (output omitted)
. predict ProbM ProbB ProbC ProbW ProbP
(option p assumed; predicted probabilities)
```

The variables created by `predict` are

```
. describe Prob*

                storage  display    value
variable name   type     format     label      variable label

ProbM           float    %9.0g                  Pr(occ==1)
ProbB           float    %9.0g                  Pr(occ==2)
ProbC           float    %9.0g                  Pr(occ==3)
ProbW           float    %9.0g                  Pr(occ==4)
ProbP           float    %9.0g                  Pr(occ==5)

. sum Prob*
     Variable |        Obs        Mean    Std. Dev.        Min        Max

        ProbM |        337    .0919881     .059396    .0010737    .3281906
        ProbB |        337    .2047478    .1450568    .0012066    .6974148
        ProbC |        337    .2492582    .1161309    .0079713     .551609
        ProbW |        337    .1216617    .0452844    .0083857    .2300058
        ProbP |        337    .3323442    .2870992    .0001935    .9597512
```

Using predict to compare mlogit and ologit

An interesting way to illustrate how predictions can be plotted is to compare predictions from ordered logit and multinomial logit when the models are applied to the same data. Recall from Chapter 5 that the range of the predicted probabilities for middle categories abruptly ended, while predictions for the end categories had a more gradual distribution. To illustrate this point, the example in Chapter 5 is estimated using `ologit` and `mlogit`, with predicted probabilities computed for each case:

```
. use http://www.stata-press.com/data/lfr/ordwarm2,clear
(77 & 89 General Social Survey)
```

```
. ologit warm yr89 male white age ed prst, nolog
  (output omitted)
. predict SDologit Dologit Aologit SAologit
(option p assumed; predicted probabilities)
. label var Dologit "ologit-D"
. mlogit warm yr89 male white age ed prst, nolog
  (output omitted)
. predict SDmlogit Dmlogit Amlogit SAmlogit
(option p assumed; predicted probabilities)
. label var Dmlogit "mlogit-D"
```

We can plot the predicted probabilities of disagreeing in the two models with the command `dotplot Dologit Dmlogit, ylabel(0(.25).75)`, which leads to

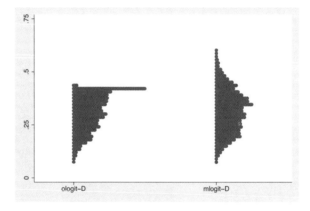

While the two sets of predictions have a correlation of .92 (computed by the command `corr Dologit Dmlogit`), the abrupt truncation of the distribution for the ordered logit model strikes us as substantively unrealistic.

6.6.3   Individual predicted probabilities with prvalue

Predicted probabilities for individuals with specified characteristics can be computed with `prvalue`. For example, we might compute the probabilities of each occupational category to compare nonwhites and whites who are average on education and experience:

```
. use http://www.stata-press.com/data/lfr/nomocc2, clear
(1982 General Social Survey)
. mlogit occ white ed exper, basecategory(5) nolog
  (output omitted)
. quietly prvalue, x(white=0) rest(mean) save
```

```
. prvalue, x(white=1) rest(mean) dif
mlogit: Change in Predictions for  occ
Predicted probabilities for each category:
                        Current      Saved  Difference
      Pr(y=Menial|x):    0.0860     0.2168     -0.1309
      Pr(y=BlueCol|x):   0.1862     0.1363      0.0498
      Pr(y=Craft|x):     0.2790     0.4387     -0.1597
      Pr(y=WhiteCol|x):  0.1674     0.0877      0.0797
      Pr(y=Prof|x):      0.2814     0.1204      0.1611

                   white        ed       exper
      Current=         1  13.094955  20.501484
        Saved=         0  13.094955  20.501484
         Diff=         1          0           0
```

This example also illustrates how to use **prvalue** to compute differences between two sets of probabilities. Our first call of **prvalue** is done **quietly**, but we **save** the results. The second call uses the **dif** option, and the output compares the results for the first and second set of values computed.

6.6.4   Tables of predicted probabilities with prtab

If you want predicted probabilities for all combinations of a set of categorical independent variables, **prtab** is useful. For example, we might want to know how white and nonwhite respondents differ in their probability of having a menial job by years of education:

```
. label def lwhite 0 NonWhite 1 White
. label val white lwhite
. prtab ed white, novarlbl outcome(1)
mlogit: Predicted probabilities of outcome 1 (Menial) for occ
```

	white	
ed	NonWhite	White
3	0.2847	0.1216
6	0.2987	0.1384
7	0.2988	0.1417
8	0.2963	0.1431
9	0.2906	0.1417
10	0.2814	0.1366
11	0.2675	0.1265
12	0.2476	0.1104
13	0.2199	0.0883
14	0.1832	0.0632
15	0.1393	0.0401
16	0.0944	0.0228
17	0.0569	0.0120
18	0.0310	0.0060
19	0.0158	0.0029
20	0.0077	0.0014

```
         white         ed       exper
x=   .91691395  13.094955  20.501484
```

Tip: `outcome()` **option** In this example, we use the `outcome()` option to restrict the output to a single outcome category. Without this option, `prtab` will produce a separate table for each outcome category.

The table produced by `prtab` shows the substantial differences between whites and nonwhites in the probabilities of having menial jobs and how these probabilities are affected by years of education. However, given the number of categories for `ed`, plotting these predicted probabilities with `prgen` is probably a more useful way to examine the results.

6.6.5   Graphing predicted probabilities with prgen

Predicted probabilities can be plotted using the same methods considered for the ordinal regression model. After fitting the model, we use `prgen` to compute the predicted probabilities for whites with average working experience as education increases from 6 years to 20 years:

```
. prgen ed, x(white=1) from(6) to(20) generate(wht) ncases(15)
mlogit: Predicted values as ed varies from 6 to 20.
        white        ed       exper
x=          1  13.094955  20.501484
```

Here is what the options specify:

`x(white=1)` sets `white` to 1. Because the `rest()` option is not included, all other variables are set to their means by default.

`from(6)` and `to(20)` set the minimum and maximum values over which `ed` is to vary. The default is to use the variable's minimum and maximum values.

`ncases(15)` indicates that 15 evenly spaced values of `ed` between 6 and 20 are to be generated. We chose 15 for the number of values from 6 to 20, inclusive.

`gen(wht)` specifies the root name for the new variables generated by `prgen`. For example, the variable `whtx` contains values of `ed`, the p-variables (e.g., `whtp2`) contain the predicted probabilities for each outcome, and the s-variables contain the summed probabilities. A list of these variables should make this clear:

(Continued on next page)

```
. describe wht*

                storage  display    value
variable name    type    format     label     variable label
─────────────────────────────────────────────────────────────────
whtx            float   %9.0g                 Changing value of ed
whtp1           float   %9.0g                 pr(Menial) [1]
whts1           float   %9.0g                 pr(y<=1)
whtp2           float   %9.0g                 pr(BlueCol) [2]
whts2           float   %9.0g                 pr(y<=2)
whtp3           float   %9.0g                 pr(Craft) [3]
whts3           float   %9.0g                 pr(y<=3)
whtp4           float   %9.0g                 pr(WhiteCol) [4]
whts4           float   %9.0g                 pr(y<=4)
whtp5           float   %9.0g                 pr(Prof) [5]
whts5           float   %9.0g                 pr(y<=5)
```

The same thing can be done to compute predicted probabilities for nonwhites:

```
. prgen ed, x(white=0) from(6) to(20) generate(nwht) ncases(15)
mlogit: Predicted values as ed varies from 6 to 20.
        white         ed        exper
x=          0   13.094955   20.501484
```

Plotting probabilities for one outcome and two groups

The variables **nwhtp1** and **whtp1** contain the predicted probabilities of having menial jobs for nonwhites and whites. Plotting these provides clearer information than the results of **prtab** given above:

```
. label var whtp1 "Whites"
. label var nwhtp1 "Nonwhites"
. graph twoway connected whtp1 nwhtp1 nwhtx,
      xtitle("Years of Education")
      ytitle("Pr(Menial Job)")
      ylabel(0(.25).50) xlabel(6 8 12 16 20)
      ysize(2.7051) xsize(4.0421)
```

(Continued on next page)

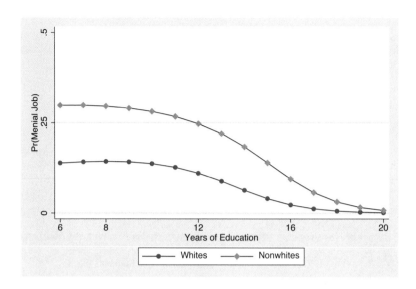

Graphing probabilities for all outcomes for one group

Even though nominal outcomes are not ordered, plotting the summed probabilities can be a useful way to show predicted probabilities for all outcome categories. To illustrate this, we construct a graph to show how education affects the probability of each occupation for whites (a similar graph could be plotted for nonwhites). This is done using the *roots#* variables created by **prgen**, which provide the probability of being in an outcome less than or equal to some value. For example, the label for **whts3** is **pr(y<=3)**, which indicates that all nominal categories coded as 3 or less are added together. To plot these probabilities, the first thing we do is change the variable labels to the name of the highest category in the sum, which makes the graph clearer (as you will see below):

```
. label var whts1 "Menial"
. label var whts2 "Blue Collar"
. label var whts3 "Craft"
. label var whts4 "White Collar"
```

To create the summed plot, we use the following commands:

```
. graph twoway connected whts1 whts2 whts3 whts4 whtx,
        xtitle("Whites: Years of Education")
        ytitle("Summed Probability")
        xlabel(6(2)20)
        ylabel(0(.25)1)
        ysize(2.6195) xsize(4.0421)
```

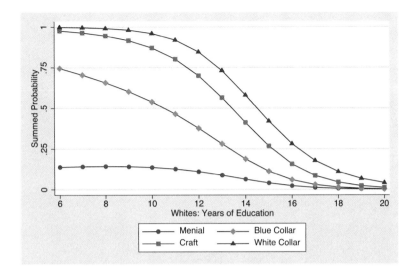

The graph plots the four summed probabilities against `whtx`, where standard options for `graph` are used. This graph is not ideal, but before revising it, let's make sure we understand what is being plotted. The lowest line with circles, labeled "Menial" in the key, plots the probability of having a menial job for a given year of education. This is the same information as plotted in our prior graph for whites. The next line with small diamonds, labeled "Blue Collar" in the key, plots the sum of the probability of having a menial job or a blue collar job. Thus, the area between the line with circles and the line with diamonds is the probability of having a blue collar job, and so on.

Because what we really want to illustrate are the regions between the curves, this graph is not as effective as we would like. In the `graph` command below, we use the `rarea` plot type to shade the regions between the curves. The syntax for an `rarea` plot[2] is

graph twoway rarea *y1var* *y2var* *xvar* [if *exp*] [in *range*] [, *rarea_options*]

where *y1var* defines the lower boundary and *y2var* defines the upper of the region for each *x*-value given in the variable *xvar*.

Continuing with our example, as the probabilities are bounded between zero and one, we begin by creating variables that hold these extreme values.

```
. gen zero = 0
. gen one  = 1
```

Now, we are ready to draw to the full graph. In the command below, we draw the `rarea` plots before the connected line plots. Because the `rarea` plots obscure whatever is previously drawn, this is important.

---

[2]Type `help twoway_rarea` for more information.

```
. graph twoway (rarea zero whts1 whtx, bc(gs1))
         (rarea whts1 whts2 whtx, bc(gs4))
         (rarea whts2 whts3 whtx, bc(gs8))
         (rarea whts3 whts4 whtx, bc(gs11))
         (rarea whts4 one whtx, bc(gs14)),
         ytitle("Summed Probability")
         legend( order( 1 2 3 4 5)
         label( 1 "Menial")
         label( 2 "Blue Collar") label( 3 "Craft")
         label(4 "White Collar") label(5 "Professional"))
         xtitle("Whites: Years of Education")
         xlabel(6 8 12 16 20) ylabel(0(.25)1)
         ysize(2.6195) xsize(4.0421)
         plotregion(margin(zero))
```

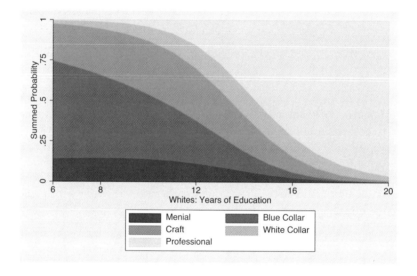

Figure 6.1: Whites: Years of Education

The changes in the shaded regions in Figure 6.1 clearly illustrate how the probability of selecting any one occupation changes as education increases.

6.6.6   Changes in predicted probabilities

Marginal and discrete change can be used in the same way as in models for ordinal outcomes. As before, both can be computed using `prchange`.

Marginal change is defined as

$$\frac{\partial \Pr (y = m \mid \mathbf{x})}{\partial x_k} = \Pr (y = m \mid \mathbf{x}) \left\{ \beta_{k,m|J} - \sum_{j=1}^{J} \beta_{k,j|J} \Pr(y = j \mid \mathbf{x}) \right\}$$

As this equation combines all of the $\beta_{k,j|J}$'s, the value of the marginal change depends on the levels of all variables in the model. Further, as the value of x_k changes, the sign of the marginal can change. For example, at one point the marginal effect of education on having a craft occupation could be positive, while at another point the marginal effect could be negative.

Discrete change is defined as

$$\frac{\Delta \Pr(y = m \mid \mathbf{x})}{\Delta x_k} = \Pr(y = m \mid \mathbf{x}, x_k = x_E) - \Pr(y = m \mid \mathbf{x}, x_k = x_S)$$

where the magnitude of the change depends on the levels of all variables and the size of the change that is being made. The J discrete change coefficients for a variable (one for each outcome category) can be summarized by computing the average of the *absolute values* of the changes across all of the outcome categories,

$$\overline{\Delta} = \frac{1}{J} \sum_{j=1}^{J} \left| \frac{\Delta \Pr(y = j \mid \overline{\mathbf{x}})}{\Delta x_k} \right|$$

where the absolute value is taken because the sum of the changes without taking the absolute value is necessarily zero.

Computing marginal and discrete change with prchange

Discrete and marginal changes are computed with **prchange** (the full syntax for which is provided in Chapter 3). For example,

```
. mlogit occ white ed exper
  (output omitted)
. prchange
mlogit: Changes in Predicted Probabilities for occ

white
             Avg|Chg|      Menial      BlueCol        Craft     WhiteCol
    0->1    .11623582   -.13085523    .04981799   -.15973434    .07971004
                  Prof
    0->1     .1610615

ed
             Avg|Chg|      Menial      BlueCol        Craft     WhiteCol
Min->Max    .39242268   -.13017954   -.70077323   -.15010394    .02425591
   -+1/2    .05855425   -.02559762   -.06831616   -.05247185    .01250795
  -+sd/2     .1640657   -.07129153   -.19310513   -.14576758    .03064777
MargEfct    .05894859   -.02579097   -.06870635   -.05287415    .01282041
                  Prof
Min->Max    .95680079
   -+1/2    .13387768
  -+sd/2    .37951647
MargEfct    .13455107
```

```
exper
            Avg|Chg|      Menial      BlueCol        Craft     WhiteCol
Min->Max    .12193559  -.11536534  -.18947365    .03115708    .09478889
   -+1/2    .00233425  -.00226997  -.00356567    .00105992     .0016944
   -+sd/2   .03253578  -.03167491  -.04966453    .01479983    .02360725
MargEfct    .00233427  -.00226997  -.00356571    .00105992    .00169442

                 Prof
Min->Max    .17889298
   -+1/2    .00308132
   -+sd/2   .04293236
MargEfct    .00308134

            Menial     BlueCol       Craft     WhiteCol        Prof
Pr(y|x)   .09426806   .18419114   .29411051    .16112968   .26630062

            white         ed     exper
    x=    .916914     13.095   20.5015
 sd(x)=   .276423    2.94643   13.9594
```

The first thing to notice is the output labeled Pr(y | x), which is the predicted prob-
abilities at the values set by x() and rest(). Marginal change is listed in the rows
MargEfct. For variables that are not binary, discrete change is reported over the entire
range of the variable (reported as Min->Max), for changes of one unit centered on the
base values (reported as -+1/2), and for changes of one standard deviation centered on
the base values (reported as -+sd/2). If the **uncentered** option is used, the changes
begin at the value specified by x() or rest() and increase one unit or one standard
deviation from there. For binary variables, the discrete change from 0 to 1 is the only
appropriate quantity and is the only quantity that is presented. Looking at the results
for white above, we can see that for someone who is average in education and experience, the predicted probability of having a professional job is .16 higher for whites than
nonwhites. The average change is listed in the column Avg | Chg |. For example, for
white, $\overline{\Delta} = 0.12$, the average absolute change in the probability of various occupational
categories for being white as opposed to nonwhite is .12.

Marginal change with mfx compute

The marginal change can also be computed using mfx compute, where the at() option
is used to set values of the independent variables. Like prchange, mfx compute sets all
values of the independent variables to their means by default. As noted in Chapter 5,
mfx compute does not allow you to compute effects only for a subset of variables in the
model. Also, we must estimate the marginal effects for one outcome at a time, using
the predict(outcome(#)) option to specify the outcome for which we want marginal
effects:

```
. mfx compute, predict(outcome(1))

Marginal effects after mlogit
      y  = Pr(occ==1) (predict, outcome(1))
         =  .09426806
```

variable	dy/dx	Std. Err.	z	P>|z|	[95% C.I.]	X
white*	-.1308552	.08915	-1.47	0.142	-.305592 .043882	.916914
ed	-.025791	.00688	-3.75	0.000	-.039269 -.012312	13.0950
exper	-.00227	.00126	-1.80	0.071	-.004737 .000197	20.5015

```
(*) dy/dx is for discrete change of dummy variable from 0 to 1
```

These results are for the Menial category (occ==1). Estimates for exper and ed match the results in the MargEfct rows of the prchange output above. Meanwhile, for the binary variable white, the discrete change from 0 to 1 is presented, which also matches the corresponding result from prchange. An advantage of mfx compute is that standard errors for the effects are also provided; a disadvantage is that mfx compute can take a long time to produce results after mlogit, especially if the number of observations and independent variables is large.

6.6.7 Plotting discrete changes with prchange and mlogview

One difficulty with nominal outcomes is the large number of coefficients that need to be considered: one for each variable times the number of outcome categories minus one. To help you sort out all of this information, discrete change coefficients can be plotted using our program mlogview. After fitting the model with mlogit and computing discrete changes with prchange, executing mlogview opens the following dialog box:[3]

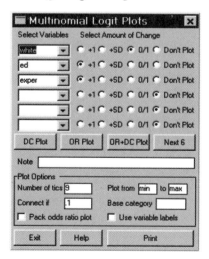

---

[3]StataCorp recently increased the limits for the number of options that can be contained in a dialog box. Accordingly, future versions of mlogview are likely to have additional options, and the dialog box you get could look different.

Dialog boxes are easier to use than to explain. So, as we describe various features, the best advice is to generate the dialog box shown above and experiment.

Selecting variables If you click and hold a 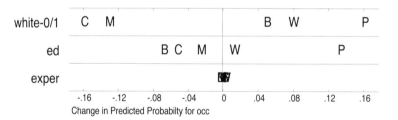 button, you can select a variable to be plotted. The same variable can be plotted more than once, for example, showing the effects of different amounts of change.

Selecting the amount of change The radio buttons allow you to select the type of discrete change coefficient to plot for each selected variable: +1 selects coefficients for a change of one unit; +SD selects coefficients for a change of one standard deviation; 0/1 selects changes from 0 to 1; and Don't Plot is self-explanatory.

Making a plot Even though there are more options to explain, you should try plotting your selections by clicking on DC Plot, which produces a graph. The command `mlogview` works by generating the syntax for the command `mlogplot`, which actually draws the plot. In the Results window, you will see the `mlogplot` command that was used to generate your graph (full details on `mlogplot` are given in Section 6.6.9). If there is an error in the options you select, the error message will appear in the Results window.

Assuming everything has worked, we generate the following graph:

The graph immediately shows how a unit increase in each variable affects the probability of each outcome. While it appears that the effects of being white are the largest, changes of one unit in education and (especially) experience are often too small to be as informative. It would make more sense to look at the effects of a standard deviation change in these variables. To do this, we return to the dialog box and click on the radio button +SD. Before we see what this does, let's consider several other options that can be used.

Adding labels The box Note allows you to enter text that will be placed at the top of the graph. Clicking the box for Use variable labels replaces the names of the variables on the left axis with the variable labels associated with each variable. When you do this, you may find that the labels are too long. If so, you can use the `label variable` command to change them.

Tick marks The values for the tick marks are determined by specifying the minimum and maximum values to plot and the number of tick marks. For example, we could specify a plot from $-.2$ to $.4$ with 7 tick marks. This will lead to labels every $.1$ units.

Using some of the features discussed above, our dialog box would look like this:

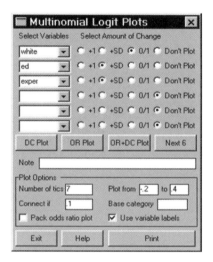

Clicking on DC Plot produces the following graph:

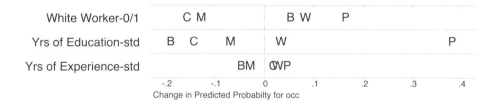

In this figure, you can see that the effects of education are largest, and that those of experience are smallest. Or, each coefficient can be interpreted individually, such as the following:

> The effects of a standard deviation change in education are largest, with an increase of over .35 in the probability of having a professional occupation.

> The effects of race are also substantial, with average blacks being less likely to enter blue collar, white collar, or professional jobs than average whites.

> Expected changes due to a standard deviation change in experience are much smaller and show that experience increases the probabilities of more highly skilled occupations.

In using these graphs, keep in mind that different values for discrete change are obtained at different levels of the variables, which are specified with the x() and rest() options for prchange.

Value labels with mlogview The value labels for the different categories of the dependent variables must begin with different letters because the plots generated with mlogview use the first letter of the value label.

6.6.8   Odds ratios using listcoef and mlogview

Discrete change does little to illuminate the dynamics among the outcomes. For example, a decrease in education increases the probability of both blue collar and craft jobs, but how does it affect the odds of a person choosing a craft job relative to a blue collar job? To deal with these issues, odds ratios (also referred to as factor change coefficients) can be used. Holding other variables constant, the factor change in the odds of outcome m versus outcome n as x_k increases by δ equals

$$\frac{\Omega_{m|n}\left(\mathbf{x}, x_k + \delta\right)}{\Omega_{m|n}\left(\mathbf{x}, x_k\right)} = e^{\beta_{k,m|n}\delta}$$

If the amount of change is $\delta = 1$, the odds ratio can be interpreted that

> for a unit change in x_k, the odds of m versus n are expected to change by a factor of $\exp(\beta_{k,m|n})$, holding all other variables constant.

If the amount of change is $\delta = s_{x_k}$, then the odds ratio can be interpreted that

> for a standard deviation change in x_k, the odds of m versus n are expected to change by a factor of $\exp(\beta_{k,m|n} \times s_k)$, holding all other variables constant.

Listing odds ratios with listcoef

The difficulty in interpreting odds ratios for the MNLM is that, to understand the effect of a variable, you need to examine the coefficients for comparisons among all pairs of outcomes. The standard output from mlogit includes only $J - 1$ comparisons with the base category. While you could estimate coefficients for all possible comparisons by rerunning mlogit with different base categories (e.g., mlogit occ white ed exper, basecategory(3)), using listcoef is much simpler. For example, to examine the effects of race, type

```
. listcoef white, help

mlogit (N=337): Factor Change in the Odds of occ

Variable: white (sd=    .28)

      Odds comparing|
     Group 1 vs Group 2|       b         z       P>|z|      e^b     e^bStdX
```

Odds comparing Group 1 vs Group 2		b	z	P>|z|	e^b	e^bStdX
Menial	-BlueCol	-1.23650	-1.707	0.088	0.2904	0.7105
Menial	-Craft	-0.47234	-0.782	0.434	0.6235	0.8776
Menial	-WhiteCol	-1.57139	-1.741	0.082	0.2078	0.6477
Menial	-Prof	-1.77431	-2.350	0.019	0.1696	0.6123
BlueCol	-Menial	1.23650	1.707	0.088	3.4436	1.4075
BlueCol	-Craft	0.76416	1.208	0.227	2.1472	1.2352
BlueCol	-WhiteCol	-0.33488	-0.359	0.720	0.7154	0.9116
BlueCol	-Prof	-0.53780	-0.673	0.501	0.5840	0.8619
Craft	-Menial	0.47234	0.782	0.434	1.6037	1.1395
Craft	-BlueCol	-0.76416	-1.208	0.227	0.4657	0.8096
Craft	-WhiteCol	-1.09904	-1.343	0.179	0.3332	0.7380
Craft	-Prof	-1.30196	-2.011	0.044	0.2720	0.6978
WhiteCol-Menial		1.57139	1.741	0.082	4.8133	1.5440
WhiteCol-BlueCol		0.33488	0.359	0.720	1.3978	1.0970
WhiteCol-Craft		1.09904	1.343	0.179	3.0013	1.3550
WhiteCol-Prof		-0.20292	-0.233	0.815	0.8163	0.9455
Prof	-Menial	1.77431	2.350	0.019	5.8962	1.6331
Prof	-BlueCol	0.53780	0.673	0.501	1.7122	1.1603
Prof	-Craft	1.30196	2.011	0.044	3.6765	1.4332
Prof	-WhiteCol	0.20292	0.233	0.815	1.2250	1.0577

```
          b = raw coefficient
          z = z-score for test of b=0
      P>|z| = p-value for z-test
        e^b = exp(b) = factor change in odds for unit increase in X
   e^bStdX = exp(b*SD of X) = change in odds for SD increase in X
```

The odds ratios of interest are in the column labeled e^b. For example, the odds ratio for the effect of race on having a professional versus a menial job is 5.90, which can be interpreted as follows:

> The odds of having a professional occupation relative to a menial occupation are 5.90 times greater for whites than for blacks, holding education and experience constant.

Plotting odds ratios

However, examining all of the coefficients for even a single variable with only five dependent categories is complicated. An *odds ratio plot* makes it easy to quickly see patterns in results for even a complex MNLM (see Long 1997, Chapter 6 for full details). To explain how to interpret an odds ratio plot, we begin with some hypothetical output from a MNLM with three outcomes and three independent variables:

		Logit Coefficient for		
Comparison		x_1	x_2	x_3
$B \mid A$	$\beta_{B\mid A}$	-0.693	0.693	0.347
	$\exp(\beta_{B\mid A})$	0.500	2.000	1.414
	p	0.04	0.01	0.42
$C \mid A$	$\beta_{C\mid A}$	0.347	-0.347	0.693
	$\exp(\beta_{C\mid A})$	1.414	0.707	2.000
	p	0.21	0.04	0.37
$C \mid B$	$\beta_{C\mid B}$	1.040	-1.040	0.346
	$\exp(\beta_{C\mid B})$	2.828	0.354	1.414
	p	0.02	0.03	0.21

These coefficients were constructed to have some fixed relationships among categories and variables:

- The effects of x_1 and x_2 on $B \mid A$ (which you can read as B versus A) are equal but of opposite size. The effect of x_3 is half as large.

- The effects of x_1 and x_2 on $C \mid A$ are half as large (and in opposite directions) as the effects on $B \mid A$, while the effect of x_3 is in the same direction but twice as large.

In the odds ratio plot, the independent variables are each represented on a separate row, and the horizontal axis indicates the relative magnitude of the β coefficients associated with each outcome. Here is the plot, where the letters correspond to the outcome categories:

The plot reveals a great deal of information, which we now summarize.

Sign of coefficients

If a letter is to the right of another letter, increases in the independent variable make the outcome to the right more likely. Thus, relative to outcome A, an increase in x_1 makes it more likely that we will observe outcome C and less likely that we will observe outcome B. This corresponds to the positive sign of the $\beta_{1,C|A}$ coefficient and the negative sign of the $\beta_{1,B|A}$ coefficient. The signs of these coefficients are reversed for x_2, and accordingly, the odds ratio plot for x_2 is a mirror image of that for x_1.

Magnitude of effects

The distance between a pair of letters indicates the magnitude of the effect. For both x_1 and x_2, the distance between A and B is twice the distance between A and C, which reflects that $\beta_{B|A}$ is twice as large as $\beta_{C|A}$ for both variables. For x_3, the distance between A and B is half the distance between A and C, reflecting that $\beta_{3,C|A}$ is twice as large as $\beta_{3,B|A}$.

The additive relationship

The additive relationships among coefficients shown in (6.1) are also fully reflected in this graph. For any of the independent variables, $\beta_{C|A} = \beta_{B|A} + \beta_{C|B}$. Accordingly, the distance from A to C is the sum of the distances from A to B and B to C.

The base category

The additive scale on the bottom axis measures the value of the $\beta_{k,m|n}$s. The multiplicative scale on the top axis measures the $\exp\left(\beta_{k,m|n}\right)$s. The reason why the As are stacked on top of one another is that the plot uses A as its base category for graphing the coefficients. The choice of base category is arbitrary. We could have used outcome B instead. If we had, the rows of the graph would be shifted to the left or right so that the Bs lined up. Doing this leads to the following graph:

	Factor Change Scale Relative to Category B						
	.35	.5	.71	1	1.41	2	2.83
x1				B		A	C
x2	C	A		B			
x3			A	B	C		
	-1.04	-.69	-.35	0	.35	.69	1.04
	Logit Coefficient Scale Relative to Category B						

Creating odds ratio plots

These graphs can be created using `mlogview` after running `mlogit`. Using our example and after changing a few options, we obtain this dialog box:

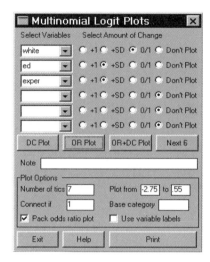

Clicking on OR Plot gives

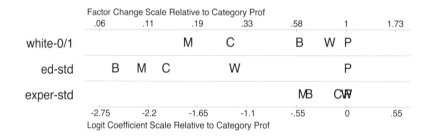

Several things are immediately apparent. The effect of experience is the smallest, although increases in experience make it more likely that one will be in a craft, white collar, or professional occupation relative to a menial or blue collar one. We also see that education has the largest effect; as expected, increases in education increase the odds of having a professional job relative to any other type.

Adding significance levels

The current graph does not reflect statistical significance. This is added by drawing a line between categories for which there is *not* a significant coefficient. The *lack* of statistical significance is shown by a connecting line, suggesting that those two outcomes are "tied together". You can add the significance level to the plot with the Connect if box on the dialog box. For example, if we enter .1 in this box and uncheck the "pack odds ratio plot" box, we obtain

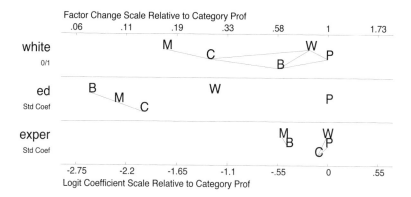

To make the connecting lines clear, vertical spacing is added to the graph. *This vertical spacing has no meaning and is only used to make the lines clearer.* The graph shows that race orders occupations from menial to craft to blue collar to white collar to professional, but the connecting lines show that none of the adjacent categories are significantly differentiated by race. Being white increases the odds of being a craft worker relative to having a menial job, but the effect is not significant. However, being white significantly increases the odds of being a blue collar worker, a white collar worker, or a professional, relative to having a menial job. The effects of ed and exper can be interpreted similarly.

Adding discrete change

In Chapter 4, we emphasized that *while the factor change in the odds is constant across the levels of all variables, the discrete change gets larger or smaller at different values of the variables.* For example, if the odds increase by a factor of 10 but the current odds are 1 in 10,000, the substantive impact is small. But, if the current odds were 1 in 5, the impact is large. Information on the discrete change in probability can be incorporated in the odds ratio graph by making the size of the letter proportional to the discrete change in the odds (specifically, the area of the letter is proportional to the size of the discrete change). This can be added to our graph very simply. First, after estimating the MNLM, run prchange at the levels of the variables that you want. Then, enter mlogview to open the dialog box. Set any of the options, and then click the OR+DC Plot button:

(*Continued on next page*)

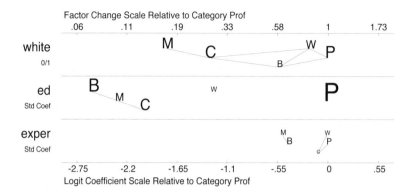

With a little practice, you can create and interpret these graphs very quickly.

6.6.9   Using mlogplot*

The dialog box `mlogview` does not actually draw the plots but only sends the options you select to `mlogplot`, which creates the graph. Once you click a plot button in `mlogview`, the necessary `mlogplot` command, including options, appears in the Results window. This is done because `mlogview` invokes a dialog box and so cannot be used effectively in a do-file. But, once you create a plot using the dialog box, you can copy the generated `mlogplot` command from the Results window and paste it into a do-file. This should be clear by looking at the following screenshot:

(*Continued on next page*)

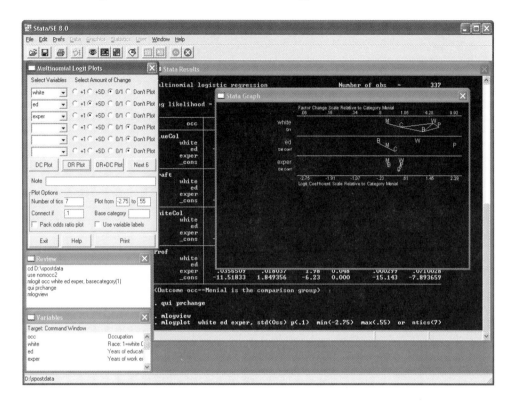

The dialog box with selected options appears in the upper left of the screen. After we clicked on the **OR Plot** button, the graph in the upper right appeared along with the following command in the Results window:

```
. mlogplot white ed exper, std(0ss) p(.1) min(-2.75) max(.55) or ntics(7)
```

If you enter this command from the Command window or run it from a do-file, the same graph will be generated. The full syntax for `mlogplot` is described in Appendix A.

6.6.10 Plotting estimates from matrices with mlogplot*

You can also use `mlogplot` to construct odds ratio plots (but not discrete change plots) using coefficients that are to be contained in matrices. For example, you can plot coefficients from published papers or generate examples like those we used above. To do this, you must construct matrices containing the information to be plotted and add the option `matrix` to the command. The easiest way to see how this is done is with an example, followed by details on each matrix. The commands

```
. matrix mnlbeta = (-.693, .693, .347   .347, -.347, .693 )
. matrix mnlsd   = (1, 2, 4)
. global mnlname = "x1 x2 x3"
```

```
. global mnlcatnm = "B C A"
. global mnldepnm "depvar"
. mlogplot, matrix std(uuu) vars(x1 x2 x3) packed
```

create the following plot:

Factor Change Scale Relative to Category A

.5	.63	.79	1	1.26	1.59	2

x1	B		A		C	
x2		C	A			B
x3			A		B		C

-.69	-.46	-.23	0	.23	.46	.69

Logit Coefficient Scale Relative to Category A

Options for using matrices with mlogplot

matrix indicates that the coefficients to be plotted are contained in matrices.

vars(*variable-list*) contains the names of the variables to be plotted. This list must contain names from mnlname, which will be described next, but does not need to be in the same order as in mnlname. The list can contain the same name more than once and can select a subset of the names from mnlname.

Global macros and matrices used by mlogplot

mnlname is a string containing the names of the variables corresponding to the columns of the matrix mnlbeta. For example, global mnlname = "x1 x2 x3".

mnlbeta is a matrix with the βs, where element (i, j) is the coefficient $\beta_{j,i|b}$. That is, rows i are for different contrasts; columns j are for variables. For example, matrix mnlbeta = (-.693, .693, .347 \ .347, -.347, .693). As constant terms are *not* plotted, they are not included in mnlbeta.

mnlsd is a vector with the standard deviations for the variables listed in mnlname. For example, matrix mnlsd = (1, 2, 4). If you do not want to view standardized coefficients, this matrix can be made all 1s.

mnlcatnm is a string with labels for the outcome categories with each label separated by a space. For example, global mnlcatnm = "B C A". The first label corresponds to the first row of mnlbeta, the second to the second, and so on. *The label for the base category is last.*

Example

Suppose that you want to compare the logit coefficients estimated from two groups, such as whites and nonwhites from the example used in this chapter. We begin by estimating the logit coefficients for whites:

```
. use http://www.stata-press.com/data/lfr/nomocc2, clear
(1982 General Social Survey)
. mlogit occ ed exper if white==1, base(5) nolog
Multinomial logistic regression                 Number of obs   =        309
                                                LR chi2(8)      =     154.60
                                                Prob > chi2     =     0.0000
Log likelihood = -388.21313                     Pseudo R2       =     0.1660
```

occ	Coef.	Std. Err.	z	P>|z|	[95% Conf. Interval]	
Menial						
ed	-.8307514	.1297238	-6.40	0.000	-1.085005	-.5764973
exper	-.0338038	.0192045	-1.76	0.078	-.071444	.0038364
_cons	10.34842	1.779603	5.82	0.000	6.860465	13.83638
BlueCol						
ed	-.9225522	.1085452	-8.50	0.000	-1.135297	-.7098075
exper	-.031449	.0150766	-2.09	0.037	-.0609987	-.0018994
_cons	12.27337	1.507683	8.14	0.000	9.318368	15.22838
Craft						
ed	-.6876114	.0952882	-7.22	0.000	-.8743729	-.50085
exper	-.0002589	.0131021	-0.02	0.984	-.0259385	.0254207
_cons	9.017976	1.36333	6.61	0.000	6.345897	11.69005
WhiteCol						
ed	-.4196403	.0956209	-4.39	0.000	-.6070539	-.2322268
exper	.0008478	.0147558	0.06	0.954	-.0280731	.0297687
_cons	4.972973	1.421146	3.50	0.000	2.187578	7.758368

```
(Outcome occ==Prof is the comparison group)
```

Next, we compute coefficients for nonwhites:

(Continued on next page)

```
. mlogit occ ed exper if white==0, base(5) nolog
Multinomial logistic regression          Number of obs   =        28
                                         LR chi2(8)      =     17.79
                                         Prob > chi2     =    0.0228
Log likelihood = -32.779416              Pseudo R2       =    0.2135
```

occ	Coef.	Std. Err.	z	P>|z|	[95% Conf. Interval]
Menial					
ed	-.7012628	.3331146	-2.11	0.035	-1.354155 -.0483701
exper	-.1108415	.0741488	-1.49	0.135	-.2561705 .0344876
_cons	12.32779	6.053743	2.04	0.042	.4626714 24.19291
BlueCol					
ed	-.560695	.3283292	-1.71	0.088	-1.204208 .0828185
exper	-.0261099	.0682348	-0.38	0.702	-.1598477 .1076279
_cons	8.063397	6.008358	1.34	0.180	-3.712768 19.83956
Craft					
ed	-.882502	.3359805	-2.63	0.009	-1.541012 -.2239924
exper	-.1597929	.0744172	-2.15	0.032	-.305648 -.0139378
_cons	16.21925	6.059753	2.68	0.007	4.342356 28.09615
WhiteCol					
ed	-.5311514	.369815	-1.44	0.151	-1.255976 .1936728
exper	-.0520881	.0838967	-0.62	0.535	-.2165227 .1123464
_cons	7.821371	6.805367	1.15	0.250	-5.516904 21.15965

```
(Outcome occ==Prof is the comparison group)
```

The two sets of coefficients for `ed` are placed in `mnlbeta`:

```
. matrix mnlbeta = (-.8307514, -.9225522, -.6876114, -.4196403 \
        -.7012628, -.560695 , -.882502 , -.5311514)
```

Notice that the rows of the matrix correspond to the variables (i.e., `ed` for whites and `ed` for nonwhites) since this was the easiest way to enter the coefficients. For `mlogplot`, the columns must correspond to variables, so we transpose the matrix:

```
. matrix mnlbeta = mnlbeta´
```

We assign names to the columns using `mnlname` and to the rows using `mnlcatnm` (where the last element is the name of the reference category):

```
. global mnlname = "White NonWhite"
. global mnlcatnm = "Menial BlueCol Craft WhiteCol Prof"
```

We named the coefficients for `ed` for whites, `White`, and the coefficients for `ed` for nonwhites, `NonWhite`, as this will make the plot clearer. Next, we compute the standard deviation of `ed`:

```
. sum ed
```

Variable	Obs	Mean	Std. Dev.	Min	Max
ed	337	13.09496	2.946427	3	20

and enter the information into `mnlsd`:

```
matrix mnlsd = (2.946427,2.946427)
```

The same value is entered twice because we want to use the overall standard deviation in education for both groups. To create the plot, we use the command

```
. mlogplot, vars(White NonWhite) packed
      or matrix std(ss)
      note("Racial Differences in Effects of Education")
```

which leads to

Racial Differences in Effects of Education

Factor Change Scale Relative to Category Prof

White-std   B M   C   W   P

NonWhite-std   C   M   BW   P

Logit Coefficient Scale Relative to Category Prof

Given the limitations of our dataset (e.g., there were only 28 cases in the logit for nonwhites) and our simple model, these results do not represent serious research on racial differences in occupational outcomes, but they do illustrate the flexibility of the `mlogplot` command.

6.7 The conditional logit model

In the multinomial logit model, we estimate how *individual-specific* variables affect the likelihood of observing a given outcome. For example, we considered how individual characteristics such as education and experience affect a person's occupation. In the conditional logit model (CLM), *alternative-specific* variables that vary by outcome and individual are used to predict the outcome that is chosen. Consider the following examples:

- The dependent variable is the mode of transportation that an individual uses to get to work: car, bus (of any color), or train (e.g., Hensher 1986). We are interested in the effect of time: we think that how long it would take a person to get to work for a given alternative might affect her probability of selecting the alternative. We want to estimate the effect of time on the respondent's choice, but the amount of time for a given mode of travel is different for each respondent.

- The dependent variable is the type of car an individual purchases: European, American, or Japanese (see [R] **clogit**). We are interested in the effect of the number of dealerships in the buyer's city: we think that the more dealerships that sell cars of a given type, the more likely it is that buyers will purchase cars of that type. We want to estimate the effect of the number of dealerships, but the number of dealerships of each type in the buyer's city varies for different buyers (because they live in different cities).

- The dependent variable is which candidate a respondent votes for in a multiparty election (see, e.g., Alvarez and Nagler 1998). For example, in 1992, the major candidates were Clinton, Bush, and Perot. We are interested in the effect of the distance between the respondent and the candidate on particular issues (e.g., taxation, defense, gun control). We want to estimate how distance on different issues affects vote choice, but the distance from each candidate to the respondent varies for each respondent.

The conditional logit model (CLM) allows us to estimate how nominal outcomes are affected by characteristics of the outcomes that vary across individuals. In the CLM, the predicted probability of observing outcome m is given by

$$\Pr\left(y_i = m \mid \mathbf{z}_i\right) = \frac{\exp\left(\mathbf{z}_{im}\gamma\right)}{\sum_{j=1}^{J} \exp\left(\mathbf{z}_{ij}\gamma\right)} \quad \text{for } m = 1 \text{ to } J$$

where $\mathbf{z}_{im}$ contains values of the independent variables for outcome m for individual i. In the example of the CLM that we use, there are three choices for transportation: train, bus, and car. Suppose that we consider a single independent variable, where z_{im} is the amount of time it would take respondent i to travel using mode of transportation m. Then, γ is a single parameter indicating the effect of time on the probability of choosing one mode over another. In general, for each variable z_k, there are J values of the variable for each individual but only the single parameter γ_k.

6.7.1   Data arrangement for conditional logit

Fitting the CLM in Stata requires that the data be arranged differently than for the other models we consider in this book, which we illustrate with an example from Greene and Hensher (1997). We have data on 152 groups of people traveling for their vacation, choosing between three modes of travel: train, bus or car. The group is indicated by the variable **id**. For each group of travelers, there are three rows of data corresponding to the three choices faced by each group. Accordingly, we have $N \times J = 152 \times 3 = 456$ observations. For each group, the first observation is for the option of taking a train, the second for taking a bus, and the third for taking a car. Two dummy variables are used to indicate the mode of travel corresponding to a given row of data. Variable **train** is 1 if the observation contains information about taking the train, else **train** is 0. **bus** is 1 if the observation contains information about taking a bus, else 0. If both **train** and **bus** are 0, the observation has information about driving a car. The actual choice made for a group is indicated with the dummy variable `choice` equal to 1 if the person took the mode of travel corresponding to a specific observation. For example, let's look at the first two groups (i.e., six records):

```
. use http://www.stata-press.com/data/lfr/travel2, clear
(Greene & Hensher 1997 data on travel mode choice)
. list id mode train bus time invc choice in 1/6, sepby(id)
```

	id	mode	train	bus	time	invc	choice
1.	1	Train	1	0	406	31	0
2.	1	Bus	0	1	452	25	0
3.	1	Car	0	0	180	10	1
4.	2	Train	1	0	398	31	0
5.	2	Bus	0	1	452	25	0
6.	2	Car	0	0	255	11	1

Both groups traveled by car, as indicated by `choice`, which equals 1 in the third observation for each group. The variable `time` indicates how long a group thinks it will take them to travel using a given mode of transportation. Thus, `time` is an alternative-specific variable. For the first group, we can see that their trip would take 406 minutes by train, 452 minutes by bus, and 180 minutes by car. We might expect that the longer the time required, the less likely a person is to choose a particular mode of transportation. Similarly, the variable `invc` contains the in-vehicle cost of the trip: we might expect that the higher the cost of traveling by some mode, the less likely a person is to choose that mode. While many datasets with alternative-specific variables are already arranged in this way, later we talk about commands for setting up your data.

6.7.2   Fitting the conditional logit model

The syntax for `clogit` is

clogit *depvar* $\left[\text{indepvars}\right]$ $\left[\text{weight}\right]$ $\left[\text{if } \text{exp}\right]$ $\left[\text{in } \text{range}\right]$, <u>gro</u>up(*varname*)
    $\left[\underline{\text{l}}\text{evel}(\#) \text{ or }\right]$

Options

group(*varname*) is required and specifies the variable that identifies the different groups of observations in the dataset. In our example, the group variable is `id`, which identifies the different respondents.

level(#) specifies the level, in percent, for confidence intervals. The default is 95 percent.

or requests that odds ratios $\exp\left(\widehat{\gamma}_k\right)$ be reported instead of $\widehat{\gamma}_k$.

Example of the clogit model

For our transportation example, the dependent variable is choice, a binary variable indicating which mode of transportation was actually chosen. The independent variables include the $J-1$ dummy variables train and bus that identify each alternative mode of transportation and the alternative-specific variables time and invc. To fit the model, we use the option group(id) to specify that the id variable identifies the groups in the sample:

```
. clogit choice train bus time invc, group(id) nolog
Conditional (fixed-effects) logistic regression   Number of obs   =       456
                                                  LR chi2(4)      =    172.06
                                                  Prob > chi2     =    0.0000
Log likelihood = -80.961135                       Pseudo R2       =    0.5152
```

choice	Coef.	Std. Err.	z	P>|z|	[95% Conf. Interval]	
train	2.671238	.453161	5.89	0.000	1.783059	3.559417
bus	1.472335	.4007151	3.67	0.000	.6869475	2.257722
time	-.0191453	.0024509	-7.81	0.000	-.0239489	-.0143417
invc	-.0481658	.0119516	-4.03	0.000	-.0715905	-.0247411

6.7.3   Interpreting results from clogit

Using odds ratios

In the results that we just obtained, the coefficients for time and invc are negative. This indicates that the longer it takes to travel by a given mode, the less likely that mode is to be chosen. Similarly, the more it costs, the less likely a mode is to be chosen. More specific interpretations are possible by using listcoef to transform the estimates into odds ratios:

```
. listcoef
clogit (N=456): Factor Change in Odds
  Odds of: 1 vs 0
```

choice	b	z	P>|z|	e^b
train	2.67124	5.895	0.000	14.4579
bus	1.47233	3.674	0.000	4.3594
time	-0.01915	-7.812	0.000	0.9810
invc	-0.04817	-4.030	0.000	0.9530

For the alternative-specific variables, time and invc, each odds ratio is the multiplicative effect of a unit change in a given independent variable on the odds of any given mode of travel. For example, increasing the time of travel by one minute for a given mode of transportation decreases the odds of using that mode of travel by a factor of .98 (2%), holding the values for the other alternatives constant.

That is, if the time it takes to travel by car increases by one minute while the time it takes to travel by train and bus remain constant, the odds of traveling by car decrease by 2 percent.

The odds ratios for the alternative-specific constants **bus** and **train** indicate the relative likelihood of selecting these alternatives versus traveling by car (the omitted category), assuming that cost and time are the same for all modes. For example,

if cost and time were equal, individuals would be 4.36 times more likely to travel by bus than by car, and they would be 14.46 times more likely to travel by train than by car.

Using predicted probabilities

While the SPost commands **prvalue**, **prtab**, **prcounts**, and **prgen** do not work with **clogit**, you can use Stata's **predict** to compute predicted probabilities for each alternative for each group in the sample, where the predicted probabilities sum to 1 for each group. For example,

```
. predict prob
(option pc1 assumed; conditional probability for single outcome within group)
```

The message in parentheses indicates that, by default, conditional probabilities are being computed. To see what was done, let's list the variables in the model along with the predicted probabilities for the first group:

```
. list train bus time invc choice prob in 1/3
```

	train	bus	time	invc	choice	prob
1.	1	0	406	31	0	.0642477
2.	0	1	452	25	0	.0107205
3.	0	0	180	10	1	.9250318

The predicted probability of traveling by car (the option chosen) is .93, while the predicted probability of traveling by train is only .06. In this case, the choice corresponds to choosing the cheapest and quickest mode of transportation. If we consider another observation where train was chosen,

```
. list train bus time invc choice prob in 16/18
```

	train	bus	time	invc	choice	prob
16.	1	0	385	20	1	.5493771
17.	0	1	452	13	0	.0643481
18.	0	0	284	12	0	.3862748

we see that the probability of choosing train was estimated to be .55, while the probability of driving was .39. In this case, the respondent chose to travel by train, even though it was neither cheapest nor fastest.

6.7.4   Fitting the multinomial logit model using clogit*

Any multinomial logit model can be fitted using `clogit` by expanding the dataset (explained below) and respecifying the independent variables as a set of interactions. This is of more than academic interest for two reasons. First, it opens up the possibility of mixed models that include both individual-specific and alternative-specific variables (see Section 6.7.5). Second, it is possible to impose constraints on parameters in `clogit` that are not possible with `mlogit` (see `mclgen` and `mclest` by Hendrickx (2000) for further details).

Setting up the data

To illustrate how this is done, we show how to use `clogit` to fit the model of occupational attainment that we used to illustrate `mlogit` earlier in the chapter. The first step in rearranging the data is to create one record for each outcome:

```
. use http://www.stata-press.com/data/lfr/nomocc2, clear
(1982 General Social Survey)
. gen id = _n
. expand 5
(1348 observations created)
```

The command `gen id = _n` creates an id number that is equal to the observation's row number. The `expand` *n* command creates *n* duplicate observations for each current observation. We need 5 observations per individual because there are 5 alternatives. Next, we sort the data so that observations with the same `id` value are adjacent. This is necessary so that we can use the `mod`(*modulo*) function to `generate` variable `alt` with values 1 through 5 corresponding to the codings for the different values of `occ` (our dependent variable):

```
. sort id
. gen alt = mod(_n, 5)
. replace alt = 5 if alt == 0
(337 real changes made)
```

The values of `alt` are then used to create the four dummy variables for the different occupational types, leaving professional as the reference category (`alt==5`).

```
. gen menial = (alt==1)
. gen bluecol = (alt==2)
. gen craft = (alt==3)
. gen whitecol = (alt==4)
```

Finally, we generate a new variable `choice` that equals 1 if `choice==alt` and equals 0 otherwise. That is, `choice` indicates the occupation attained:

```
. gen choice = (occ==alt)
```

For the first two individuals (which is 10 observations),

```
. list id menial bluecol craft whitecol choice in 1/10, sepby(id)
```

	id	menial	bluecol	craft	whitecol	choice
1.	1	1	0	0	0	1
2.	1	0	1	0	0	0
3.	1	0	0	1	0	0
4.	1	0	0	0	1	0
5.	1	0	0	0	0	0
6.	2	1	0	0	0	1
7.	2	0	1	0	0	0
8.	2	0	0	1	0	0
9.	2	0	0	0	1	0
10.	2	0	0	0	0	0

Creating interactions

Next, we create interactions by multiplying each of the four dummy variables by each of the independent variables `white`, `ed`, and `exper`:

```
. gen whiteXm = white*menial
. gen whiteXbc = white*bluecol
. gen whiteXc = white*craft
. gen whiteXwc = white*whitecol
. gen edXm = ed*menial
. gen edXbc = ed*bluecol
. gen edXc = ed*craft
. gen edXwc = ed*whitecol
. gen experXm = exper*menial
. gen experXbc = exper*bluecol
. gen experXc = exper*craft
. gen experXwc = exper*whitecol
```

To see what this does, we list the interactions with `ed`:

```
. list menial bluecol craft whitecol edXm edXbc edXc edXwc in 1/5
```

	menial	bluecol	craft	whitecol	edXm	edXbc	edXc	edXwc
1.	1	0	0	0	11	0	0	0
2.	0	1	0	0	0	11	0	0
3.	0	0	1	0	0	0	11	0
4.	0	0	0	1	0	0	0	11
5.	0	0	0	0	0	0	0	0

The trick is that the interaction of `ed` with the indicator variable for a given outcome is only equal to `ed` in the record corresponding to that outcome (see Long 1997, 181 for details on the mathematics involved).

Fitting the model

These interactions are then included as independent variables for `clogit`, where we order the terms in the same way as the output from `mlogit` on page 196.

```
. clogit choice whiteXm edXm experXm menial whiteXbc edXbc experXbc
      bluecol whiteXc edXc experXc craft whiteXwc edXwc experXwc
      whitecol, group(id) nolog
```

```
Conditional (fixed-effects) logistic regression    Number of obs   =      1685
                                                   LR chi2(16)     =    231.16
                                                   Prob > chi2     =    0.0000
Log likelihood = -426.80048                        Pseudo R2       =    0.2131
```

choice	Coef.	Std. Err.	z	P>|z|	[95% Conf. Interval]	
whiteXm	-1.774306	.7550518	-2.35	0.019	-3.254181	-.2944322
edXm	-.7788519	.1146287	-6.79	0.000	-1.00352	-.5541839
experXm	-.0356509	.018037	-1.98	0.048	-.0710027	-.0002991
menial	11.51833	1.849346	6.23	0.000	7.89368	15.14298
whiteXbc	-.5378027	.7996015	-0.67	0.501	-2.104993	1.029387
edXbc	-.8782767	.1005441	-8.74	0.000	-1.075339	-.6812139
experXbc	-.0309296	.0144086	-2.15	0.032	-.0591699	-.0026894
bluecol	12.25956	1.668135	7.35	0.000	8.990079	15.52905
whiteXc	-1.301963	.6474136	-2.01	0.044	-2.57087	-.0330555
edXc	-.6850365	.089299	-7.67	0.000	-.8600593	-.5100138
experXc	-.0079671	.0127054	-0.63	0.531	-.0328693	.0169351
craft	10.42698	1.517934	6.87	0.000	7.451883	13.40207
whiteXwc	-.2029212	.8693059	-0.23	0.815	-1.906729	1.500887
edXwc	-.4256943	.0922188	-4.62	0.000	-.6064398	-.2449487
experXwc	-.001055	.0143582	-0.07	0.941	-.0291966	.0270865
whitecol	5.279722	1.683999	3.14	0.002	1.979146	8.580299

As the estimated parameters are identical to those produced by `mlogit` earlier, their interpretation is also the same.

6.7.5   Using clogit to fit mixed models*

The MNLM has individual-specific variables, for example, an individual's income. For individual-specific variables, the value of a variable does not differ across outcomes, but we want to estimate $J - 1$ parameters for each individual-specific variable. The CLM has alternative-specific variables, such as the time it takes to get to work with a given mode of transportation. For alternative-specific variables, values varied across alternatives, but we estimate a single parameter for the effect of the variable. An interesting possibility is combining the two in a single model, referred to as a *mixed model*. For example, in explaining the choices people make about mode of transportation, we might want to know if wealthier people are more likely to drive than take the bus.

To create a mixed model, we combine the formulas for the MNLM and the CLM (see Long 1997, 178–182 and Powers and Xie 2000, 242–245):

$$\Pr(y_i = m \mid \mathbf{x}_i, \mathbf{z}_i) = \frac{\exp(\mathbf{z}_{im}\gamma + \mathbf{x}_i\beta_m)}{\sum_{j=1}^{J} \exp(\mathbf{z}_{ij}\gamma + \mathbf{x}_i\beta_j)} \qquad \text{where } \beta_1 = 0$$

As in the CLM, $\mathbf{z}_{im}$ contains values of the alternative-specific variables for outcome m and individual i, and γ contains the effects of the alternative-specific variables. As in the multinomial logit model, $\mathbf{x}_i$ contains individual-specific independent variables for individual i, and β_m contains coefficients for the effects on outcome m relative to the base category.

This mixed model can be fitted using `clogit`. For the alternative-specific variables the data are set up in the same way as for the conditional logit model above. For individual-specific variables, interaction terms are created as illustrated in the last section. To illustrate this approach, we add two individual-specific variables to our model of travel demand: `hhinc` is household income and `psize` is the number of people who will be traveling together. First we create the interactions:

```
. use http://www.stata-press.com/data/lfr/travel2, clear
(Greene & Hensher 1997 data on travel mode choice)
. gen hincXbus = hinc*bus
. gen hincXtrn = hinc*train
. gen sizeXbus = psize*bus
. gen sizeXtrn = psize*train
```

Then, we fit the model with `clogit`:

(*Continued on next page*)

```
. clogit choice train bus time invc hincXbus hincXtrn sizeXbus sizeXtrn,
     group(id) nolog
Conditional (fixed-effects) logistic regression    Number of obs   =        456
                                                    LR chi2(8)      =     178.97
                                                    Prob > chi2     =     0.0000
Log likelihood = -77.504846                         Pseudo R2       =     0.5359
```

choice	Coef.	Std. Err.	z	P>|z|	[95% Conf. Interval]	
train	3.499641	.7579659	4.62	0.000	2.014055	4.985227
bus	2.486465	.8803643	2.82	0.005	.7609827	4.211947
time	-.0185035	.0025035	-7.39	0.000	-.0234103	-.0135966
invc	-.0402791	.0134851	-2.99	0.003	-.0667095	-.0138488
hincXbus	-.0080174	.0200322	-0.40	0.689	-.0472798	.031245
hincXtrn	-.0342841	.0158471	-2.16	0.031	-.0653438	-.0032243
sizeXbus	-.5141037	.4007012	-1.28	0.199	-1.299464	.2712563
sizeXtrn	-.0038421	.3098074	-0.01	0.990	-.6110533	.6033692

To interpret these results, we can again transform the coefficients into odds ratios using listcoef:

```
. listcoef
clogit (N=456): Factor Change in Odds
   Odds of: 1 vs 0
```

choice	b	z	P>|z|	e^b
train	3.49964	4.617	0.000	33.1036
bus	2.48647	2.824	0.005	12.0187
time	-0.01850	-7.391	0.000	0.9817
invc	-0.04028	-2.987	0.003	0.9605
hincXbus	-0.00802	-0.400	0.689	0.9920
hincXtrn	-0.03428	-2.163	0.031	0.9663
sizeXbus	-0.51410	-1.283	0.199	0.5980
sizeXtrn	-0.00384	-0.012	0.990	0.9962

The interpretation for the individual-specific variables is the same as the interpretation of odds ratios in the MNLM. For example, a unit increase in income decreases the odds of traveling by train versus traveling by car by a factor of .97. Similarly, each additional member of the traveling party decreases the odds of traveling by bus versus traveling by car by a factor of .60.

Note We have only considered the conditional logit model in the context of choices among an unordered set of alternatives. The possible uses of clogit are much broader. The *Stata Base Reference Manual* entry for clogit contains additional examples and references.

---

7 Models for Count Outcomes

Count variables indicate how many times something has happened. While the use of regression models for counts is relatively recent, even a brief survey of recent applications illustrates how common these outcomes are and the importance of this class of models. Examples include the number of patients, hospitalizations, daily homicides, international conflicts, beverages consumed, industrial injuries, new companies, and arrests by police, to name only a few.

While the linear regression model has often been applied to count outcomes, this can result in inefficient, inconsistent, and biased estimates. Even though there are situations in which the LRM provides reasonable results, it is much safer to use models specifically designed for count outcomes. Four such models are considered in this chapter: Poisson regression (PRM), negative binomial regression (NBRM), and variations of these models for zero-inflated counts (ZIP and ZINB). As with earlier chapters, we begin with a quick review of the statistical model, consider issues of testing and fit, and then discuss methods of interpretation. These discussions are intended as a review for those who are familiar with the models. For further details, see Long (1997) or Cameron and Trivedi (1998, the definitive work in this area). As always, you can obtain sample do-files and data files by downloading the `spostst4` package (see Chapter 1 for details).

7.1 The Poisson distribution

The univariate Poisson distribution is fundamental to understanding regression models for counts. Accordingly, we start by exploring this distribution. Let μ be the rate of occurence or the expected number of times an event will occur over a given period of time. Let y be a random variable indicating the number of times an event did occur. Sometimes the event will occur fewer times than the average rate or even not at all, and other times it will occur more often. The relationship between the expected count μ and the probability of observing any observed count y is specified by the Poisson distribution

$$\Pr(y \mid \mu) = \frac{e^{-\mu}\mu^y}{y!} \quad \text{for } y = 0, 1, 2, \ldots$$

where $\mu > 0$ is the sole parameter defining the distribution. The easiest way to get a sense of this distribution is to compare the plot of the predicted probability for different values of the rate parameter μ (labeled as mu in the graph):

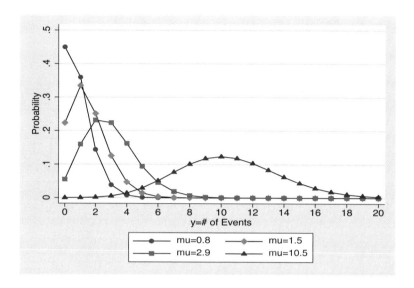

The plot illustrates four characteristics of the Poisson distribution that are important for understanding regression models for counts:

1. μ is the mean of the distribution. As μ increases, the mass of the distribution shifts to the right.

2. μ is also the variance. Thus, $\mathrm{Var}(y) = \mu$, which is known as *equidispersion*. In real data, many count variables have a variance greater than the mean, which is called *overdispersion*.

3. As μ increases, the probability of a zero count decreases. For many count variables, there are more observed zeros than predicted by the Poisson distribution.

4. As μ increases, the Poisson distribution approximates a normal distribution. This is shown by the distribution for $\mu = 10.5$.

7.1.1   Fitting the Poisson distribution with the poisson command

To illustrate the models in this chapter, we use data from Long (1990) on the number of publications produced by Ph.D. biochemists. The variables considered are

```
. use http://www.stata-press.com/data/lfr/couart2, clear
(Academic Biochemists / S Long)
```

```
. describe
Contains data from http://www.stata-press.com/data/lfr/couart2.dta
  obs:           915                        Academic Biochemists / S Long
  vars:            6                        30 Jan 2001 10:49
  size:       11,895 (99.9% of memory free)  (_dta has notes)
```

variable name	storage type	display format	value label	variable label
art	byte	%9.0g		Articles in last 3 yrs of PhD
fem	byte	%9.0g	sexlbl	Gender: 1=female 0=male
mar	byte	%9.0g	marlbl	Married: 1=yes 0=no
kid5	byte	%9.0g		Number of children < 6
phd	float	%9.0g		PhD prestige
ment	byte	%9.0g		Article by mentor in last 3 yrs

```
Sorted by:   art
. summarize
```

Variable	Obs	Mean	Std. Dev.	Min	Max
art	915	1.692896	1.926069	0	19
fem	915	.4601093	.4986788	0	1
mar	915	.6622951	.473186	0	1
kid5	915	.495082	.76488	0	3
phd	915	3.103109	.9842491	.755	4.62
ment	915	8.767213	9.483916	0	77

A useful place to begin when analyzing a count outcome is to compare the observed distribution with a Poisson distribution that has the same mean. The command poisson fits the Poisson regression model that is presented in Section 7.2. Here we use poisson without any independent variables in order to fit a univariate Poisson distribution with a mean equal to that of our outcome variable art. That is, we estimate the parameter of the model:

$$\mu = \exp\left(\beta_0\right)$$

The results are

```
. poisson art, nolog
Poisson regression                          Number of obs   =         915
                                            LR chi2(0)      =        0.00
                                            Prob > chi2     =           .
Log likelihood = -1742.5735                 Pseudo R2       =      0.0000
```

art	Coef.	Std. Err.	z	P>|z|	[95% Conf. Interval]
_cons	.5264408	.0254082	20.72	0.000	.4766416 .57624

Because $\widehat{\beta}_0 = .5264$, $\widehat{\mu} = \exp\left(.5264\right) = 1.6929$, which is the same as the estimated mean of art obtained with summarize earlier. To compute the observed probabilities for each count and the predictions from counts drawn from a Poisson distribution with this mean, we use prcounts, which is part of SPost.

7.1.2   Computing predicted probabilities with prcounts

For `poisson` and other models for count outcomes, `prcounts` extends the features of `predict` by computing the predicted rate and predicted probabilities of each count from 0 to the specified maximum for every observation. Optionally, `prcounts` creates variables with observed probabilities and sample averages of predicted probabilities for each count; these variables can be used to construct plots to assess the fit of count models, as shown in Section 7.5.1.

Syntax

> `prcounts` *name* [if *exp*] [in *range*] [, max(*max*) plot]

where *name* is a prefix to the new variables that are generated. *name* cannot be the name of an existing variable.

Options

max(*max*) is the maximum count for which predicted probabilities should be computed. The default is 9. For example, with `max(2)` predictions for $\Pr(y=0)$, $\Pr(y=1)$, and $\Pr(y=2)$ are computed.

plot specifies that variables for plotting expected counts should be generated. If this option is not used, only predictions for individual observations are computed.

if and in restrict the sample for which predictions are made. By default, `prcounts` computes predicted values for all observations in memory. To restrict the computations to the estimation sample, you should add the condition if `e(sample)==1`.

Variables generated

The following variables are generated, where *name* represents the prefix specified with `prcounts`. y is the count variable and each prediction is conditional on the independent variables in the regression. If there are no independent variables, as in our current example, the values are unconditional.

*name*rate The predicted rate or count $E(y)$.

*name*prk The predicted probability $\Pr(y=k)$ for $k=0$ to *max*. By default, $max=9$.

*name*prgt The predicted probability $\Pr(y>max)$.

*name*cuk The predicted cumulative probability $\Pr(y \leq k)$ for $k=0$ to *max*. By default, $max=9$.

When the `plot` option is specified, *max*+1 observations (for counts 0 through *max*) are generated for the following variables:

*name*val The value k of the count y ranging from 0 to *max*.

*name*obeq The *observed* probability $\Pr(y=k)$. These values are the same as the ones you could obtain by running `tabulate` on the count variable (e.g., `tabulate art`).

*name*oble The *observed* cumulative probability $\Pr(y\le k)$.

*name*preq The average *predicted* probability $\Pr(y=k)$.

*name*prle The average *predicted* cumulative probability $\Pr(y\le k)$.

Which observations are used to compute the averages? By default, `prcounts` computes averages for all observations in memory, which could include observations that were not used in the estimation. For example, if your model was `poisson art if fem==1`, then the averages computed by `prcounts` would be based on all observations, including those where `fem` is not 1. To restrict the averages to the sample used in estimation, you need to add the condition `if e(sample)==1`. For example, `prcounts isfem if e(sample)==1, plot`.

7.1.3   Comparing observed and predicted counts with prcounts

If the `plot` option was used with `prcounts`, it is simple to construct a graph that compares the observed probabilities for each value of the count variable to the predicted probabilities from fitting the Poisson distribution. For example,

```
. prcounts psn, plot max(9)
. label var psnobeq "Observed Proportion"
. label var psnpreq "Poisson Prediction"
. label var psnval "# of Articles"
. list psnval psnobeq psnpreq in 1/10
```

	psnval	psnobeq	psnpreq
1.	0	.3005464	.1839859
2.	1	.2688525	.311469
3.	2	.1945355	.2636423
4.	3	.0918033	.148773
5.	4	.073224	.0629643
6.	5	.0295082	.0213184
7.	6	.0185792	.006015
8.	7	.0131148	.0014547
9.	8	.0010929	.0003078
10.	9	.0021858	.0000579

The listed values are the observed and predicted probabilities for observing scientists with 0 through 9 publications. These can be plotted with `graph`:

```
. graph twoway connected psnobeq psnpreq psnval,
        ytitle("Probability") ylabel(0(.1).4)
        xlabel(0(1)9)
        ysize(2.7051) xsize(4.0421)
```

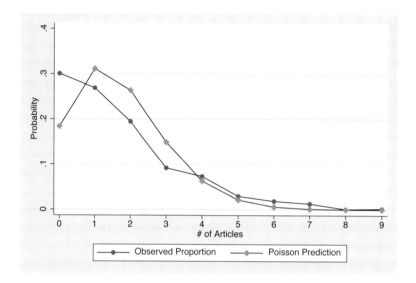

The graph clearly shows that the fitted Poisson distribution (represented by ◆) un-derpredicts 0s and over-predicts counts 1, 2, and 3. This pattern of over- and under-prediction is characteristic of fitting a count model that does not take into account *heterogeneity* among sample members in their rate μ. As fitting the univariate Pois-son distribution assumes that all scientists have the same rate of productivity, which is clearly unrealistic, our next step is to allow heterogeneity in μ based on observed characteristics of the scientists.

Advanced: plotting Poisson distributions Earlier we plotted the Poisson distribu-tion for four values of μ. The trick to doing this is to construct artificial data with a given mean rate of productivity. Here are the commands we used to generate the graph on page 246:

```
. clear
. set obs 25
. gen ya = .8
. poisson ya, nolog
. prcounts pya, plot max(20)
. gen yb = 1.5
. poisson yb, nolog
. prcounts pyb, plot max(20)
```

```
. gen yc = 2.9
. poisson yc, nolog
. prcounts pyc, plot max(20)
. gen yd = 10.5
. poisson yd, nolog
. prcounts pyd, plot max(20)
. label var pyapreq "mu=0.8"
. label var pybpreq "mu=1.5"
. label var pycpreq "mu=2.9"
. label var pydpreq "mu=10.5"
. label var pyaval "y=# of Events"
. graph twoway connected pyapreq pybpreq pycpreq pydpreq pyaval,
         ytitle("Probability") ylabel(0(.1).5) xlabel(0(2)20)
```

7.2   The Poisson regression model

The Poisson regression model (PRM) extends the Poisson distribution by allowing each observation to have a different value of μ. More formally, the PRM assumes that the observed count for observation i is drawn from a Poisson distribution with mean μ_i, where μ_i is estimated from observed characteristics. This is sometimes referred to as incorporating *observed heterogeneity* and leads to the structural equation

$$\mu_i = E\left(y_i \mid \mathbf{x}_i\right) = \exp\left(\mathbf{x}_i\beta\right)$$

Taking the exponential of $\mathbf{x}\beta$ forces μ to be positive, which is necessary because counts can only be 0 or positive. To see how this works, consider the PRM with a single independent variable, $\mu = \exp\left(\alpha + \beta x\right)$, which can be plotted as

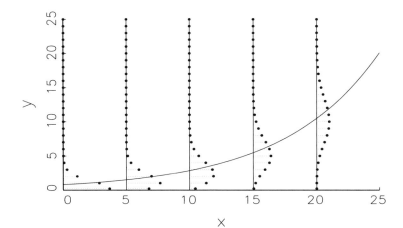

In this graph the mean μ, shown by the curved line, increases as x increases. For each value of μ, the distribution around the mean is shown by the dots, which should be thought of as coming out of the page and which represent the probability of each count. Interpretation of the model involves assessing how changes in the independent variables affect the conditional mean and the probabilities of various counts. Details on interpretation are given after we consider estimation.

7.2.1   Fitting the PRM with poisson

The Poisson regression model is fitted with the command

poisson *depvar* [*indepvars*] [*weight*] [if *exp*] [in *range*] [, irr level(#)
    exposure(*varname*) robust cluster(*varname*) score(*newvar*) nolog]

In our experience, poisson converges quickly, and difficulties are rarely encountered.

Variable lists

depvar is the dependent variable. poisson does not require this to be an integer. But, if you have noninteger values, you obtain the warning

Note: you are responsible for interpretation of noncount dep variable.

indepvars is a list of independent variables. If *indepvars* is not included, a model with only an intercept is fitted, which corresponds to fitting a univariate Poisson distribution, as shown in the last section.

Specifying the estimation sample

if and in qualifiers can be used to restrict the estimation sample. For example, if you want to fit a model for only women, you could specify poisson art mar kid5 phd ment if fem==1.

Listwise deletion Stata excludes observations in which there are missing values for any of the variables in the model. Accordingly, if two models are fitted using the same data but have different independent variables, it is possible to have different samples. We recommend that you use mark and markout (discussed in Chapter 3) to explicitly remove observations with missing data.

Weights

poisson can be used with fweights, pweights, and iweights. See Chapter 3 for details.

Options

irr reports estimated coefficients transformed to incidence rate ratios defined as $\exp(\beta)$. These are discussed in Section 7.2.3.

level(#) specifies the level of the confidence interval for estimated parameters. By default, a 95% interval is used. You can change the default level, say, to a 90% interval, with the command set level 90.

exposure(*varname*) specifies a variable indicating the amount of time during which an observation was "at risk" of the event occurring. Details are given in an example below.

robust requests that robust variance estimates be used. See Chapter 3 for details.

cluster(*varname*) specifies that the observations are independent across the groups specified by unique values of *varname* but not necessarily within the groups. When cluster() is specified, robust standard errors are automatically used. See Chapter 3 for details.

score(*newvar*) creates *newvar* containing $u_j = \partial \ln L_j / \partial(\mathbf{x}_j \mathbf{b})$ for each observation j in the sample. The score vector is $\sum \partial \ln L_j / \partial \mathbf{b} = \sum u_j \mathbf{x}_j$; i.e., the product of *newvar* with each covariate summed over observations. See [U] **23.15 Obtaining scores**.

nolog suppresses the iteration history.

7.2.2   Example of fitting the PRM

If scientists who differ in their rates of productivity are combined, the univariate distribution of articles will be overdispersed (i.e., the variance is greater than the mean). Differences among scientists in their rates of productivity could be due to factors such as gender, marital status, number of young children, prestige of the graduate program, and the number of articles written by a scientist's mentor. To account for these differences, we add these variables as independent variables:

```
. use http://www.stata-press.com/data/lfr/couart2, clear
(Academic Biochemists / S Long)
. poisson art fem mar kid5 phd ment, nolog
```

```
Poisson regression                              Number of obs   =        915
                                                LR chi2(5)      =     183.03
                                                Prob > chi2     =     0.0000
Log likelihood = -1651.0563                     Pseudo R2       =     0.0525
```

art	Coef.	Std. Err.	z	P>|z|	[95% Conf. Interval]	
fem	-.2245942	.0546138	-4.11	0.000	-.3316352	-.1175532
mar	.1552434	.0613747	2.53	0.011	.0349512	.2755356
kid5	-.1848827	.0401272	-4.61	0.000	-.2635305	-.1062349
phd	.0128226	.0263972	0.49	0.627	-.038915	.0645601
ment	.0255427	.0020061	12.73	0.000	.0216109	.0294746
_cons	.3046168	.1029822	2.96	0.003	.1027755	.5064581

The way in which you interpret a count model depends on whether you are interested in the expected value of the count variable or in the distribution of counts. If your interest is in the expected count, several methods can be used to compute the change in the expectation for a change in an independent variable. If your interest is in the distribution of counts or perhaps just the probability of a specific count, the probability of a count for a given level of the independent variables can be computed. Each of these methods is now considered.

7.2.3   Interpretation using the rate μ

In the PRM,
$$\mu = E\left(y \mid \mathbf{x}\right) = \exp\left(\mathbf{x}\beta\right)$$
Changes in μ for changes in the independent variable can be interpreted in a variety of ways.

Factor change in E(y | x)

Perhaps the most common method of interpretation is the factor change in the rate. If we define $E\left(y \mid \mathbf{x}, x_k\right)$ as the expected count for a given $\mathbf{x}$, where we explicitly note the value of x_k, and define $E\left(y \mid \mathbf{x}, x_k + \delta\right)$ as the expected count after increasing x_k by δ units, then
$$\frac{E\left(y \mid \mathbf{x}, x_k + \delta\right)}{E\left(y \mid \mathbf{x}, x_k\right)} = e^{\beta_k \delta} \tag{7.1}$$
Therefore, the parameters can be interpreted that

> for a change of δ in x_k, the expected count increases by a factor of $\exp(\beta_k \times \delta)$, holding all other variables constant.

For example,

> *Factor change*: For a unit change in x_k, the expected count changes by a factor of $\exp(\beta_k)$, holding all other variables constant.

> *Standardized factor change*: For a standard deviation change in x_k, the expected count changes by a factor of $\exp(\beta_k \times s_k)$, holding all other variables constant.

Incidence rate ratio

In some discussions of count models, μ is referred to as the *incidence rate*, and (7.1) for $\delta = 1$ is called the *incidence rate ratio*. These coefficients can be computed by adding the option `irr` to the estimation command. Alternatively, they are computed with our `listcoef`, which is illustrated below.

Percent change in E(y | x)

Alternatively, the percentage change in the expected count for a δ unit change in x_k, holding other variables constant, can be computed as

$$100 \times \frac{E\left(y \mid \mathbf{x},x_k + \delta\right) - E\left(y \mid \mathbf{x},x_k\right)}{E\left(y \mid \mathbf{x},x_k\right)} = 100 \times \left\{\exp\left(\beta_k \times \delta\right) - 1\right\}$$

Example of factor and percent change

Factor change coefficients can be computed using `listcoef`:

```
. poisson art fem mar kid5 phd ment, nolog
  (output omitted)
. listcoef fem ment, help
poisson (N=915): Factor Change in Expected Count
  Observed SD: 1.926069
```

art	b	z	P>|z|	e^b	e^bStdX	SDofX
fem	-0.22459	-4.112	0.000	0.7988	0.8940	0.4987
ment	0.02554	12.733	0.000	1.0259	1.2741	9.4839

```
      b = raw coefficient
      z = z-score for test of b=0
  P>|z| = p-value for z-test
    e^b = exp(b) = factor change in expected count for unit increase in X
e^bStdX = exp(b*SD of X) = change in expected count for SD increase in X
  SDofX = standard deviation of X
```

For example, the coefficients for `fem` and `ment` can be interpreted as follows:

Being a female scientist decreases the expected number of articles by a factor of .80, holding all other variables constant.

For a standard deviation increase in the mentor's productivity, roughly 9.5 articles, a scientist's mean productivity increases by a factor of 1.27, holding other variables constant.

(Continued on next page)

To compute *percent change*, we add the option `percent`:

```
. listcoef fem ment, percent help
poisson (N=915): Percentage Change in Expected Count
  Observed SD: 1.926069
```

art	b	z	P>|z|	%	%StdX	SDofX
fem	-0.22459	-4.112	0.000	-20.1	-10.6	0.4987
ment	0.02554	12.733	0.000	2.6	27.4	9.4839

```
      b = raw coefficient
      z = z-score for test of b=0
  P>|z| = p-value for z-test
      % = percent change in expected count for unit increase in X
  %StdX = percent change in expected count for SD increase in X
  SDofX = standard deviation of X
```

For example, the percent change coefficients for `fem` and `ment` can be interpreted as follows:

> Being a female scientist decreases the expected number of articles by 20 percent, holding all other variables constant.

> For every additional article by the mentor, a scientist's predicted mean productivity increases by 2.6 percent, holding other variables constant.

The standardized percent change coefficient can be interpreted as follows:

> For a standard deviation increase in the mentor's productivity, a scientist's mean productivity increases by 27 percent, holding all other variables constant.

Marginal change in E(y | x)

Another method of interpretation is the marginal change in $E(y \mid \mathbf{x})$:

$$\frac{\partial E(y \mid \mathbf{x})}{\partial x_k} = E(y \mid \mathbf{x}) \beta_k$$

For $\beta_k > 0$, the larger the current value of $E(y \mid \mathbf{x})$, the larger the rate of change; for $\beta_k < 0$, the smaller the rate of change. The marginal with respect of x_k depends on both β_k and $E(y \mid \mathbf{x})$. Thus, the value of the marginal depends on the levels of all variables in the model. In practice, this measure is often computed with all variables held at their means.

Example of marginal change using prchange

As the marginal is not appropriate for binary independent variables, we only request the change for the continuous variables phd and ment. The marginal effects are in the column that is labeled MargEfct:

```
. prchange phd ment, rest(mean)

poisson: Changes in Predicted Rate for art
            min->max        0->1      -+1/2      -+sd/2   MargEfct
  phd       0.0794        0.0200     0.0206      0.0203    0.0206
  ment      7.9124        0.0333     0.0411      0.3910    0.0411

exp(xb):    1.6101

                fem         mar        kid5        phd       ment
       x=   .460109     .662295     .495082    3.10311    8.76721
    sd(x)=  .498679     .473186      .76488    .984249    9.48392
```

Example of marginal change using mfx compute

By default, mfx compute computes the marginal change with variables held at their means:

```
. mfx compute

Marginal effects after poisson
      y  = predicted number of events (predict)
         =  1.6100936
```

variable	dy/dx	Std. Err.	z	P>|z|	[	95% C.I.	]	X
fem*	-.3591461	.08648	-4.15	0.000	-.528643	-.189649		.460109
mar*	.2439822	.09404	2.59	0.009	.059671	.428293		.662295
kid5	-.2976785	.06414	-4.64	0.000	-.423393	-.171964		.495082
phd	.0206455	.04249	0.49	0.627	-.062635	.103926		3.10311
ment	.0411262	.00317	12.97	0.000	.034912	.04734		8.76721

(*) dy/dx is for discrete change of dummy variable from 0 to 1

The estimated marginal effects for phd and ment match those given above. For dummy variables, mfx compute, by default, computes the discrete change, as the variable changes from 0 to 1, a topic we will now consider.

Discrete change in E(y | x)

It is also possible to compute the discrete change in the expected count for a change in x_k from x_S to x_E,

$$\frac{\Delta E\left(y \mid \mathbf{x}\right)}{\Delta x_k} = E\left(y \mid \mathbf{x}, x_k = x_E\right) - E\left(y \mid \mathbf{x}, x_k = x_S\right)$$

which can be interpreted as follows:

> For a change in x_k from x_S to x_E, the expected count changes by $\Delta E\left(y \mid \mathbf{x}\right)/\Delta x_k$, holding all other variables at the specified values.

As was the case in earlier chapters, the discrete change can be computed in a variety of ways depending on your purpose:

1. The total possible effect of x_k is found by letting x_k change from its minimum to its maximum.

2. The effect of a binary variable x_k is computed by letting x_k change from 0 to 1. This is the quantity computed by mfx compute for binary variables.

3. The *uncentered* effect of a unit change in x_k at the mean is computed by changing from $\bar{x}_k$ to $\bar{x}_k + 1$. The *centered* discrete change is computed by changing from $(\bar{x}_k - 1/2)$ to $(\bar{x}_k + 1/2)$.

4. The *uncentered* effect of a standard deviation change in x_k at the mean is computed by changing from $\bar{x}_k$ to $\bar{x}_k + s_k$. The *centered* change is computed by changing from $(\bar{x}_k - s_k/2)$ to $(\bar{x}_k + s_k/2)$.

5. The *uncentered* effect of a change of δ units in x_k from $\bar{x}_k$ to $\bar{x}_k + \delta$. The *centered* change is computed by changing from $(\bar{x}_k - \delta/2)$ to $(\bar{x}_k + \delta/2)$.

Discrete changes are computed with prchange. By default, changes are computed centered on the values specified with x() and rest(). To compute changes that begin at the specified values, such as a change from $\bar{x}_k$ to $\bar{x}_k + 1$, you must specify the uncentered option. By default, prchange computes results for changes in the independent variables of 1 unit and a standard deviation. With the delta(#) option, you can request changes of # units. When using discrete change, remember that the magnitude of the change in the expected count depends on the levels of all variables in the model.

Example of discrete change using prchange

In this example, we set all variables to their mean:

```
. prchange fem ment, rest(mean)

poisson: Changes in Predicted Rate for art
        min->max      0->1     -+1/2    -+sd/2  MargEfct
  fem   -0.3591   -0.3591   -0.3624   -0.1804   -0.3616
  ment   7.9124    0.0333    0.0411    0.3910    0.0411

exp(xb):    1.6101

               fem       mar      kid5       phd      ment
      x=   .460109   .662295   .495082   3.10311   8.76721
  sd(x)=   .498679   .473186    .76488   .984249   9.48392
```

Examples of interpretation are the following:

Being a female scientist decreases the expected productivity by .36 articles, holding all other variables at their means.

A standard deviation increase in the mentor's articles increases the scientist's rate of productivity by .39, holding all other variables at their mean.

To illustrate the use of the `uncentered` option, suppose that we want to know the effect of a change from 1 to 2 young children:

```
. prchange kid5, uncentered x(kid5=1)
poisson: Changes in Predicted Rate for art
          min->max        0->1         +1        +sd  MargEfct
kid5      -0.7512     -0.2978    -0.2476    -0.1934   -0.2711

exp(xb):    1.4666

                fem         mar        kid5        phd       ment
     x=    .460109     .662295           1    3.10311    8.76721
 sd(x)=    .498679     .473186     .76488    .984249    9.48392
```

The rate of productivity decreases by .25 as the number of young children increases from 1 to 2. To examine the effect of a change from 1 to 3 children, we add the `delta()` option:

```
. prchange kid5, uncentered x(kid5=1) delta(2)
poisson: Changes in Predicted Rate for art
(Note: delta = 2)
          min->max        0->1     +delta        +sd  MargEfct
kid5      -0.7512     -0.2978    -0.4533    -0.1934   -0.2711

exp(xb):    1.4666

                fem         mar        kid5        phd       ment
     x=    .460109     .662295           1    3.10311    8.76721
 sd(x)=    .498679     .473186     .76488    .984249    9.48392
```

The results show a decrease of .45 in the expected number of articles as the number of young children increases from 1 to 3.

7.2.4   Interpretation using predicted probabilities

The estimated parameters can also be used to compute predicted probabilities using the following formula:

$$\widehat{\Pr}(y = m \mid \mathbf{x}) = \frac{e^{-\mathbf{x}\widehat{\beta}}\left(\mathbf{x}\widehat{\beta}\right)^m}{m!}$$

Predicted probabilities at specified values can be computed using `prvalue`. Predictions at the observed values for all observations can be made using `prcounts`, or `prgen` can be used to compute predictions that can be plotted. These commands are now illustrated.

Example of predicted probabilities using prvalue

`prvalue` computes predicted probabilities for values of the independent variables specified with `x()` and `rest()`. For example, to compare the predicted probabilities for married and unmarried women without young children, we first compute the predicted counts for single women without children by specifying `x(mar=0 fem=1 kid5=0)` and `rest(mean)`. We suppress the output with `quietly` but `save` the results for later use:

```
. * single women without children
. quietly prvalue, x(mar=0 fem=1 kid5=0) rest(mean) save
```

Next, we compute the predictions for married women without children and use the `dif`
option to compare these results with those we just saved:

```
. * compared to married women without children
. prvalue, x(mar=1 fem=1 kid5=0) rest(mean) dif

poisson: Change in Predictions for   art
Predicted rate: 1.65      95% CI [1.49, 1.82]
        Saved: 1.41
   Difference: .237

Predicted probabilities:
                        Current     Saved  Difference
    Pr(y=0|x):           0.1926    0.2441     -0.0515
    Pr(y=1|x):           0.3172    0.3442     -0.0270
    Pr(y=2|x):           0.2613    0.2427      0.0186
    Pr(y=3|x):           0.1434    0.1141      0.0293
    Pr(y=4|x):           0.0591    0.0402      0.0188
    Pr(y=5|x):           0.0195    0.0113      0.0081
    Pr(y=6|x):           0.0053    0.0027      0.0027
    Pr(y=7|x):           0.0013    0.0005      0.0007
    Pr(y=8|x):           0.0003    0.0001      0.0002
    Pr(y=9|x):           0.0000    0.0000      0.0000

                fem       mar      kid5        phd       ment
    Current=      1         1         0  3.1031093  8.7672131
      Saved=      1         0         0  3.1031093  8.7672131
       Diff=      0         1         0          0          0
```

The results show that married women are less likely than unmarried women to have one
or no publications, and are more likely to have two or more publications. Overall, their
rate of productivity is .24 publications higher.

To examine the effects of the number of young children, we can use a series of calls
to `prvalue`, where the `brief` option limits the amount of output:

```
. prvalue, x(mar=1 fem=1 kid5=0) rest(mean) brief
Predicted rate: 1.65
Predicted probabilities:
   Pr(y=0|x):    0.1926  Pr(y=1|x):    0.3172
   Pr(y=2|x):    0.2613  Pr(y=3|x):    0.1434
   Pr(y=4|x):    0.0591  Pr(y=5|x):    0.0195
   Pr(y=6|x):    0.0053  Pr(y=7|x):    0.0013
   Pr(y=8|x):    0.0003  Pr(y=9|x):    0.0000
. prvalue, x(mar=1 fem=1 kid5=1) rest(mean) brief
Predicted rate: 1.37
Predicted probabilities:
   Pr(y=0|x):    0.2544  Pr(y=1|x):    0.3482
   Pr(y=2|x):    0.2384  Pr(y=3|x):    0.1088
   Pr(y=4|x):    0.0372  Pr(y=5|x):    0.0102
   Pr(y=6|x):    0.0023  Pr(y=7|x):    0.0005
   Pr(y=8|x):    0.0001  Pr(y=9|x):    0.0000
```

```
. prvalue, x(mar=1 fem=1 kid5=2) rest(mean) brief

Predicted rate: 1.14
Predicted probabilities:
  Pr(y=0|x):    0.3205  Pr(y=1|x):    0.3647
  Pr(y=2|x):    0.2075  Pr(y=3|x):    0.0787
  Pr(y=4|x):    0.0224  Pr(y=5|x):    0.0051
  Pr(y=6|x):    0.0010  Pr(y=7|x):    0.0002
  Pr(y=8|x):    0.0000  Pr(y=9|x):    0.0000

. prvalue, x(mar=1 fem=1 kid5=3) rest(mean) brief

Predicted rate: .946
Predicted probabilities:
  Pr(y=0|x):    0.3883  Pr(y=1|x):    0.3673
  Pr(y=2|x):    0.1737  Pr(y=3|x):    0.0548
  Pr(y=4|x):    0.0130  Pr(y=5|x):    0.0025
  Pr(y=6|x):    0.0004  Pr(y=7|x):    0.0001
  Pr(y=8|x):    0.0000  Pr(y=9|x):    0.0000
```

These values could be presented in a table or plotted, but overall it is clear that the probabilities of a zero count increase as the number of young children increases.

Example of predicted probabilities using prgen

The command **prgen** computes a series of predictions by holding all variables but one constant and allowing that variable to vary. The resulting predictions can then be plotted. In this example, we plot the predicted probability of not publishing for married men and married women with different numbers of children. First, we compute the predictions for women using the prefix **fprm** to indicate predictions for women from the PRM:

```
. prgen kid5, x(fem=1 mar=1) rest(mean) from(0) to(3) gen(fprm) n(4)

poisson: Predicted values as kid5 varies from 0 to 3.
          fem       mar      kid5       phd      ment
x=          1         1  .49508197 3.1031093 8.7672125
```

Next, we compute predictions for men, using the prefix **mprm**:

```
. prgen kid5, x(fem=0 mar=1) rest(mean) from(0) to(3) gen(mprm) n(4)

poisson: Predicted values as kid5 varies from 0 to 3.
          fem       mar      kid5       phd      ment
x=          0         1  .49508197 3.1031093 8.7672125
```

In both calls of **prgen**, we requested four values with the **n(4)** option. This creates predictions for 0, 1, 2, and 3 children. To plot these predictions, we begin by adding value labels to the newly generated variables. Then, we use the now familiar **graph** command:

```
. label var fprmp0 "Married Women"

. label var mprmp0 "Married Men"

. label var mprmx  "Number of Children"
```

```
. graph twoway connected fprmp0 mprmp0 mprmx,
    ylabel(0(.1).4) yline(.1 .2 .3) xlabel(0(1)3)
    ytitle("Probability of No Articles")
    ysize(2.5051) xsize(4.0421)
```

This leads to the following graph, where the points marked with circles and diamonds
are placed at the tick marks for the number of children:

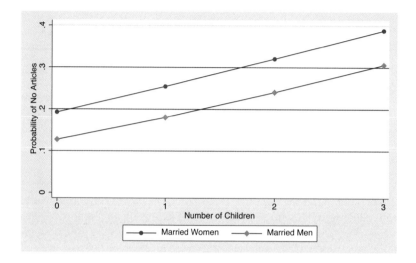

If you compare the values plotted for women with those computed with `prvalue` in the
prior section, you will see that they are exactly the same, just computed in a different
way.

Example of predicted probabilities using prcounts

`prcounts` computes predictions for all observations in the dataset. In addition, the
predictions are averaged across observations:

$$\overline{\Pr}(y = m) = \frac{1}{N} \sum_{i=1}^{N} \widehat{\Pr}(y_i = m \mid \mathbf{x}_i)$$

To illustrate how this command can be used to compare predictions from different
models, we begin by fitting a univariate Poisson distribution and computing predictions
with `prcounts`:

```
. poisson art, nolog
Poisson regression                          Number of obs   =        915
                                            LR chi2(0)      =       0.00
                                            Prob > chi2     =          .
Log likelihood = -1742.5735                 Pseudo R2       =     0.0000
```

art	Coef.	Std. Err.	z	P>|z|	[95% Conf. Interval]	
_cons	.5264408	.0254082	20.72	0.000	.4766416	.57624

```
. prcounts psn, plot max(9)
. label var psnpreq "Univariate Poisson Dist."
```

Because we specified the `plot` option and the prefix `psn`, the command `prcounts` created a new variable called `psnpreq` that contains the average predicted probabilities of counts 0 through 9 from a univariate Poisson distribution. We then estimate the PRM with independent variables and again compute predictions with `prcounts`:

```
. poisson art fem mar kid5 phd ment, nolog
Poisson regression                          Number of obs   =        915
                                            LR chi2(5)      =     183.03
                                            Prob > chi2     =     0.0000
Log likelihood = -1651.0563                 Pseudo R2       =     0.0525
```

art	Coef.	Std. Err.	z	P>|z|	[95% Conf. Interval]	
fem	-.2245942	.0546138	-4.11	0.000	-.3316352	-.1175532
mar	.1552434	.0613747	2.53	0.011	.0349512	.2755356
kid5	-.1848827	.0401272	-4.61	0.000	-.2635305	-.1062349
phd	.0128226	.0263972	0.49	0.627	-.038915	.0645601
ment	.0255427	.0020061	12.73	0.000	.0216109	.0294746
_cons	.3046168	.1029822	2.96	0.003	.1027755	.5064581

```
. prcounts prm, plot max(9)
. label var prmpreq "PRM"
. label var prmobeq "Observed"
```

In addition to the new variable `prmpreq`, `prcounts` also generates `prmobeq`, which contains the observed probability of counts 0 through 9. Another new variable, `prmval`, contains the value of the count. We now plot the values of `psnpreq`, `prmpreq`, and `prmobeq` with `prmval` on the x-axis:

```
. graph twoway connected prmobeq psnpreq prmpreq prmval,
      ytitle("Probability of Count") ylabel(0(.1).4)
      xlabel(0(1)9)
      ysize(2.7051) xsize(4.0413)
```

This produces the following graph:

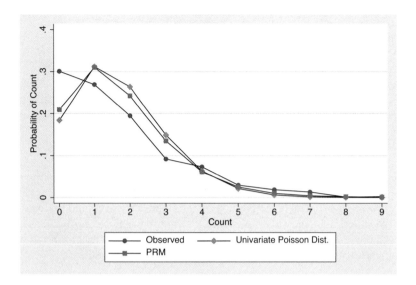

This graph shows that, even though many of the independent variables have significant effects on the number of articles published, there is only a modest improvement in the predictions made by the PRM over the univariate Poisson distribution, with somewhat more 0s predicted and slightly fewer 2s and 3s. While this suggests the need for an alternative model, we will first discuss how different periods of exposure can be incorporated into count models.

7.2.5   Exposure time[*]

So far we have implicitly assumed that each observation was "at risk" of an event occurring for the same amount of time. In terms of our example, this means that, for each person in the sample, we counted their articles over the same period of time. Often when collecting data, however, different observations have different *exposure* times. For example, the sample of scientists might have received their degrees in different years, and our outcome might have been total publications from Ph.D. to the date of the survey. Clearly the amount of time in the career affects the total number of publications.

Different exposure times can be incorporated quite simply into count models. Let t_i be the amount of time that observation i is at risk. If the rate (i.e., the expected number of observations for a single unit of time) for that case is μ_i, then we would expect $t_i\mu_i$ to be the expected count over a period of length t_i. Then, assuming only two independent variables for simplicity, our count equation becomes

$$\mu_i t_i = \{\exp\left(\beta_0 + \beta_1 x_1 + \beta_2 x_2\right)\} \times t_i$$

Because $t = \exp\left(\ln t\right)$, the equation can be rewritten as

$$\mu_i t_i = \exp\left(\beta_0 + \beta_1 x_1 + \beta_2 x_2 + \ln t_i\right)$$

This shows that the effect of different exposure times can be included as the log of the exposure time with a regression coefficient constrained to equal 1. While we do not have data with different exposure times, we have artificially constructed three variables to illustrate this issue. `profage` is a scientist's professional age, which corresponds to the time a scientist has been "exposed" to the possibility of publishing; `lnage` is the natural log of `profage`; and `totalarts` is the total number of articles during the career (to see how these were created, you can examine the sample file `st4ch7.do`). To fit the model including exposure time, we use the `exposure()` option:

```
. poisson totalarts fem mar kid5 phd ment, nolog exposure(profage)
```

Poisson regression				Number of obs	=	915
				LR chi2(5)	=	912.15
				Prob > chi2	=	0.0000
Log likelihood = -6408.8645				Pseudo R2	=	0.0664

totalarts	Coef.	Std. Err.	z	P>|z|	[95% Conf. Interval]	
fem	-.1779011	.0218925	-8.13	0.000	-.2208096	-.1349925
mar	.1490357	.0247031	6.03	0.000	.1006186	.1974528
kid5	-.1484304	.0158661	-9.36	0.000	-.1795274	-.1173334
phd	.0053858	.0104624	0.51	0.607	-.01512	.0258917
ment	.0234991	.000818	28.73	0.000	.0218959	.0251023
_cons	.340734	.0405566	8.40	0.000	.2612446	.4202234
profage	(exposure)					

The results can be interpreted using the same methods discussed above.

To show you what the `exposure()` option is doing, we can obtain the same results by adding `lnage` as an independent variable and constraining the coefficient for `lnage` to 1:

```
. constraint define 1 lnage=1

. poisson totalarts fem mar kid5 phd ment lnage, nolog constraint(1)
```

Poisson regression				Number of obs	=	915
				LR chi2(6)	=	3611.18
				Prob > chi2	=	0.0000
Log likelihood = -6408.8645				Pseudo R2	=	0.2198

(1)   [totalarts]lnage = 1

totalarts	Coef.	Std. Err.	z	P>|z|	[95% Conf. Interval]	
fem	-.1779011	.0218925	-8.13	0.000	-.2208096	-.1349925
mar	.1490357	.0247031	6.03	0.000	.1006186	.1974528
kid5	-.1484304	.0158661	-9.36	0.000	-.1795274	-.1173334
phd	.0053859	.0104624	0.51	0.607	-.01512	.0258917
ment	.0234991	.000818	28.73	0.000	.0218959	.0251023
lnage	1	.	.	.	.	.
_cons	.340734	.0405566	8.40	0.000	.2612445	.4202234

You can also obtain the same result with `offset()` instead of `exposure()`, except that with `offset()` you specify a variable that is equal to the log of the exposure time. For example,

```
. poisson totalarts fem mar kid5 phd ment, nolog offset(lnage)
```

While the `exposure()` and `offset()` are not considered further in this chapter, they can be used with the other models we discuss.

7.3   The negative binomial regression model

The PRM accounts for observed heterogeneity (i.e., observed differences among sample members) by specifying the rate μ_i as a function of observed x_ks. In practice, the PRM rarely fits, due to *overdispersion*. That is, the model underestimates the amount of dispersion in the outcome. The negative binomial regression model (NBRM) addresses the failure of the PRM by adding a parameter α that reflects *unobserved* heterogeneity among observations.[1] For example, with three independent variables, the PRM is

$$\mu_i = \exp\left(\beta_0 + \beta_1 x_{i1} + \beta_2 x_{i2} + \beta_3 x_{i3}\right)$$

The NBRM adds an error ε that is assumed to be uncorrelated with the xs,

$$\begin{aligned}
\widetilde{\mu}_i &= \exp\left(\beta_0 + \beta_1 x_{i1} + \beta_2 x_{i2} + \beta_3 x_{i3} + \varepsilon_i\right) \\
&= \exp\left(\beta_0 + \beta_1 x_{i1} + \beta_2 x_{i2} + \beta_3 x_{i3}\right)\exp\left(\varepsilon_i\right) \\
&= \exp\left(\beta_0 + \beta_1 x_{i1} + \beta_2 x_{i2} + \beta_3 x_{i3}\right)\delta_i
\end{aligned}$$

where the second step follows by basic algebra, and the last step simply defines $\delta \equiv \exp\left(\varepsilon\right)$. To identify the model, we assume that

$$E\left(\delta\right) = 1$$

which corresponds to the assumption $E\left(\varepsilon\right) = 0$ in the LRM. With this assumption, it is easy to show that

$$E\left(\widetilde{\mu}\right) = \mu E\left(\delta\right) = \mu$$

Thus, *the PRM and the NBRM have the same mean structure*. That is, if the assumptions of the NBRM are correct, the expected rate for a given level of the independent variables will be the same in both models. However, the standard errors in the PRM will be biased downward, resulting in spuriously large z-values and spuriously small p-values (Cameron and Trivedi 1986, 31).

The distribution of observations given both the values of the xs *and* δ is still Poisson in the NBRM. That is,

$$\Pr(y_i \mid \mathbf{x}_i, \delta_i) = \frac{e^{-\widetilde{\mu}_i}\widetilde{\mu}_i^{y_i}}{y_i!}$$

---

[1]The NBRM can also be derived through a process of contagion where the occurrence of an event changes the probability of further events. That approach is not considered further here.

Because δ is unknown, we cannot compute $\Pr(y \mid \mathbf{x})$. This is resolved by assuming that δ is drawn from a gamma distribution (see Long 1997, 231–232 or Cameron and Trivedi 1998, 70–79 for details). Then we can compute $\Pr(y \mid \mathbf{x})$ as a weighted combination of $\Pr(y \mid \mathbf{x}, \delta)$ for all values of δ, where the weights are determined by $\Pr(\delta)$. The mathematics for this mixing of values of $\Pr(y \mid \mathbf{x}, \delta)$ are complex (and not particularly helpful for understanding the interpretation of the model) but lead to the negative binomial distribution

$$\Pr(y \mid \mathbf{x}) = \frac{\Gamma(y + \alpha^{-1})}{y!\Gamma(\alpha^{-1})} \left(\frac{\alpha^{-1}}{\alpha^{-1} + \mu}\right)^{\alpha^{-1}} \left(\frac{\mu}{\alpha^{-1} + \mu}\right)^{y}$$

where $\Gamma()$ is the gamma function.

In the negative binomial distribution, the parameter α determines the degree of dispersion in the predictions, as illustrated by the following figure:

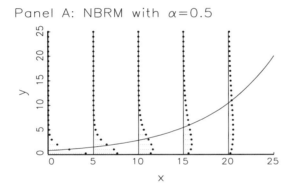

Panel A: NBRM with $\alpha = 0.5$

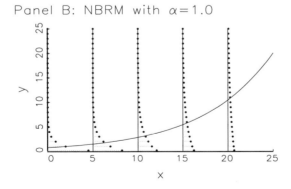

Panel B: NBRM with $\alpha = 1.0$

In both panels, the dispersion of predicted counts for a given value of x is larger than in the PRM. In particular, note the greater probability of a 0 count. Further, the larger value of α in Panel B results in greater spread in the data. Indeed, if $\alpha = 0$, the NBRM reduces to the PRM, which turns out to be the key to testing for overdispersion. This is discussed in Section 7.3.3.

7.3.1   Fitting the NBRM with nbreg

The NBRM is fitted with the following command:

nbreg *depvar* [*indepvars*] [*weight*] [if *exp*] [in *range*] [, irr level(*#*)

exposure(*varname*) robust cluster(*varname*) score(*newvar*) nolog]

where the options are the same as those for poisson. Because of differences in how
poisson and nbreg are implemented in Stata, models fitted with nbreg take substantially longer to converge.

7.3.2   Example of fitting the NBRM

Here, we use the same example as for the PRM above:

```
. use http://www.stata-press.com/data/lfr/couart2, clear
(Academic Biochemists / S Long)

. nbreg art fem mar kid5 phd ment, nolog
Negative binomial regression                    Number of obs   =        915
                                                LR chi2(5)      =      97.96
                                                Prob > chi2     =     0.0000
Log likelihood = -1560.9583                     Pseudo R2       =     0.0304
```

art	Coef.	Std. Err.	z	P>|z|	[95% Conf. Interval]	
fem	-.2164184	.0726724	-2.98	0.003	-.3588537	-.0739832
mar	.1504895	.0821063	1.83	0.067	-.0104359	.3114148
kid5	-.1764152	.0530598	-3.32	0.001	-.2804105	-.07242
phd	.0152712	.0360396	0.42	0.672	-.0553652	.0859075
ment	.0290823	.0034701	8.38	0.000	.0222811	.0358836
_cons	.256144	.1385604	1.85	0.065	-.0154294	.5277174
/lnalpha	-.8173044	.1199372			-1.052377	-.5822318
alpha	.4416205	.0529667			.3491069	.5586502

```
Likelihood-ratio test of alpha=0:  chibar2(01) =  180.20 Prob>=chibar2 = 0.000
```

The output is similar to that of poisson, with the exception of the results at the
bottom of the output, which initially can be confusing. While the model was defined
in terms of the parameter α, nbreg estimates $\ln(\alpha)$ with the estimate given in the line
/lnalpha. This is done because estimating $\ln(\alpha)$ forces the estimated α to be positive.
The value of $\widehat{\alpha}$ is given on the next line. z-values are not given because they require
special treatment, as discussed in Section 7.3.3.

Comparing the PRM and NBRM using estimates table

We can use `estimates table` to combine the results from `poisson` and `nbreg`:

```
. poisson art fem mar kid5 phd ment, nolog
(output omitted)
. estimates store PRM
. nbreg art fem mar kid5 phd ment, nolog
(output omitted)
. estimates store NBRM
. estimates table PRM NBRM, b(%9.3f) t label varwidth(32) drop(lnalpha:_cons)
    stats(alpha N)
```

Variable	PRM	NBRM
Gender: 1=female 0=male	-0.225	-0.216
	-4.11	-2.98
Married: 1=yes 0=no	0.155	0.150
	2.53	1.83
Number of children < 6	-0.185	-0.176
	-4.61	-3.32
PhD prestige	0.013	0.015
	0.49	0.42
Article by mentor in last 3 yrs	0.026	0.029
	12.73	8.38
Constant	0.305	0.256
	2.96	1.85
alpha		0.442
N	915.000	915.000

```
legend: b/t
```

The estimates of the corresponding parameters from the PRM and the NBRM are close, but the z-values for the NBRM are consistently smaller than those for the PRM. This is the expected consequence of overdispersion.

7.3.3 Testing for overdispersion

If there is overdispersion, estimates from the PRM are inefficient with standard errors that are biased downward, even if the model includes the correct variables. Accordingly, it is important to test for overdispersion. Because the NBRM reduces to the PRM when $\alpha = 0$, we can test for overdispersion by testing H_0: $\alpha = 0$. There are two points to keep in mind in making this test. First, `nbreg` estimates $\ln(\alpha)$ rather than α. A test of H_0: $\ln(\alpha) = 0$ corresponds to testing H_0: $\alpha = 1$, which is *not* the test we want. Second, as α must be greater than or equal to 0, the asymptotic distribution of $\widehat{\alpha}$ when $\alpha = 0$ is only half of a normal distribution. That is, all values less than 0 have a probability of 0. This requires an adjustment to the usual significance level of the test.

To test the hypothesis H_0: $\alpha = 0$, Stata provides a LR test that is listed after the estimates of the parameters:

```
Likelihood-ratio test of alpha=0:   chibar2(01) =   180.20 Prob>=chibar2 = 0.000
```

As this output is different than that from `lrtest`, it is worth clarifying what it means. The test statistic `chibar2(01)` is computed by the same formula shown in Chapter 3:

$$G^2 = 2\left(\ln L_{\text{NBRM}} - \ln L_{\text{PRM}}\right)$$
$$= 2\left(-1560.96 - -1651.06\right) = 180.2$$

The significance level of the test is adjusted to account for the truncated sampling distribution of $\widehat{\alpha}$. For details, you can click on `chibar2(01)`, which will be listed in blue in the Results window (recall that blue means that you can click for further information). In our example, the results are very significant and provide strong evidence of overdispersion. You can summarize this by saying that

> because there is significant evidence of overdispersion ($G^2 = 180.2, p < .01$), the negative binomial regression model is preferred to the Poisson regression model.

7.3.4   Interpretation using the rate μ

As the mean structure for the NBRM is identical to that for the PRM, the same methods of interpretation based on $E\left(y \mid \mathbf{x}\right)$ can be used based on the equation

$$\frac{E\left(y \mid \mathbf{x}, x_k + \delta\right)}{E\left(y \mid \mathbf{x}, x_k\right)} = e^{\beta_k \delta}$$

This leads to the interpretation that

> for a change of δ in x_k, the expected count increases by a factor of $\exp(\beta_k \times \delta)$, holding all other variables constant.

(Continued on next page)

Factor and percent change coefficients can be obtained using `listcoef`. For example,

```
. listcoef fem ment, help
nbreg (N=915): Factor Change in Expected Count
 Observed SD: 1.926069
```

art	b	z	P>|z|	e^b	e^bStdX	SDofX
fem	-0.21642	-2.978	0.003	0.8054	0.8977	0.4987
ment	0.02908	8.381	0.000	1.0295	1.3176	9.4839
ln alpha	-0.81730					
alpha	0.44162	SE(alpha) = 0.05297				

```
 LR test of alpha=0: 180.20   Prob>=LRX2 = 0.000
```

```
        b = raw coefficient
        z = z-score for test of b=0
    P>|z| = p-value for z-test
      e^b = exp(b) = factor change in expected count for unit increase in X
 e^bStdX = exp(b*SD of X) = change in expected count for SD increase in X
    SDofX = standard deviation of X
. listcoef fem ment, help percent
nbreg (N=915): Percentage Change in Expected Count
 Observed SD: 1.926069
```

art	b	z	P>|z|	%	%StdX	SDofX
fem	-0.21642	-2.978	0.003	-19.5	-10.2	0.4987
ment	0.02908	8.381	0.000	3.0	31.8	9.4839
ln alpha	-0.81730					
alpha	0.44162	SE(alpha) = 0.05297				

```
 LR test of alpha=0: 180.20   Prob>=LRX2 = 0.000
```

```
        b = raw coefficient
        z = z-score for test of b=0
    P>|z| = p-value for z-test
        % = percent change in expected count for unit increase in X
    %StdX = percent change in expected count for SD increase in X
    SDofX = standard deviation of X
```

These coefficients can be interpreted as follows:

Being a female scientist decreases the expected number of articles by a factor of .81, holding all other variables constant. Equivalently, being a female scientist decreases the expected number of articles by 19.5 percent, holding all other variables constant.

For every additional article by the mentor, a scientist's expected mean productivity increases by 3.0 percent, holding other variables constant.

For a standard deviation increase in the mentor's productivity, a scientist's expected mean productivity increases by 32 percent, holding all other variables constant.

Interpretations for marginal and discrete change can be computed and interpreted using
the methods discussed for the PRM.

7.3.5 Interpretation using predicted probabilities

The methods from the PRM can also be used for interpreting predicted probabilities.
The *only* difference is that the predicted probabilities are computed with the formula

$$\widehat{\Pr}\left(y \mid \mathbf{x}\right) = \frac{\Gamma(y + \alpha^{-1})}{y!\Gamma(\alpha^{-1})} \left(\frac{\widehat{\alpha}^{-1}}{\widehat{\alpha}^{-1} + \widehat{\mu}}\right)^{\widehat{\alpha}^{-1}} \left(\frac{\widehat{\mu}}{\widehat{\alpha}^{-1} + \widehat{\mu}}\right)^{y}$$

where $\widehat{\mu} = \exp\left(\mathbf{x}\widehat{\beta}\right)$. As before, predicted probabilities can be computed using prgen,
prchange, prcounts, and prvalue. As there is nothing new in how to use these com-
mands, we provide only two examples that are designed to illustrate key differences
and similarities between the PRM and the NBRM. First, we use prvalue to compute
predicated values for an "average" respondent. For the PRM,

```
. quietly poisson art fem mar kid5 phd ment
. prvalue

poisson: Predictions for art
Predicted rate: 1.61      95% CI [1.53, 1.7]
Predicted probabilities:
  Pr(y=0|x):   0.1999  Pr(y=1|x):   0.3218
  Pr(y=2|x):   0.2591  Pr(y=3|x):   0.1390
  Pr(y=4|x):   0.0560  Pr(y=5|x):   0.0180
  Pr(y=6|x):   0.0048  Pr(y=7|x):   0.0011
  Pr(y=8|x):   0.0002  Pr(y=9|x):   0.0000
            fem        mar       kid5        phd       ment
x=   .46010929  .66229508  .49508197  3.1031093  8.7672131
```

and for the NBRM,

```
. quietly nbreg art fem mar kid5 phd ment
. prvalue

nbreg: Predictions for art
Predicted rate: 1.6
Predicted probabilities:
  Pr(y=0|x):   0.2978  Pr(y=1|x):   0.2794
  Pr(y=2|x):   0.1889  Pr(y=3|x):   0.1113
  Pr(y=4|x):   0.0607  Pr(y=5|x):   0.0315
  Pr(y=6|x):   0.0158  Pr(y=7|x):   0.0077
  Pr(y=8|x):   0.0037  Pr(y=9|x):   0.0018
            fem        mar       kid5        phd       ment
x=   .46010929  .66229508  .49508197  3.1031093  8.7672131
```

The first thing to notice is that the predicted rate is nearly identical for both models: 1.610 versus 1.602. This illustrates that even with overdispersion (which there is in this example), the estimates from the PRM are consistent. But, substantial differences emerge when we examine predicted probabilities: $\widehat{\text{Pr}}_{\text{PRM}}(y = 0 \mid \overline{\mathbf{x}}) = 0.200$ compared with $\widehat{\text{Pr}}_{\text{NBRM}}(y = 0 \mid \overline{\mathbf{x}}) = 0.298$. We also find higher probabilities in the NBRM for larger counts. For example, $\widehat{\text{Pr}}_{\text{NBRM}}(y = 5 \mid \overline{\mathbf{x}}) = 0.0315$ compared with $\widehat{\text{Pr}}_{\text{PRM}}(y = 5 \mid \overline{\mathbf{x}}) = 0.0180$. These probabilities reflect the greater dispersion in the NBRM compared with the PRM.

Another way to see the greater probability for 0 counts in the NBRM is to plot the probability of 0s as values of an independent variable change. This is done with `prgen`:

```
. quietly nbreg art fem mar kid5 phd ment, nolog
. prgen ment, rest(mean) f(0) t(50) gen(nb) n(20)
nbreg: Predicted values as ment varies from 0 to 50.
         fem        mar       kid5        phd       ment
x=   .46010929  .66229508  .49508197  3.1031093  8.7672131
. quietly poisson art fem mar kid5 phd ment
. prgen ment, rest(mean) f(0) t(50) gen(psn) n(20)
poisson: Predicted values as ment varies from 0 to 50.
         fem        mar       kid5        phd       ment
x=   .46010929  .66229508  .49508197  3.1031093  8.7672131
. label var psnp0 "Pr(0) for PRM"
. label var nbp0 "Pr(0) for NBRM"
. graph twoway connected psnp0 nbp0 nbx,
      ylabel(0(.1).4) yline(.1 .2 .3)
      ytitle("Probability of a Zero Count")
      ysize(2.7051) xsize(4.0421)
```

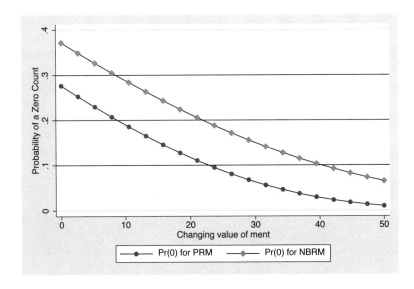

The probability of having zero publications is computed when each variable except the mentor's number of articles is held at its mean. For both models, the probability of a zero decreases as the mentor's articles increase. But, the proportion of predicted zeros is significantly higher for the NBRM. As both models have the same expected number of publications, the higher proportion of predicted zeros for the NBRM is offset by the higher proportion of larger counts that are also predicted by this model.

7.4   Zero-inflated count models

The NBRM improves upon the underprediction of zeros in the PRM by increasing the conditional variance without changing the conditional mean, which was illustrated by the output from `prvalue` in the prior section. Zero-inflated count models, introduced by Lambert (1992), respond to the failure of the PRM model to account for dispersion and excess zeros by changing the mean structure to allow zeros to be generated by two distinct processes. To make this clearer, consider our example of scientific productivity. The PRM and NBRM assume that *every* scientist has a positive probability of publishing any given number of papers. The probability differs across individuals according to their characteristics, but *all* scientists have some probability of publishing. Substantively, this is unrealistic because some scientists are not potential publishers. For example, they could hold positions, perhaps in industry, where publishing is not allowed. Zero-inflated models allow for this possibility, and in the process, they increase the conditional variance and the probability of zero counts.

The zero-inflated model assumes that there are two *latent* (i.e., unobserved) groups. An individual in the *Always-0 group* (Group A) has an outcome of 0 with a probability of 1, while an individual in the *Not Always-0 group* (Group ~A) might have a zero count, but there is a nonzero probability that she has a positive count. This process is developed in three steps: 1) Model membership into the latent groups; 2) Model counts for those in Group ~A; and 3) Compute observed probabilities as a mixture of the probabilities for the two groups.

Step 1: Membership in Group A

Let $A = 1$ if someone is in Group A, else $A = 0$. Group membership is a binary outcome that can be modeled using the logit or probit model of Chapter 4,

$$\psi_i = \Pr\left(A_i = 1 \mid \mathbf{z}_i\right) = F\left(\mathbf{z}_i \gamma\right) \qquad (7.2)$$

where ψ_i is the probability of being in Group A for individual i. The z-variables are referred to as *inflation* variables because they serve to inflate the number of 0s as shown below. To illustrate (7.2), assume that two variables affect the probability of an individual being in Group A and that we model this with a logit equation:

$$\psi_i = \frac{\exp\left(\gamma_0 + \gamma_1 z_1 + \gamma_2 z_2\right)}{1 + \exp\left(\gamma_0 + \gamma_1 z_1 + \gamma_2 z_2\right)}$$

If we had an observed variable indicating group membership, this would be a standard, binary regression model. But, because group membership is a latent variable, we do not know whether an individual is in Group A or Group ~A.

Step 2: Counts for those in Group ~A

Among those who are *not* always zero, the probability of each count (including zeros) is determined by either a Poisson or a negative binomial regression. Notice that in the equations that follow, we are conditioning both on the x_ks and on $A = 0$. Also note that the x_ks are not necessarily the same as the inflation variables z_k in the first step (although the two sets of variables can be the same). For the *zero-inflated Poisson* (ZIP) *model*, we have

$$\Pr(y_i \mid \mathbf{x}_i, \; A_i = 0) = \frac{e^{-\mu_i} \mu_i^{y_i}}{y_i!}$$

or, for the *zero-inflated negative binomial* (ZINB) *model*,

$$\Pr(y_i \mid \mathbf{x}_i, \; A_i = 0) = \frac{\Gamma(y_i + \alpha^{-1})}{y_i!\Gamma(\alpha^{-1})} \left(\frac{\alpha^{-1}}{\alpha^{-1} + \mu_i}\right)^{\alpha^{-1}} \left(\frac{\mu_i}{\alpha^{-1} + \mu_i}\right)^{y_i}$$

In both equations, $\mu_i = \exp(\mathbf{x}_i\beta)$. If we knew which observations were in Group ~A, these equations would define the PRM and the NBRM. But, here the equations only apply to those observations in Group ~A, and we do not have an observed variable indicating group membership.

Step 3: Mixing Groups A and ~A

The simplest way to understand the mixing is to start with an example. Suppose that retirement status is indicated by $r = 1$ for retired folks and $r = 0$ for those not retired, where

$$\Pr(r = 1) = .2$$
$$\Pr(r = 0) = 1 - .2 = .8$$

Let y indicate living in a warm climate, with $y = 1$ for yes and $y = 0$ for no. Suppose that the conditional probabilities are

$$\Pr(y = 1 \mid r = 1) = .5$$
$$\Pr(y = 1 \mid r = 0) = .3$$

so that people are more likely to live in a warm climate if they are retired. What is the probability of living in a warm climate for the population as a whole? The answer is a mixture of the probabilities for the two groups weighted by the proportion in each group:

$$\begin{aligned}
\Pr\left(y=1\right) &= \{\Pr\left(r=1\right) \times \Pr\left(y=1 \mid r=1\right)\} \\
&\quad + \{\Pr\left(r=0\right) \times \Pr\left(y=1 \mid r=0\right)\} \\
&= [.2 \times .5] + [.8 \times .3] = .34
\end{aligned}$$

In other words, the two groups are mixed according to their proportions in the population to determine the overall rate. The same thing is done for the zero-inflated models.

The proportion in each group is defined by

$$\begin{aligned}
\Pr\left(A_i = 1\right) &= \psi_i \\
\Pr\left(A_i = 0\right) &= 1 - \psi_i
\end{aligned}$$

and the probabilities of a zero within each group are

$$\begin{aligned}
\Pr\left(y_i = 0 \mid A_i = 1, \mathbf{x}_i, \mathbf{z}_i\right) &= 1 \quad \text{by definition of the } A \text{ group} \\
\Pr\left(y_i = 0 \mid A_i = 0, \mathbf{x}_i, \mathbf{z}_i\right) &= \text{outcome of \textsc{prm} or \textsc{nbrm}.}
\end{aligned}$$

Then, the overall probability of a 0 count is

$$\begin{aligned}
\Pr\left(y_i = 0 \mid \mathbf{x}_i, \mathbf{z}_i\right) &= \{\psi_i \times 1\} + [(1 - \psi_i) \times \Pr\left(y_i = 0 \mid \mathbf{x}_i, A_i = 0\right)\} \\
&= \psi_i + \{(1 - \psi_i) \times \Pr\left(y_i = 0 \mid \mathbf{x}_i, A_i = 0\right)\}
\end{aligned}$$

For outcomes other than 0,

$$\begin{aligned}
\Pr\left(y_i = k \mid \mathbf{x}_i, \mathbf{z}_i\right) &= [\psi_i \times 0] + \{(1 - \psi_i) \times \Pr\left(y_i = k \mid \mathbf{x}_i, A_i = 0\right)\} \\
&= (1 - \psi_i) \times \Pr\left(y_i = k \mid \mathbf{x}_i, A_i = 0\right)
\end{aligned}$$

where we use the assumption that the probability of a positive count in Group A is 0.

Expected counts are computed in a similar fashion:

$$\begin{aligned}
E\left(y \mid \mathbf{x}, \mathbf{z}\right) &= [0 \times \psi] + \{\mu \times (1 - \psi)\} \\
&= \mu\left(1 - \psi\right)
\end{aligned}$$

Because $0 \le \psi \le 1$, the expected value will be smaller than μ, which shows that the mean structure in zero-inflated models differs from that in the \textsc{prm} or \textsc{nbrm}.

7.4.1   Fitting zero-inflated models with zinb and zip

The ZIP and ZINB models are fitted with the `zip` and `zinb` commands. The syntax is

```
zip depvar [indepvars] [weight] [if exp] [in range] [, inflate(indepvars2)
    irr level(#) robust cluster(varname) score(newvarlist | stub*)
    exposure(varname) probit vuong nolog ]
```

```
zinb depvar [indepvars] [weight] [if exp] [in range] [, inflate(indepvars2)
    irr level(#) robust cluster(varname) score(newvarlist | stub*)
    exposure(varname) probit vuong nolog ]
```

Variable lists

depvar is the dependent variable, which must be a count variable.

indepvars is a list of independent variables that determine counts among those who are not always zeros. If *indepvars* is not included, a model with only an intercept is fitted.

indepvars2 is a list of inflation variables that determine if you are in the Always-0 group or the Not Always-0 group.

indepvars and *indepvars2* can be the same variables, but they do not have to be.

Options

Here we only consider options that differ from those in earlier models for this chapter.

`probit` specifies that the model determining the probability of being in the Always-0 group versus the Not Always-0 group is to be a binary probit model. By default, a binary logit model is used.

`vuong` requests a Vuong (1989) test of the ZIP model versus the PRM, or of the ZINB versus the NBRM. Details are given in Section 7.5.2.

7.4.2 Example of fitting the ZIP and ZINB models

The output from `zip` and `zinb` is very similar, so here we show only the output for `zinb`:

(Continued on next page)

```
. zinb art fem mar kid5 phd ment, inf(fem mar kid5 phd ment) nolog
Zero-inflated negative binomial regression      Number of obs    =        915
                                                Nonzero obs      =        640
                                                Zero obs         =        275

Inflation model = logit                         LR chi2(5)       =      67.97
Log likelihood  = -1549.991                     Prob > chi2      =     0.0000
```

art	Coef.	Std. Err.	z	P>|z|	[95% Conf. Interval]	
art						
fem	-.1955068	.0755926	-2.59	0.010	-.3436655	-.0473481
mar	.0975826	.084452	1.16	0.248	-.0679402	.2631054
kid5	-.1517325	.054206	-2.80	0.005	-.2579744	-.0454906
phd	-.0007001	.0362696	-0.02	0.985	-.0717872	.0703869
ment	.0247862	.0034924	7.10	0.000	.0179412	.0316312
_cons	.4167466	.1435962	2.90	0.004	.1353032	.69819
inflate						
fem	.6359328	.8489175	0.75	0.454	-1.027915	2.299781
mar	-1.499469	.9386701	-1.60	0.110	-3.339228	.3402909
kid5	.6284274	.4427825	1.42	0.156	-.2394105	1.496265
phd	-.0377153	.3080086	-0.12	0.903	-.641401	.5659705
ment	-.8822932	.3162276	-2.79	0.005	-1.502088	-.2624984
_cons	-.1916865	1.322821	-0.14	0.885	-2.784368	2.400995
/lnalpha	-.9763565	.1354679	-7.21	0.000	-1.241869	-.7108443
alpha	.3766811	.0510282			.288844	.4912293

The top set of coefficients, labeled `art` at the left margin, correspond to the NBRM for those in the Not Always-0 group. The lower set of coefficients, labeled `inflate`, correspond to the binary model predicting group membership.

7.4.3   Interpretation of coefficients

When interpreting zero-inflated models, it is easy to be confused by the direction of the coefficients. `listcoef` makes interpretation simpler. For example, consider the results for the ZINB:

```
. zinb art fem mar kid5 phd ment, inf(fem mar kid5 phd ment) nolog
  (output omitted)
. listcoef, help
zinb (N=915): Factor Change in Expected Count
Observed SD: 1.926069
Count Equation: Factor Change in Expected Count for Those Not Always 0
```

art	b	z	P>|z|	e^b	e^bStdX	SDofX
fem	-0.19551	-2.586	0.010	0.8224	0.9071	0.4987
mar	0.09758	1.155	0.248	1.1025	1.0473	0.4732
kid5	-0.15173	-2.799	0.005	0.8592	0.8904	0.7649
phd	-0.00070	-0.019	0.985	0.9993	0.9993	0.9842
ment	0.02479	7.097	0.000	1.0251	1.2650	9.4839
ln alpha	-0.97636					
alpha	0.37668	SE(alpha) = 0.05103				

```
    b = raw coefficient
    z = z-score for test of b=0
 P>|z| = p-value for z-test
   e^b = exp(b) = factor change in expected count for unit increase in X
e^bStdX = exp(b*SD of X) = change in expected count for SD increase in X
 SDofX = standard deviation of X
Binary Equation: Factor Change in Odds of Always 0
```

Always0	b	z	P>|z|	e^b	e^bStdX	SDofX
fem	0.63593	0.749	0.454	1.8888	1.3732	0.4987
mar	-1.49947	-1.597	0.110	0.2232	0.4919	0.4732
kid5	0.62843	1.419	0.156	1.8747	1.6172	0.7649
phd	-0.03772	-0.122	0.903	0.9630	0.9636	0.9842
ment	-0.88229	-2.790	0.005	0.4138	0.0002	9.4839

```
    b = raw coefficient
    z = z-score for test of b=0
 P>|z| = p-value for z-test
   e^b = exp(b) = factor change in odds for unit increase in X
e^bStdX = exp(b*SD of X) = change in odds for SD increase in X
 SDofX = standard deviation of X
```

The top half of the output, labeled Count Equation, contains coefficients for the factor change in the expected count for those in the Not Always-0 group. This group comprises those scientists who have the opportunity to publish. The coefficients can be interpreted in the same way as coefficients from the PRM or the NBRM. For example,

among those who have the opportunity to publish, being a woman decreases the expected rate of publication by a factor of .91, holding all other factors constant.

The bottom half, labeled `Binary Equation`, contains coefficients for the factor change in the odds of being in the Always-0 group compared to the Not Always-0 group. These can be interpreted just as the coefficients for a binary logit model. For example,

> being a woman increases the odds of not having the opportunity to publish by a factor of 1.89, holding all other variables constant.

As we found in this example, when the same variables are included in both equations, the signs of the corresponding coefficients from the binary equation are often in the opposite direction of the coefficients for the count equation. This often makes substantive sense because the binary process is predicting membership in the group that always has zero counts, so a positive coefficient implies lower productivity. The count process predicts number of publications so that a negative coefficient would indicate lower productivity.

7.4.4   Interpretation of predicted probabilities

For the ZIP model,

$$\widehat{\Pr}\left(y = 0 \mid \mathbf{x}, \mathbf{z}\right) = \widehat{\psi} + \left(1 - \widehat{\psi}\right) e^{-\widehat{\mu}}$$

where $\widehat{\mu} = \exp\left(\mathbf{x}\widehat{\beta}\right)$ and $\widehat{\psi} = F\left(\mathbf{z}\widehat{\gamma}\right)$. The predicted probability of a positive count applies only to the $1 - \widehat{\psi}$ observations in the Not Always-0 Group:

$$\widehat{\Pr}(y \mid \mathbf{x}) = \left(1 - \widehat{\psi}\right) \frac{e^{-\widehat{\mu}_i} \widehat{\mu}^y}{y!}$$

Similarly, for the ZINB model,

$$\widehat{\Pr}\left(y = 0 \mid \mathbf{x}, \mathbf{z}\right) = \widehat{\psi} + \left(1 - \widehat{\psi}\right) \left(\frac{\widehat{\alpha}^{-1}}{\widehat{\alpha}^{-1} + \widehat{\mu}_i}\right)^{\widehat{\alpha}^{-1}}$$

and the predicted probability for a positive count is

$$\widehat{\Pr}(y \mid \mathbf{x}) = \left(1 - \widehat{\psi}\right) \frac{\Gamma\left(y + \widehat{\alpha}^{-1}\right)}{y!\,\Gamma(\widehat{\alpha}^{-1})} \left(\frac{\widehat{\alpha}^{-1}}{\widehat{\alpha}^{-1} + \widehat{\mu}}\right)^{\widehat{\alpha}^{-1}} \left(\frac{\widehat{\mu}}{\widehat{\alpha}^{-1} + \widehat{\mu}}\right)^y$$

The probabilities can be computed with `prvalue`, `prcounts`, and `prgen`.

Predicted probabilities with prvalue

`prvalue` works in the same way for `zip` and `zinb` as it did for earlier count models, although the output is slightly different. Suppose that we want to compare the predicted probabilities for a married female scientist with young children who came from a weak graduate program with those for a married male from a strong department who had a productive mentor:

```
. quietly prvalue, x(fem=0 mar=1 kid5=3 phd=3 ment=10) save
. prvalue, x(fem=1 mar=1 kid5=3 phd=1 ment=0) dif

zinb: Change in Predictions for  art
Predicted rate: .272              Saved: 1.36
    Difference: -1.08
Predicted probabilities:
                        Current       Saved  Difference
    Pr(y=0|x,z):         0.9290      0.3344      0.5945
    Pr(y=1|x):           0.0593      0.3001     -0.2408
    Pr(y=2|x):           0.0101      0.1854     -0.1754
    Pr(y=3|x):           0.0015      0.0973     -0.0958
    Pr(y=4|x):           0.0002      0.0465     -0.0463
    Pr(y=5|x):           0.0000      0.0209     -0.0209
    Pr(y=6|x):           0.0000      0.0090     -0.0090
    Pr(y=7|x):           0.0000      0.0038     -0.0038
    Pr(y=8|x):           0.0000      0.0015     -0.0015
    Pr(y=9|x):           0.0000      0.0006     -0.0006
Pr(Always0|z):           0.6883      0.0002      0.6882

x values for count equation
            fem   mar  kid5   phd  ment
Current=     1     1     3     1     0
  Saved=     0     1     3     3    10
   Diff=     1     0     0    -2   -10

z values for binary equation
            fem   mar  kid5   phd  ment
Current=     1     1     3     1     0
  Saved=     0     1     3     3    10
   Diff=     1     0     0    -2   -10
```

There are two major differences in the output of **prvalue** for **zip** and **zinb** compared to other count models. First, levels of both the x variables from the count equation and the z variables from the binary equation are listed. In this example, they are the same variables, but they could be different. Second, there are two probabilities of 0 counts. For example, for our female scientists, **prvalue** lists **Pr(y=0 | x,z): 0.9290**, which is the probability of having no publications, either because a scientist does not have the opportunity to publish or because a scientist is a potential publisher who by chance did not publish. The quantity **Pr(Always0 | z): 0.6883** is the probability of not having the opportunity to publish. Thus, most of the 0s for women are due to being in the group that never publishes. The remaining probabilities listed are the probabilities of observing each count of publications for the specified set of characteristics.

Predicted probabilities with prgen

prgen is used to plot predictions. In this case, we examine the two sources of 0s. First, we call **prgen** to compute the predicted values to be plotted:

```
. prgen ment, rest(mean) f(0) t(20) gen(zinb) n(21)
zinb: Predicted values as ment varies from 0 to 20.
base x values for count equation:
           fem         mar        kid5         phd        ment
x=   .46010929   .66229508   .49508197   3.1031093   8.7672131
base z values for binary equation:
           fem         mar        kid5         phd        ment
z=   .46010929   .66229508   .49508197   3.1031093   8.7672131
```

prgen created two probabilities for 0 counts: **zinbp0** contains the probability of a 0 count from both the count and the binary equation. **zinball0** is the probability due to observations being in the Always-0 group. We use **generate zinbnb0 = zinbp0 - zinball0** to compute the probability of 0s from the count portion of the model:

```
. gen zinbnb0 = zinbp0 - zinball0
(894 missing values generated)
. label var zinbp0 "0s from Both Equations"
. label var zinball0 "0s from Binary Equation"
. label var zinbnb0 "0s from Count Equation"
. label var zinbx "Mentor's publications"
```

These are plotted with the command

```
. graph twoway connected zinball0 zinbnb0 zinbp0 zinbx,
      xlabel(0(5)20) ylabel(0(.1).7)
      ytitle(Probability of zero) msymbol(Oh Sh O)
      ysize(2.6541) xsize(3.9678)
```

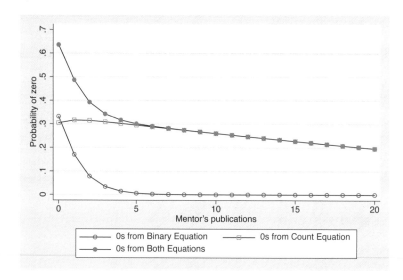

The curve marked with ○s is a probability curve just like those shown in Chapter 4 for binary models. The curve marked with □s shows the probability of 0s from a series of negative binomial distributions, each with different rate parameters μ determined by the level of mentor's publications. The overall probability of a zero count is the sum of the two curves, which is shown by the line with ●s.

7.5   Comparisons among count models

There are two methods that can be used to compare the results of the PRM, NBRM, ZIP, and ZINB models.

7.5.1   Comparing mean probabilities

One way to compare these models is to compare predicted probabilities across models. First, we compute the mean predicted probability. For example, in the PRM,

$$\overline{\Pr}_{\mathrm{PRM}}(y=m) = \frac{1}{N}\sum_{i=1}^{N}\widehat{\Pr}_{\mathrm{PRM}}(y_i = m \mid \mathbf{x}_i)$$

This is simply the average across all observations of the probability of each count. The difference between the observed probabilities and the mean prediction can be computed as

$$\Delta\overline{\Pr}_{\mathrm{PRM}}(y=m) = \widehat{\Pr}_{\mathrm{Observed}}(y=m) - \overline{\Pr}_{\mathrm{PRM}}(y=m)$$

This can be done for each model and then plotted. The commands are

```
. use http://www.stata-press.com/data/lfr/couart2, clear
(Academic Biochemists / S Long)
. quietly poisson art fem mar kid5 phd ment, nolog
. prcounts prm, plot max(9)
. label var prmpreq "Predicted: PRM"
. label var prmobeq "Observed"
. quietly nbreg art fem mar kid5 phd ment, nolog
. prcounts nbrm, plot max(9)
. label var nbrmpreq "Predicted: NBRM"
. quietly zip art fem mar kid5 phd ment,
      inf(fem mar kid5 phd ment) vuong nolog
. prcounts zip, plot max(9)
. label var zippreq "Predicted: ZIP"
. quietly zinb art fem mar kid5 phd ment,
      inf(fem mar kid5 phd ment) vuong nolog
. prcounts zinb, plot max(9)
. label var zinbpreq "Predicted: ZINB"
. * create deviations
. gen obs = prmobeq
(905 missing values generated)
. gen dprm = obs - prmpreq
(905 missing values generated)
```

```
. label var dprm "PRM"
. gen dnbrm = obs - nbrmpreq
(905 missing values generated)
. label var dnbrm "NBRM"
. gen dzip = obs - zippreq
(905 missing values generated)
. label var dzip "ZIP"
. gen dzinb = obs - zinbpreq
(905 missing values generated)
. label var dzinb "ZINB"
. * plot deviations
. graph twoway connected dprm dnbrm dzip dzinb prmval,
      ytitle(Observed-Predicted) ylabel(-.10(.05).10)
      xlabel(0(1)9) msymbol(Oh Sh O S)
      ysize(2.7051) xsize(4.0413)
```

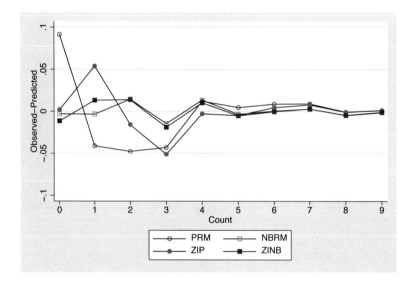

Points above 0 on the y-axis indicate more observed counts than predicted; those below 0 indicate more predicted counts than observed. The graph shows that only the PRM has a problem predicting the average number of 0s. Among the other models, the ZIP does less well, predicting too many 1s and too few 2s and 3s. The NBRM and ZINB do about equally well. Based on these results, we might prefer the NBRM because it is simpler.

7.5.2 Tests to compare count models

Plotting predictions is only an informal method of assessing the fit of a count model. More formal testing can be done with an LR test of overdispersion and a Vuong test to compare two models.

LR tests of α

Because the NBRM reduces to the PRM when $\alpha=0$, the PRM and NBRM can be compared by testing H_0: $\alpha = 0$. As shown in Section 7.3.3, we find that

```
Likelihood-ratio test of alpha=0:  chibar2(01) =  180.20 Prob>=chibar2 = 0.000
```

which provides strong evidence for preferring the NBRM over the PRM.

Because the ZIP and ZINB models are also nested, the same LR test can be applied to compare them. While Stata does not compute this for you, it is simple to do. First, we fit the ZIP model:

```
. quietly zip art fem mar kid5 phd ment, inf(fem mar kid5 phd ment) vuong nolog
. scalar llzip = e(ll)
```

The command `scalar llzip = e(ll)` saves the log likelihood that was left in memory by `zip`. Next, we do the same thing for `zinb` and compute the difference between the two log likelihoods:

```
. quietly zinb art fem mar kid5 phd ment, inf(fem mar kid5 phd ment) nolog
. scalar llzinb = e(ll)
. scalar lr = -2*(llzip-llzinb)
```

The following commands can be used to compute the p-value. Note that if you do this with your own model, you need to substitute the value of `lnalpha`, which is listed as part of the output for `zinb`:

```
. scalar pvalue = chiprob(1,lr)/2
. scalar lnalpha = -.9763565
. if (lnalpha < -20)  scalar pvalue= 1
. di as text "Likelihood-ratio test comparing ZIP to ZINB: " as res %8.3f
     lr as text " Prob>=" as res %5.3f pvalue
```

The first line is the standard way to compute the p-value for a chi-squared test with one degree of freedom, except that we divide by 2. This is because α cannot be negative, as we discussed earlier with regard to comparing the `poisson` and `nbreg` models (Gutierrez, Carter, and Drukker 2001). The next line assigns the estimated value of $\ln \alpha$ to a scalar. If this value is very close to 0, we conclude that the p-value is 1. The last line simply prints the result:

```
Likelihood-ratio test comparing ZIP to ZINB:  109.564 Prob>=0.000
```

We conclude that the ZINB significantly improves the fit over the ZIP model.

Vuong test of non-nested models

Greene (1994) points out that the PRM and the ZIP are not nested. For the ZIP model to reduce to the PRM, it is necessary for ψ to equal zero. This does *not* occur when $\gamma = \mathbf{0}$ because $\psi = F(\mathbf{z0}) = .5$. Similarly, the NBRM and the ZINB are not nested. Consequently, Greene proposes using a test by Vuong (1989, 319) for non-nested models. This test considers two models, where $\widehat{\Pr}_1 (y_i \mid \mathbf{x}_i)$ is the predicted probability of observing y in the first model, and $\widehat{\Pr}_2 (y_i \mid \mathbf{x}_i)$ is the predicted probability for the second model. Defining

$$m_i = \ln \left\{ \frac{\widehat{\Pr}_1 (y_i \mid \mathbf{x}_i)}{\widehat{\Pr}_2 (y_i \mid \mathbf{x}_i)} \right\}$$

let $\overline{m}$ be the mean and let s_m be the standard deviation of m_i. The Vuong statistic to test the hypothesis that $E(m) = 0$ equals

$$V = \frac{\sqrt{N}\,\overline{m}}{s_m}$$

V has an asymptotic normal distribution. If $V > 1.96$, the first model is favored; if $V < -1.96$, the second model is favored.

For zip, the vuong option computes the Vuong statistic, comparing the ZIP model with the PRM; for zinb it compares ZINB with NBRM. For example,

```
. zip art fem mar kid5 phd ment, inf(fem mar kid5 phd ment) vuong nolog
  (output omitted)
Vuong Test of Zip vs. Poisson: Std. Normal   =     4.18   Pr> Z      = 0.0000
```

The significant, positive value of V supports the ZIP model over the PRM. If you use listcoef, you get more guidance in interpreting the result:

```
. listcoef, help
zip (N=915): Factor Change in Expected Count
  (output omitted)
  Vuong Test =  4.18 (p=0.000) favoring ZIP over PRM.
```

For the ZINB,

```
. listcoef, help
zinb (N=915): Factor Change in Expected Count
  (output omitted)
  Vuong Test =  2.24 (p=0.012) favoring ZINB over NBRM.
```

While it is possible to compute a Vuong statistic to compare other pairs of models, such as ZIP and NBRM, these are currently not available in Stata.

Overall, these tests provide evidence that the ZINB model fits the data best. However, when fitting a series of models without any theoretical rationale, it is easy to overfit the data. In our example, the most compelling evidence for the ZINB is that it makes substantive sense. Within science, there are some scientists who for structural reasons cannot publish, but for other scientists, the failure to publish in an given period is a matter of chance. This is the basis of the zero-inflated models. The negative binomial version of the model seems preferable to the Poisson version, since it is likely that there are unobserved sources of heterogeneity that differentiate the scientists. In sum, the ZINB makes substantive sense and fits the data well.

8 Additional Topics

In this final chapter, we discuss some disparate topics that were not covered in the preceding chapters. We begin by considering complications on the right-hand side of the model: nonlinearities, interactions, and nominal or ordinal variables coded as a set of dummy variables. While the same principles of interpretation apply in these cases, several tricks are necessary for computing the appropriate quantities. Next, we discuss briefly what is required if you want to modify SPost to work with other estimation commands. The final section discusses a menagerie of Stata "tricks" that we find useful for working more efficiently in Stata.

8.1 Ordinal and nominal independent variables

When an independent variable is categorical, it should be entered into the model as a set of binary indicator variables. While our example uses an ordinal variable, the discussion applies equally to nominal independent variables, with one exception that is clearly noted.

8.1.1 Coding a categorical independent variable as a set of dummy variables

A categorical independent variable with J categories can be included in a regression model as a set of $J-1$ dummy variables. In this section, we use a binary logit model to analyze factors affecting whether a scientist has published. The outcome is a dummy variable hasarts that is equal to 1 if the scientist has one or more publications and equals 0 otherwise. In our analysis in the last chapter, we included the independent variable ment, which we treated as continuous. But, suppose instead that the data were from a survey in which the mentor was asked to indicate whether he or she had 0 articles (none), 1 to 3 articles (few), 4 to 9 (some), 10 to 20 (many), or more than 20 articles (lots). The resulting variable, which we call mentord, has the following frequency distribution:[1]

(Continued on next page)

---

[1] Details on creating mentord from the data in couart2.dta are located in st4ch8.do, which is part of the spostrm4 package. For details, when you are in Stata and online, type net search spostrm4.

```
. tab mentord, missing
   Ordinal │
measure of │
  mentor's │
   articles │      Freq.      Percent        Cum.
───────────┼──────────────────────────────────────
      None │        90         9.84         9.84
       Few │       201        21.97        31.80
      Some │       324        35.41        67.21
      Many │       213        23.28        90.49
      Lots │        87         9.51       100.00
───────────┼──────────────────────────────────────
     Total │       915       100.00
```

We can convert `mentord` into a set of dummy variables using a series of `generate` commands. Because the dummy variables are used to indicate in which category an observation belongs, they are often referred to as *indicator variables*. First, we construct `none` to indicate that the mentor had no publications:

```
. gen none = (mentord == 0) if mentord < .
```

Expressions in Stata equal 1 if true and 0 if false. Accordingly, `gen none = (mentord==0)` creates `none` equal to 1 for scientists whose mentor had no publications and equal to 0 otherwise. Although we do not have any missing values for `mentord`, it is a good habit to always add an `if` condition so that missing values continue to be missing. This is done by adding `if mentord < .` to the command (remember that missing values are treated by Stata as larger than all nonmissing values when evaluating expressions). We use `tab` to verify that `none` was constructed correctly:

```
. tab none mentord, missing
          │       Ordinal measure of mentor's articles
     none │     None      Few     Some     Many     Lots │     Total
──────────┼─────────────────────────────────────────────┼──────────
        0 │        0      201      324      213       87 │       825
        1 │       90        0        0        0        0 │        90
──────────┼─────────────────────────────────────────────┼──────────
    Total │       90      201      324      213       87 │       915
```

In the same way, we create indicator variables for the other categories of `mentord`:

```
. gen few = (mentord == 1) if mentord < .
. gen some = (mentord == 2) if mentord < .
. gen many = (mentord == 3) if mentord < .
. gen lots = (mentord == 4) if mentord < .
```

Note You can also construct indicator variables using `xi` or `tabulate`'s `gen()` option. For further information, type `help xi` and `help tabulate`.

8.1.2 Estimation and interpretation with categorical independent variables

Since `mentord` has $J = 5$ categories, so we must include $J - 1 = 4$ indicator variables as independent variables in our model. To see why one of the indicators must be dropped, consider our example. If you know that `none`, `few`, `some`, and `many` are all 0, it must be the case that `lots` equals 1 because a person has to be in one of the five categories. Another way to think of this is to note that `none` + `few` + `some` + `many` + `lots` = 1 so that including all J categories would lead to perfect collinearity. If you include all five indicator variables, Stata automatically drops one of them. For example,

```
. logit hasarts fem mar kid5 phd none few some many lots, nolog
note: lots dropped due to collinearity
Logit estimates                        Number of obs   =      915
   (output omitted )
```

The category that is excluded is the *reference category*, as the coefficients for the included indicators are interpreted relative to the excluded category, which serves as a point of reference. Which category you exclude is arbitrary, but with an ordinal independent variable, it is generally easier to interpret the results when you exclude an extreme category. For nominal categories, it is often useful to exclude the most important category. For example, we estimate a binary logit, excluding the indicator variable `none`:

```
. logit hasarts fem mar kid5 phd few some many lots, nolog
Logit estimates                        Number of obs   =      915
                                       LR chi2(8)      =    73.80
                                       Prob > chi2     =   0.0000
Log likelihood = -522.46467            Pseudo R2       =   0.0660
```

hasarts	Coef.	Std. Err.	z	P>|z|	[95% Conf. Interval]	
fem	-.2579293	.1601187	-1.61	0.107	-.5717562	.0558976
mar	.3300817	.1822141	1.81	0.070	-.0270514	.6872147
kid5	-.2795751	.1118578	-2.50	0.012	-.4988123	-.0603379
phd	.0121703	.0802726	0.15	0.879	-.145161	.1695017
few	.3859147	.2586461	1.49	0.136	-.1210223	.8928517
some	.9602176	.2490498	3.86	0.000	.4720889	1.448346
many	1.463606	.2829625	5.17	0.000	.9090099	2.018203
lots	2.335227	.4368715	5.35	0.000	1.478975	3.19148
_cons	-.0521187	.3361977	-0.16	0.877	-.7110542	.6068167

Logit models can be interpreted in terms of factor changes in the odds, which we compute using `listcoef`:

(*Continued on next page*)

```
. listcoef
logit (N=915): Factor Change in Odds
  Odds of: Arts vs NoArts
```

hasarts	b	z	P>|z|	e^b	e^bStdX	SDofX
fem	-0.25793	-1.611	0.107	0.7726	0.8793	0.4987
mar	0.33008	1.812	0.070	1.3911	1.1690	0.4732
kid5	-0.27958	-2.499	0.012	0.7561	0.8075	0.7649
phd	0.01217	0.152	0.879	1.0122	1.0121	0.9842
few	0.38591	1.492	0.136	1.4710	1.1734	0.4143
some	0.96022	3.856	0.000	2.6123	1.5832	0.4785
many	1.46361	5.172	0.000	4.3215	1.8568	0.4228
lots	2.33523	5.345	0.000	10.3318	1.9845	0.2935

The effect of an indicator variable can be interpreted in the same way that we interpreted dummy variables in Chapter 4, but with comparisons being relative to the reference category. For example, the odds ratio of 10.33 for lots can be interpreted as

> The odds of a scientist publishing are 10.3 times larger if his or her mentor had lots of publications compared with no publications, holding other variables constant.

Or, you can say equivalently:

> If a scientist's mentor has lots of publications as opposed to no publications, the odds of a scientist publishing are 10.3 times larger, holding other variables constant.

The odds ratios for the other indicators can be interpreted in the same way.

8.1.3 Tests with categorical independent variables

The basic ideas and commands for tests that involve categorical independent variables are the same as those used in prior chapters. But, because the tests involve some special considerations, we will review them here.

Testing the effect of membership in one category versus the reference category

When a set of indicator variables is included in a regression, a test of the significance of the coefficient for any indicator variable is a test of whether being in that category compared with being in the reference category affects the outcome. For example, the coefficient for few can be used to test whether having a mentor with few publications compared with having a mentor with no publications significantly affects the scientist's publishing. In our example, $z = 1.492$ and $p = 0.136$, so we conclude that

> the effect of having a mentor with a few publications compared with none is not significant using a two-tailed test ($z = 1.492$, $p = 0.14$).

Often, the significance of an indicator variable is reported without mentioning the reference category. For example, the test of **many** could be reported as follows:

> Having a mentor with a many publications significantly affects a scientist's productivity ($z = 5.17$, $p < .01$).

Here, the comparison is implicitly being made to mentors with no publications. Such interpretations should only be used if you are confident that the implicit comparison will be apparent to the reader.

Testing the effect of membership in two nonreference categories

What if neither of the categories that we wish to compare is the reference category? A simple solution is to refit the model with a different reference category. For example, to test the effect of having a mentor with some articles compared with a mentor with many publications, we can refit the model using **some** as the reference category:

```
. logit hasarts fem mar kid5 phd none few many lots, nolog
Logit estimates                              Number of obs   =         915
                                             LR chi2(8)      =       73.80
                                             Prob > chi2     =      0.0000
Log likelihood = -522.46467                  Pseudo R2       =      0.0660
```

hasarts	Coef.	Std. Err.	z	P>|z|	[95% Conf. Interval]	
fem	-.2579293	.1601187	-1.61	0.107	-.5717562	.0558976
mar	.3300817	.1822141	1.81	0.070	-.0270514	.6872147
kid5	-.2795751	.1118578	-2.50	0.012	-.4988123	-.0603379
phd	.0121703	.0802726	0.15	0.879	-.145161	.1695017
none	-.9602176	.2490498	-3.86	0.000	-1.448346	-.4720889
few	-.5743029	.1897376	-3.03	0.002	-.9461818	-.2024241
many	.5033886	.2143001	2.35	0.019	.0833682	.9234091
lots	1.37501	.3945447	3.49	0.000	.6017161	2.148303
_cons	.9080989	.3182603	2.85	0.004	.2843202	1.531878

The z-statistics for the mentor indicator variables are now tests comparing a given category with that of the mentor having some publications.

Advanced: lincom Notice that for the model that excludes **some**, the estimated coefficient for **many** equals the difference between the coefficients for **many** and **some** in the earlier model that excluded **none**. This suggests that instead of refitting the model, we could have used **lincom** to fit $\beta_{\text{many}} - \beta_{\text{some}}$:

(Continued on next page)

```
. lincom many-some

( 1) - some + many = 0
```

hasarts	Coef.	Std. Err.	z	P>|z|	[95% Conf. Interval]
(1)	.5033886	.2143001	2.35	0.019	.0833682	.9234091

The result is identical to that obtained by refitting the model with a different base category.

Testing that a categorical independent variable has no effect

For an omnibus test of a categorical variable, our null hypothesis is that the coefficients for all of the indicator variables are zero. In our model where **none** is the excluded variable, the hypothesis to test is

$$H_0 \colon \beta_{\texttt{few}} = \beta_{\texttt{some}} = \beta_{\texttt{lots}} = \beta_{\texttt{many}} = 0$$

This hypothesis can be tested with an LR test by comparing the model with the four indicators with the model that drops the four indicator variables:

```
. logit hasarts fem mar kid5 phd few some many lots, nolog
  (output omitted )
. estimates store fmodel
. logit hasarts fem mar kid5 phd, nolog
  (output omitted )
. lrtest fmodel
likelihood-ratio test                               LR chi2(4)  =     58.32
(Assumption: . nested in fmodel)                    Prob > chi2 =    0.0000
```

The key result is the change as **few** changes from 0->1 (which, because it is a dummy variable, is also the change from **min->max**).

The effect of the mentor's productivity is significant at the .01 level ($LRX^2 = 58.32$, $df = 4, p < .01$).

Alternatively, a Wald test can be used, although the LR test is generally preferred:

```
. logit hasarts fem mar kid5 phd few some many lots, nolog
  (output omitted )
. test few some many lots
( 1)  few = 0
( 2)  some = 0
( 3)  many = 0
( 4)  lots = 0

         chi2(  4) =     51.60
       Prob > chi2 =     0.0000
```

which leads to the same conclusion as the LR test.

Note that *exactly* the same results would be obtained for either test if we had used a different reference category and tested, for example,

$$H_0\text{: } \beta_{\text{none}} = \beta_{\text{few}} = \beta_{\text{lots}} = \beta_{\text{many}} = 0$$

Testing whether treating an ordinal variable as interval loses information

Ordinal independent variables are often treated as interval in regression models. For example, rather than include the four indicator variables that were created from `mentord`, we might simply include only `mentord` in our model:

```
. logit hasarts fem mar kid5 phd mentord, nolog
Logit estimates                                  Number of obs   =        915
                                                 LR chi2(5)      =      72.73
                                                 Prob > chi2     =     0.0000
Log likelihood = -522.99932                      Pseudo R2       =     0.0650
```

hasarts	Coef.	Std. Err.	z	P>|z|	[95% Conf. Interval]	
fem	-.266308	.1598617	-1.67	0.096	-.5796312	.0470153
mar	.3329119	.1823256	1.83	0.068	-.0244397	.6902635
kid5	-.2812119	.1118409	-2.51	0.012	-.500416	-.0620078
phd	.0100783	.0802174	0.13	0.900	-.147145	.1673016
mentord	.5429222	.0747143	7.27	0.000	.3964848	.6893595
_cons	-.1553251	.3050814	-0.51	0.611	-.7532736	.4426234

The advantage of this approach is that interpretation is simpler, but to take advantage of this simplicity you must make the strong assumption that successive categories of the ordinal independent variable are equally spaced. For example, it implies that an increase from no publications by the mentor to a few publications involves an increase of the same amount of productivity as an increase from a few to some, from some to many, and from many to lots of publications.

Accordingly, before treating an ordinal independent variable as if it were interval, you should test whether this leads to a loss of information about the association between the independent and dependent variable. A likelihood-ratio test can be computed by comparing the model with only `mentord` to the model that includes both the ordinal variable (`mentord`) and all but two of the indicator variables. In the example below, we add `some`, `many`, and `lots`, but including any three of the indicators leads to the same results. If the categories of the ordinal variables are equally spaced, the coefficients of the $J-2$ indicator variables should all be 0. For example,

```
. logit hasarts fem mar kid5 phd mentord some many lots, nolog
 (output omitted )
. estimates store fmodel
. logit hasarts fem mar kid5 phd mentord, nolog
 (output omitted )
. lrtest fmodel
likelihood-ratio test                            LR chi2(3)  =       1.07
(Assumption: . nested in fmodel)                 Prob > chi2 =     0.7845
```

We conclude that the indicator variables do not add additional information to the model ($LRX^2 = 1.07, df = 3, p = .78$). If the test was significant, we would have evidence that the categories of `mentord` are not evenly spaced, and so we should not treat `mentord` as interval. A Wald test can also be computed, leading to the same conclusion:

```
. logit hasarts fem mar kid5 phd mentord some many lots, nolog
  (output omitted)

. test some many lots

 ( 1)   some = 0
 ( 2)   many = 0
 ( 3)   lots = 0

           chi2( 3) =     1.03
         Prob > chi2 =    0.7950
```

8.1.4   Discrete change for categorical independent variables

There are a few tricks that you must be aware of when computing discrete change for categorical independent variables. To show how this is done, we will compute the change in the probability of publishing for those with a mentor with few publications compared with a mentor with no publications. There are two ways to compute this discrete change. The first way is easier, but the second is more flexible.

Computing discrete change with prchange

The easy way is to use `prchange`, where we set all the indicator variables to 0:

```
. logit hasarts fem mar kid5 phd few some many lots, nolog
  (output omitted)

. prchange few, x(some=0 many=0 lots=0)

logit: Changes in Predicted Probabilities for hasarts

           min->max      0->1      -+1/2    -+sd/2  MargEfct
few        0.0957      0.0957     0.0962    0.0399   0.0965

             NoArts      Arts
Pr(y|x)    0.4920     0.5080

                 fem       mar       kid5       phd        few     some      many      lots
       x=    .460109   .662295   .495082   3.10311   .219672        0         0         0
    sd(x)=   .498679   .473186   .76488    .984249   .414251   .478501   .422839   .293489
```

We conclude that

> having a mentor with a few publications compared with none increases a scientist's probability of publishing by .10, holding all other variables at their mean.

Even though we say "holding all other variables at their mean", which is clear within the context of reporting substantive results, the key to getting the right answer from `prchange` is holding all of the indicator variables at 0, not at their mean. It does not make sense to change `few` from 0 to 1 when `some`, `many`, and `lots` are at their means.

Computing discrete change with prvalue

A second approach to computing discrete change is to use a pair of calls to `prvalue`. The advantage of this approach is that it works in situations where `prchange` does not. For example, how does the predicted probability change if we compare a mentor with a few publications to a mentor with some publications, holding all other variables constant? This involves computing probabilities as we move from `few=1` and `some=0`, to `few=0` and `some=1`. We cannot compute this with `prchange` because two variables are changing at the same time. Instead, we use two calls to `prvalue`:[2]

```
. quietly prvalue, x(few=1 some=0 many=0 lots=0) save
. prvalue, x(few=0 some=1 many=0 lots=0) dif

logit: Change in Predictions for  hasarts

                        Current      Saved  Difference
    Pr(y=Arts|x):        0.7125     0.5825      0.1300
    Pr(y=NoArts|x):      0.2875     0.4175     -0.1300

                    fem         mar        kid5         phd         few        some
   Current=   .46010929   .66229508   .49508197   3.1031093           0           1
     Saved=   .46010929   .66229508   .49508197   3.1031093           1           0
      Diff=           0           0           0           0          -1           1

                   many        lots
   Current=           0           0
     Saved=           0           0
      Diff=           0           0
```

Because we have used the `save` and `dif` options, the difference in the predicted probability (i.e., the discrete change) is reported. When we use the `save` and `dif` options, we usually add `quietly` to the first `prvalue` because all the information is listed by the second `prvalue`.

8.2 Interactions

Interaction terms are commonly included in regression models when the effect of an independent variable is thought to vary, depending on the value of another independent variable. To illustrate how interactions are used, we extend the example from Chapter 5, where the dependent variable is a respondent's level of agreement that a working mother can establish as warm a relationship with her children as mothers who do not work.

It is possible that the effect of education on attitudes towards working mothers varies by gender. To allow this possibility, we add the interaction of education (`ed`) and gender (`male`) by adding the variable `maleXed = malex*ed`. In fitting this model, we find that

```
. use http://www.stata-press.com/data/lfr/ordwarm2, clear
(77 & 89 General Social Survey)
. gen maleXed = male*ed
```

---

[2]Alternatively, we could have refitted the model adding **none** and excluding either **few** or **some**, and then used `prchange`.

```
. ologit warm age prst yr89 white male ed maleXed, nolog
Ordered logit estimates                          Number of obs   =       2293
                                                 LR chi2(7)      =     305.30
                                                 Prob > chi2     =     0.0000
Log likelihood = -2843.1198                      Pseudo R2       =     0.0510
```

warm	Coef.	Std. Err.	z	P>|z|	[95% Conf. Interval]	
age	-.0212523	.0024775	-8.58	0.000	-.0261082	-.0163965
prst	.0052597	.0033198	1.58	0.113	-.001247	.0117664
yr89	.5238686	.0799287	6.55	0.000	.3672111	.680526
white	-.3908743	.1184189	-3.30	0.001	-.622971	-.1587776
male	-.1505216	.3176105	-0.47	0.636	-.7730268	.4719836
ed	.0976341	.0226886	4.30	0.000	.0531651	.142103
maleXed	-.047534	.0251183	-1.89	0.058	-.0967649	.001697
_cut1	-2.107903	.3043008	(Ancillary parameters)			
_cut2	-.2761098	.2992857				
_cut3	1.621787	.3018749				

The interaction is marginally significant ($p = .06$) for a two-tailed Wald test. Alternatively, we can compute an LR test

```
. ologit warm age prst yr89 white male ed maleXed, nolog
  (output omitted)
. estimates store fmodel
. ologit warm age prst yr89 white male ed, nolog
  (output omitted)
. lrtest fmodel
likelihood-ratio test                            LR chi2(1)   =       3.59
(Assumption: . nested in fmodel)                 Prob > chi2  =     0.0583
```

which leads to the same conclusion.

8.2.1   Computing gender differences in predictions with interactions

What if we want to compute the difference between men and women in the predicted probabilities for the outcome categories? Gender differences are reflected in two ways in the model. First, we want to change `male` from 0 to 1 to indicate women versus men. If this was the only variable affected by changing the value of `male`, we could use `prchange`. But, when the value of `male` changes, this necessarily changes the value of `maleXed` (except in the case when ed is 0). For women, `maleXed` = male∗ed = 0∗ed = 0, while for men, `maleXed` = male ∗ ed = 1 ∗ ed = ed. Accordingly, we must examine the change in the outcome probabilities when two variables change, so `prvalue` must be used. We start by computing the predicted values for women, which requires fixing `male=0` and `maleXed=0`:

```
. prvalue, x(male=0 maleXed=0) rest(mean) save

ologit: Predictions for warm
    Pr(y=SD|x):         0.0816
    Pr(y=D|x):          0.2754
    Pr(y=A|x):          0.4304
    Pr(y=SA|x):         0.2126
            age        prst       yr89       white       male        ed     maleXed
    x=  44.935456  39.585259  .39860445   .8765809         0  12.218055         0
```

Next, we compute the predicted probability for men, where male=1 and maleXed equals the average value of education (because for men, maleXed=male*ed= 1*ed=ed). The value for maleXed can be obtained by computing the mean of ed:

```
. sum ed

    Variable |       Obs        Mean   Std. Dev.       Min        Max

          ed |      2293    12.21805    3.160827         0         20
. global meaned = r(mean)
```

summarize returns the mean to r(mean). The command global meaned = r(mean) assigns the mean of ed to the global macro meaned. In the prvalue command, we specify x(male=1 maleXed=$meaned), where $meaned tells Stata to substitute the value contained in the global macro:

```
. prvalue, x(male=1 maleXed=$meaned) dif

ologit: Change in Predictions for  warm
                    Current      Saved  Difference
    Pr(y=SD|x):      0.1559     0.0816      0.0743
    Pr(y=D|x):       0.3797     0.2754      0.1044
    Pr(y=A|x):       0.3494     0.4304     -0.0811
    Pr(y=SA|x):      0.1150     0.2126     -0.0976
                 age        prst       yr89       white    male         ed    maleXed
Current=  44.935456  39.585259  .39860445   .8765809       1  12.218055  12.218055
  Saved=  44.935456  39.585259  .39860445   .8765809       0  12.218055          0
   Diff=          0          0          0          0       1          0  12.218055
```

Warning The mean of maleXed does not equal the mean of ed. That is why we could not use the option x(male=1 maleXed=mean) and instead had to compute the mean with summarize.

While the trick of using maleXed=$meaned may seem like a lot of trouble to avoid having to type maleXed=12.21805, it can help you avoid errors, and in some cases (illustrated below), it saves a lot of time.

Substantively, we conclude that the probability of strongly agreeing that working mothers can be good mothers is .10 higher for woman than men, taking the interaction with education into account and holding other variables constant at their means. The probability of strongly disagreeing is .07 higher for men than women.

8.2.2   Computing gender differences in discrete change with interactions

We might also be interested in how the predicted outcomes are affected by a change in education from having a high school diploma (12 years of education) to having a college degree (16 years). The interaction term suggests that the effect of education varies by gender, so we must look at the discrete change separately for men and women. Again, repeated calls to `prvalue` using the `save` and `dif` options allow us to do this. For women, we hold both `male` and `maleXed` to 0 and allow `ed` to vary. For men, we hold `male` to 1 and allow both `ed` and `maleXed` to vary. For women, we find that

```
. quietly prvalue, x(male=0 maleXed=0 ed=12) rest(mean) save
. prvalue, x(male=0 maleXed=0 ed=16) rest(mean) dif

ologit: Change in Predictions for  warm
                     Current    Saved  Difference
     Pr(y=SD|x):      0.0579   0.0833    -0.0254
     Pr(y=D|x):       0.2194   0.2786    -0.0592
     Pr(y=A|x):       0.4418   0.4291     0.0127
     Pr(y=SA|x):      0.2809   0.2090     0.0718

                   age        prst       yr89      white     male       ed
   Current=  44.935456  39.585259  .39860445  .8765809        0       16
     Saved=  44.935456  39.585259  .39860445  .8765809        0       12
      Diff=          0          0          0          0        0        4

               maleXed
   Current=          0
     Saved=          0
      Diff=          0
```

For men,

```
. quietly prvalue, x(male=1 maleXed=12 ed=12) rest(mean) save
. prvalue, x(male=1 maleXed=16 ed=16) rest(mean) dif

ologit: Change in Predictions for  warm
                     Current    Saved  Difference
     Pr(y=SD|x):      0.1326   0.1574    -0.0248
     Pr(y=D|x):       0.3558   0.3810    -0.0252
     Pr(y=A|x):       0.3759   0.3477     0.0282
     Pr(y=SA|x):      0.1357   0.1139     0.0218

                   age        prst       yr89      white     male       ed
   Current=  44.935456  39.585259  .39860445  .8765809        1       16
     Saved=  44.935456  39.585259  .39860445  .8765809        1       12
      Diff=          0          0          0          0        0        4

               maleXed
   Current=         16
     Saved=         12
      Diff=          4
```

The largest difference in the discrete change between the sexes is for the probability of answering "strongly agree". For both men and women, an increase in education from 12 years to 16 years increases the probability of strong agreement, but the increase is .07 for women and only .02 for men.

8.3 Nonlinear nonlinear models

The models that we consider in this book are nonlinear models in that the effect of a change in an independent variable on the predicted probability or predicted count depends on the values of all of the independent variables. However, the right-hand side of the model includes a linear combination of variables just like the linear regression model. For example,

$$\text{Linear regression:} \qquad y = \beta_0 + \beta_1 x_1 + \beta_2 x_2 + \varepsilon$$

$$\text{Binary logit:} \qquad \Pr\left(y = 1 \mid \mathbf{x}\right) = \frac{\exp\left(\beta_0 + \beta_1 x_1 + \beta_2 x_2 + \varepsilon\right)}{1 + \exp\left(\beta_0 + \beta_1 x_1 + \beta_2 x_2 + \varepsilon\right)}$$

In the terminology of the generalized linear model, we would say that both models have the same *linear predictor*: $\beta_0 + \beta_1 x_1 + \beta_2 x_2$. In the linear regression model, this leads to predictions that are linear surfaces. For example, with one independent variable the predictions are a line, with two a plane, and so on. In the binary logit model, the prediction is a curved surface, as illustrated in Chapter 4.

8.3.1 Adding nonlinearities to linear predictors

Nonlinearities in the LRM can be introduced by adding transformations on the right-hand side. For example, in the model

$$y = \alpha + \beta_1 x + \beta_2 x^2 + \varepsilon$$

we include x and x^2 to allow predictions that are a quadratic form. For example, if the fitted model is $\widehat{y} = 1 + -.1x + .1x^2$, the plot is far from linear:

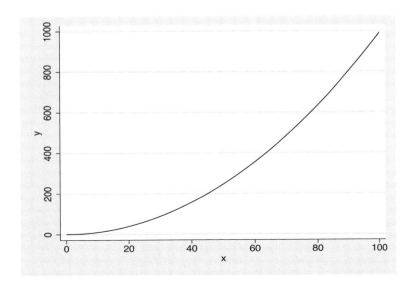

In the same fashion, nonlinearities can be added to the right-hand side of the models for categorical outcomes that we have been considering. What may seem odd is that adding nonlinearities to a nonlinear model can sometimes make the predictions *more* linear.

8.3.2   Discrete change in nonlinear nonlinear models

In the model of labor force participation from Chapter 4, we included a woman's age as an independent variable. Often when age is used in a model, terms for both the age and age-squared are included to allow for diminishing (or increasing) effects of an additional year of age. First, we fit the model *without* age squared and compute the effect of a change in age from 30 to 50 for an average respondent:

```
. use http://www.stata-press.com/data/lfr/binlfp2, clear
(Data from 1976 PSID-T Mroz)

. logit lfp k5 k618 wc hc lwg inc age, nolog
  (output omitted)

. prchange age, x(age=30) delta(20) uncentered
logit: Changes in Predicted Probabilities for lfp

(Note: delta = 20)

        min->max      0->1     +delta      +sd  MargEfct
age     -0.4372   -0.0030    -0.2894   -0.1062   -0.0118

           NotInLF      inLF
Pr(y|x)     0.2494    0.7506

             k5      k618        wc        hc       lwg       inc       age
    x=  .237716   1.35325   .281541   .391766   1.09711   20.129        30
 sd(x)=  .523959   1.31987   .450049   .488469   .587556   11.6348   8.07257
```

Notice that we have taken advantage of the `delta()` and `uncentered` options (see Chapter 3). We find that the predicted probability of a woman working decreases by .29 as age increases from 30 to 50, with all other variables at the mean. Now we add age-squared to the model:

```
. gen age2 = age*age

. logit lfp k5 k618 wc hc lwg inc age age2, nolog
Logit estimates                          Number of obs   =        753
                                         LR chi2(8)      =     125.67
                                         Prob > chi2     =     0.0000
Log likelihood = -452.03836              Pseudo R2       =     0.1220
```

lfp	Coef.	Std. Err.	z	P>|z|	[95% Conf. Interval]	
k5	-1.411597	.2001829	-7.05	0.000	-1.803948	-1.019246
k618	-.0815087	.0696247	-1.17	0.242	-.2179706	.0549531
wc	.8098626	.2299065	3.52	0.000	.3592542	1.260471
hc	.1340998	.207023	0.65	0.517	-.2716579	.5398575
lwg	.5925741	.1507807	3.93	0.000	.2970495	.8880988
inc	-.0355964	.0083188	-4.28	0.000	-.0519009	-.0192919
age	.0659135	.1188199	0.55	0.579	-.1669693	.2987962
age2	-.0014784	.0013584	-1.09	0.276	-.0041408	.001184
_cons	.511489	2.527194	0.20	0.840	-4.44172	5.464698

To test for the joint significance of age and age2, we use a likelihood-ratio test:

```
. quietly logit lfp k5 k618 wc hc lwg inc age age2, nolog
. estimates store fmodel
. quietly logit lfp k5 k618 wc hc lwg inc, nolog
. lrtest fmodel
likelihood-ratio test                               LR chi2(2)  =     26.79
(Assumption: . nested in fmodel)                    Prob > chi2 =    0.0000
```

We can no longer use prchange to compute the discrete change because we need to change two variables at the same time. Once again, we use a pair of prvalue commands, where we change age from 30 to 50 and change age2 from 30^2 (=900) to 50^2 (=2500). First, we compute the prediction with age at 30:

```
. global age30 = 30
. global age30sq = $age30*$age30
. quietly prvalue, x(age=$age30 age2=$age30sq) rest(mean) save
```

Then, we let age equal 50 and compute the difference:

```
. global age50 = 50
. global age50sq = $age50*$age50
. prvalue, x(age=$age50 age2=$age50sq) rest(mean) dif

logit: Change in Predictions for  lfp
                    Current      Saved  Difference
    Pr(y=inLF|x):    0.4699     0.7164     -0.2465
    Pr(y=NotInLF|x): 0.5301     0.2836      0.2465

               k5        k618         wc          hc         lwg        inc
Current=  .2377158   1.3532537   .2815405   .39176627   1.0971148   20.128965
  Saved=  .2377158   1.3532537   .2815405   .39176627   1.0971148   20.128965
   Diff=         0           0          0           0           0           0

              age        age2
Current=       50        2500
  Saved=       30         900
   Diff=       20        1600
```

We conclude that

> an increase in age from 30 to 50 years decreases the probability of being in the labor force by .25, holding other variables at their mean.

By adding the squared term, we have decreased our estimate of the change. While in this case the difference is not large, the example illustrates the general point of how to add nonlinearities to the model.

8.4   Using praccum and forvalues to plot predictions

In prior chapters, we used prgen to generate predicted probabilities over the range of one variable, while holding other variables constant. While prgen is a relatively simple

way of generating predictions for graphs, it can be used only when the specification of
the right-hand side of the model is straightforward. When interactions or polynomials
are included in the model, graphing the effects of a change in an independent variable
often requires computing changes in the probabilities as more than one of the variables in
the model changes (e.g., `age` and `age2`). We created `praccum` to handle such situations.
The user calculates each of the points to be plotted through a series of calls to `prvalue`.
Executing `praccum` immediately after `prvalue` accumulates these predictions.

The first time `praccum` is run, the predicted values are saved in a new matrix. Each
subsequent call to `praccum` adds new predictions to this matrix. When all of the calls
to `prvalue` have been completed, the accumulated predictions in the matrix can be
added as new variables to the dataset in an arrangement ideal for plotting, just like
with `prgen`. The syntax of `praccum` is

praccum , { <u>us</u>ing(*matrixname*) | <u>s</u>aving(*matrixname*) } [xis(*value*)
    <u>gen</u>erate(*prefix*)]

where either `using()` or `saving()` is required.

Options

using(*matrixname*) specifies the name of the matrix where the predictions from the
    previous call to `prvalue` should be added. An error is generated if the matrix does
    not have the correct number of columns. This can happen if you try to append values
    to a matrix generated from calls to `praccum` based on a different model.   Matrix
    matrixname will be created if it does not already exist.

saving(*matrixname*) specifies that a new matrix should be generated to contain the
    predicted values from the previous call to `prvalue`. You only use this option when
    you initially create the matrix. After the matrix is created, you add to it with
    `using()`. The difference between `saving()` and `using()` is that `saving()` will
    overwrite *matrixname* if it exists, while `using()` will append results to it.

xis(*value*) indicates the value of the x variable associated with the predicted values
    that are accumulated. For example, this could be the value of age if you wish to plot
    changes in predicted values as age changes. You do *not* need to include the values of
    variables created as transformations of this variable. To continue the example, you
    would not include the value of age squared.

generate(*prefix*) indicates that new variables are to be added to the current dataset.
    These variables begin with *prefix* and contain the values accumulated in the matrix
    in prior calls to `praccum`.

The generality of `praccum` requires it to be more complicated to use than `prgen`.

8.4.1   Example using age and age-squared

To illustrate the command, we use `praccum` to plot the effects of age on labor force participation for a model in which both age and age-squared are included. First, we compute the predictions from the model without `age2`:

```
. use http://www.stata-press.com/data/lfr/binlfp2,clear
(Data from 1976 PSID-T Mroz)
. quietly logit lfp k5 k618 age wc hc lwg inc
. prgen age, from(20) to(60) gen(prage) ncases(9)
logit: Predicted values as age varies from 20 to 60.
           k5       k618        age         wc         hc        lwg        inc
x=   .2377158  1.3532537  42.537849   .2815405  .39176627  1.0971148  20.128965
. label var pragep1 "Pr(lpf | age)"
```

This is the same thing we did using `prgen` in earlier chapters. Next, we fit the model with `age2` added:

```
. logit lfp k5 k618 age age2 wc hc lwg inc
  (output omitted )
```

To compute the predictions from this model, we use a series of calls to `prvalue`. For these predictions, we let `age` change by five-year increments from 20 to 60 and `age2` increase from 20^2 (= 400) to 60^2 (= 3600). In the first call of `praccum`, we use the `saving()` option to declare that `mat_age` is the matrix that will hold the results. The `xis()` option is required because it specifies the value for the x-axis of the graph that will plot these probabilities:

```
. quietly prvalue, x(age=20 age2=400) rest(mean)
. praccum, saving(mat_age) xis(20)
```

We execute `prvalue` quietly to suppress the output, because we are only generating these predictions in order to save them with `praccum`. The next set of calls adds new predictions to `mat_age`, as indicated by the option `using()`:

```
. quietly prvalue, x(age=25 age2=625) rest(mean)
. praccum, using(mat_age) xis(25)
. quietly prvalue, x(age=30 age2=900) rest(mean)
. praccum, using(mat_age) xis(30)
  (and so on )
. quietly prvalue, x(age=55 age2=3025) rest(mean)
. praccum, using(mat_age) xis(55)
```

The last call includes not only the `using()` option but also `gen()`, which tells `praccum` to save the predicted values from the matrix to variables that begin with the specified root, in this case `agesq`:

```
. quietly prvalue, x(age=60 age2=3600) rest(mean)
```

```
. praccum, using(mat_age) xis(60) gen(agesq)
New variables created by praccum:
        Variable │    Obs        Mean    Std. Dev.        Min        Max
─────────────────┼──────────────────────────────────────────────────────
          agesqx │      9          40    13.69306         20         60
         agesqp0 │      9    .4282142    .1752595   .2676314   .7479599
         agesqp1 │      9    .5717858    .1752595   .2520402   .7323686
```

To understand what has been done, it helps to look at the new variables that were
created:

```
. list agesqx agesqp0 agesqp1 in 1/10

     ┌─────────────────────────────────┐
     │ agesqx    agesqp0    agesqp1     │
     │                                 │
  1. │     20   .2676314   .7323686     │
  2. │     25   .2682353   .7317647     │
  3. │     30   .2836163   .7163837     │
  4. │     35   .3152536   .6847464     │
  5. │     40   .3656723   .6343277     │
     │                                 │
  6. │     45   .4373158   .5626842     │
  7. │     50   .5301194   .4698806     │
  8. │     55   .6381241   .3618759     │
  9. │     60   .7479599   .2520402     │
 10. │      .          .          .     │
     └─────────────────────────────────┘
```

The tenth observation is all missing values because we only made nine calls to praccum.
Each value of agesqx reproduces the value specified in xis(). The values of agesqp0
and agesqp1 are the probabilities of $y = 0$ and $y = 1$ that were computed by prvalue.
We see that the probability of observing a 1, that is, being in the labor force, was .73
the first time we executed prvalue with age at 20; the probability was .25 the last time
we executed prvalue with age at 60. Now that these predictions have been added to
the dataset, we can use graph to show how the predicted probability of being in the
labor force changes with age:

```
. label var agesqp1 "Pr(lpf | age,age2)"
. label var agesqx  "Age"
. graph twoway connected pragep1 agesqp1 agesqx,
      msymbol(Sh Dh) xlabel(20(5)60)
      ytitle("Pr(Being in the Labor Force)")  ylabel(0(.2)1)
      ysize(2.7051) xsize(4.0421)
```

We are also plotting pragep1, which was computed earlier in this section using prgen.
The graph command leads to the following plot:

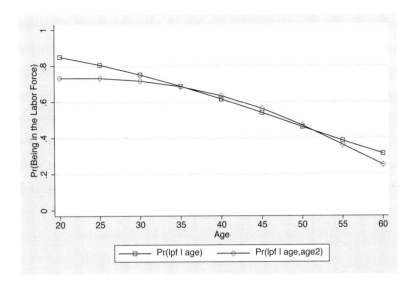

The graph shows that, as age increases from 20 to 60, a woman's probability of being in the labor force declines. In the model with only age, the decline is from .85 to .31, while in the model with age-squared, the decrease is from .73 to .25. Overall, the changes are smaller during younger years and larger after age 50.

8.4.2   Using forvalues with praccum

The use of praccum is often greatly simplified by Stata's forvalues command (which was introduced in Stata 7). The forvalues command allows you to repeat a set of commands where the only thing that you vary between successive repetitions is the value of some key number. As a trivial example, we can use forvalues to have Stata count from 0 to 100 by fives. Enter the following three lines either interactively or in a do-file:

```
forvalues count = 0(5)100 {
    display 'count'
}
```

In the forvalues statement, count is the name of a local macro that will contain the successive values of interest (see Chapter 2 if you are unfamiliar with local macros). The combination 0(5)100 indicates that Stata should begin by setting the value of count at 0 and should increase its value by 5 with each repetition until it reaches 100. The { }s enclose the commands that will be repeated for each value of count. In this case, all we want to do is to display the value of count. This is done with the command display 'count'. To indicate that count is a local macro, we use the pair of single quote marks (i.e., 'count'). The output produced is

```
0
5
10
  (and so on)
95
100
```

In our earlier example, we graphed the effect of age as it increased from 20 to 60 by five-year increments. If we specify `forvalues count = 20(5)60`, Stata will repeatedly execute the code we enclose in brackets with the value of `count` updated from 20 to 60 by increments of 5. The following lines reproduce the results we obtained earlier:

```
capture matrix drop mage
forvalues count = 20(5)60 {
    local countsq = 'count'^2
    prvalue, x(age='count' age2='countsq') rest(mean) brief
    praccum, using(mage) xis('count')
}
praccum, using(mage) gen(agsq)
```

The command `capture matrix drop mage` at the beginning will drop the matrix `mage` if it exists, but the do-file will not halt with an error if the matrix does not exist. Within the `forvalues` loop, `count` is set to the appropriate value of `age`, and we use the `local` command to create the local macro `countsq` that contains the square of `count`. After all the predictions have been computed and accumulated to matrix `mage`, we make a last call to `praccum` in which we use the `generate()` option to specify the stem of names of the new variables to be generated.

8.4.3   Using praccum for graphing a transformed variable

`praccum` can also be used when an independent variable is a transformation of the original variable. For example, you might want to include the natural log of `age` as independent variable rather than `age`. Such a model can be easily fitted:

```
. gen ageln = ln(age)
. logit lfp k5 k618 ageln wc hc lwg inc
  (output omitted)
```

As in the last example, we use `forvalues` to execute a series of calls to `prvalue` and `praccum` to generate predictions:

```
capture matrix drop mat_ln
forvalues count = 20(5)60 {
    local countln = ln('count')
    prvalue, x(ageln='countln') rest(mean) brief
    praccum, using(mat_ln) xis('count')
}
praccum, using(mat_ln) gen(ageln)
```

We use a local to compute the log of `age`, the value of which is passed to `prvalue` with the option `x(ageln='countln')`. But, in `praccum` we specify `xis('count')` not

xis(`countln'). This is because we want to plot the probability against age in its
original units. The saved values can then be plotted:

```
. label var agelnp1 "Pr(lpf | log of age)"
. graph twoway connected pragep1 agesqp1 agelnp1 agesqx,
     xlabel(20(5)60) msymbol(Sh Dh Th)
     ytitle("Pr(Being in the Labor Force)")
     ylabel(0(.2)1) ysize(2.7051) xsize(4.0421)
```

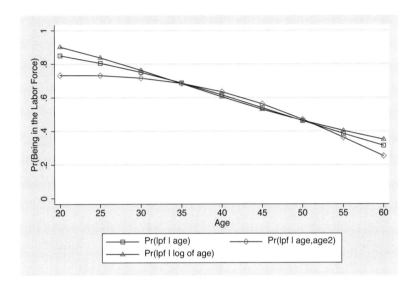

8.4.4   Using praccum to graph interactions

Earlier in this chapter, we examined an ordinal regression model of support for working
mothers that included an interaction between a respondent's sex and education. Another
way to examine the effects of the interaction is to plot the effect of education on the
predicted probability of strongly agreeing for men and women separately. First, we fit
the model:

```
. use http://www.stata-press.com/data/lfr/ordwarm2, clear
(77 & 89 General Social Survey)
. gen maleXed = male*ed
. ologit warm age prst yr89 white male ed maleXed
  (output omitted)
```

Next, we compute the predicted values of strongly agreeing as education increases for
women who are average on all other characteristics. This is done using forvalues to
make a series of calls to prvalue and praccum. For women, maleXed is always 0 because
male is 0:

```
forvalues count = 8(2)20 {
    quietly prvalue, x(male=0 ed='count' maleXed=0) rest(mean)
    praccum, using(mat_f) xis('count')
}
praccum, using(mat_f) gen(pfem)
```

In the successive calls to `prvalue`, only the variable `ed` is changing. Accordingly, we could have used `prgen`. For the men, however, we must use `praccum` because both `ed` and `maleXed` change together:

```
. forvalues count = 8(2)20 {
 2.      quietly prvalue, x(male=0 ed='count' maleXed=0) rest(mean)
 3.      praccum, using(mat_f) xis('count')
 4. }
. praccum, using(mat_f) gen(pfem)
New variables created by praccum:
```

Variable	Obs	Mean	Std. Dev.	Min	Max
pfemx	7	14	4.320494	8	20
pfemp1	7	.0737801	.0283039	.0399257	.1183256
pfemp2	7	.2501559	.0622868	.1662394	.3376414
pfemp3	7	.425898	.0176338	.3923302	.4418006
pfemp4	7	.250166	.0775179	.1517029	.3659343
pfems1	7	.0737801	.0283039	.0399257	.1183256
pfems2	7	.323936	.0904461	.2061651	.4559669
pfems3	7	.749834	.0775179	.6340657	.8482972
pfems4	7	1	0	1	1

Years of education, as specified with `xis()`, are stored in `pfemx` and `pmalx`. These variables are identical because we used the same levels for both men and women. The probabilities for women are contained in the variables `pfemp`k, where k is the category value; for models for ordered or count data, the variables `pfems`k store the cumulative probabilities $\Pr(y \leq k)$. The corresponding predictions for men are contained in `pmalp`k and `pmals`k. All that remains is to clean up the variable labels and plot the predictions:

```
. label var pfemp4 "Pr(SA | female)"
. label var pmalp4 "Pr(SA | male)"
. label var pfemx "Education in Years"
. graph twoway connected pfemp4 pmalp4 pfemx,
    msymbol(Sh Dh) xlabel(8(2)20) ylabel(0(.1).4)
    ytitle("Pr(Strongly Agreeing)")
    ysize(2.7051) xsize(4.0421)
```

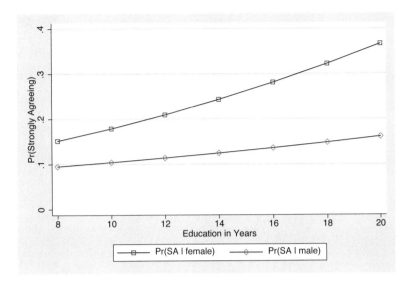

For all levels of education, women are more likely than men to strongly agree that working mothers can be good mothers, holding other variables to their mean. This difference between men and women is much larger at higher levels of education than at lower levels.

8.5 Extending SPost to other estimation commands

The commands in SPost only work with some of the many estimation commands available in Stata. If you try to use our commands after fitting other types of models, you will be told that the SPost command does not work for the last model fitted. Over the past year as we developed these commands, we have received numerous inquiries about whether we can modify SPost to work with additional estimation commands. While we would like to accommodate such requests, extensions are likely to be made mainly to estimation commands that we are using in our own work. There are two reasons for this. First, our time is limited. Second, we want to be sure that we fully understand the specifics of each model before we incorporate it into SPost. Still, users who know how to program in Stata are welcome to extend our programs to work with other models. Keep in mind, however, that we can only provide limited support. While we have attempted to write each command to make it as simple as possible to expand, some of the programs are complex, and you will need to be adept at programming in Stata.[3]

Here are some brief points that may be useful for a programmer wishing to modify our commands. First, our commands make use of ancillary programs that we have also written, all of which begin with _pe (e.g., _pebase). As will be apparent as you trace through the logic of one of our ado-files, extending a command to a new model

---

[3]StataCorp offers both introductory and advanced NetCourses in programming; more information on this can be obtained from *http://www.stata.com*.

might require modifications to these ancillary programs as well. As the _pe*.ado files
are used by many different commands, be careful that you do not make changes that
break other commands. Second, our programs use information returned in e() by the
estimation command. Some user-written estimation commands, especially older ones,
do not return the appropriate information in e(), and extending programs to work after
these estimation commands will be extremely difficult.

8.6   Using Stata more efficiently

Our introduction to Stata in Chapter 2 focused on the basics. But, as you use Stata,
you will discover various tricks that make your use of Stata more enjoyable and efficient.
While what constitutes a "good trick" depends on the needs and skills of the particular
users, in this section we describe some things that we have found useful.

8.6.1   profile.do

When Stata is launched, it looks for a do-file called profile.do in the directories listed
when you type sysdir.[4] If profile.do is found in one of these directories, Stata runs
it. Accordingly, you can customize Stata by including commands in profile.do. While
you should consult the *Getting Started with Stata* manual for full details or enter the
command help profile, the following examples show you some things that we find
useful. We have added detailed comments after the //s. The comments do not need to
be included in profile.do.

```
//  In Stata, all data is kept in memory. If you get memory errors when
//  loading a dataset or while fitting a model, you need more memory.
//  While you can change the amount of memory from the Command window,
//  we find it easier to set it here. Type -help memory- for details.

set memory 30m

//  Many programs in official Stata and many of our commands use matrices.
//  Some of our commands, such as -prchange-, use a lot of memory. So, we
//  suggest setting the amount of space for matrices to the largest value
//  allowed. Type -help matsize- for details.
//  The limit is 11,000 for Stata SE and 800 for Intercooled

set matsize 800

//  Starting with Stata 7, output in log files can be written either as text
//  (as with earlier versions of Stata) or in SMCL. We find it
//  easier to save logs as text since they can be more easily printed, copied
//  to a word processor, and so on. Type  -help log- for details.
```

---

[4]The preferred place for the file is in your default data directory (e.g., c:\data).

```
set logtype text

// You can assign commands to function keys F2 through F9. After assigning
// a text string to a key, when you press that key, the string is
// inserted into the Command window.

global F8 "set trace on"
global F9 "set trace off"

// You can tell Stata what you want your default working directory
// to be.

cd d:\statastart

// You can also add notes to yourself. Here we post a reminder that
// the command -spost- will change the working directory to the directory
// where we have the files for this book.

noisily di "spost == cd d:\spost\examples"
```

In Stata 8, much of the work that used to be done with `profile.do` files can also be done using the `permanently` option to the `set` command. Specifying this option means that the new setting will remain on subsequent occasions when Stata is launched. The `set memory`, `set matsize`, and `set logtype` commands can be invoked using the `permanently` option.

8.6.2   Changing screen fonts and window preferences

In Windows, the default font for the Results window works well on a VGA monitor with 640 by 480 resolution. But, with higher resolution monitors, we prefer a larger font. To change the font, right-click in the Results window. Select the Fonts option and choose a font you like. You do not need to select one of the fonts that are named "Stata ..." as any fixed-width font will work.

You can also change the size and position of the windows using the usual methods of clicking and dragging. After the font is selected and any new placement of windows is done, you can save your new options to be the defaults with the Preference menu and the Save Windowing Preferences option. Your windowing preferences are automatically saved when you exit Stata.

8.6.3   Using ado-files for changing directories

One of the things we like best about Stata is that you can create your own commands using ado-files. These commands work just like the commands that are part of official Stata, and indeed many commands in Stata are written as ado-files. If you are like us, at any one time you are working on several different projects. We like to keep each project in a different directory. For example, `d:\nas` includes research for the National Academy of Sciences, `d:\kinsey` is a project associated with the Kinsey Institute, and `d:\spost\examples` is (you guessed it) for this book. While you can change to these directories with the `cd` command, one of us keeps forgetting the names of directories. So, he writes a simple ado-file

```
program define spost
    cd d:\spost\examples
end
```

and saves this in his `PERSONAL` directory as `spost.ado`. Type `sysdir` to see what directory is assigned as the `PERSONAL` directory. Then, whenever he types `spost`, his working directory is immediately changed:

```
. spost
d:\spost\examples
```

8.6.4   me.hlp file

Help files in Stata are plain text or SMCL files that end with the `.hlp` extension. When you type `help` *command*, Stata searches in the same directories used for ado-files until it finds a file called `command.hlp`. We have a file called `me.hlp` that contains information on things we often use but seldom remember. For example,

```
.-
help for ^me^
.-

Reset everything              ^clear^
----------------              ^discard^

List installed packages       ^ado dir^
-----------------------

Axes options                  ^x/ylabel()^
------------                  ^x/yline()^

.-
Author: Scott Long
```

This file is saved in your `PERSONAL` directory; typing `sysdir` will tell you what your `PERSONAL` directory is. Then, whenever we are stuck and want to recall this information, we just need to type `help me`, and it is displayed on our screen.

8.6.5   Scrolling in the Results window in Windows

After you run a command whose output scrolls off the Results window, you will notice that a scroll bar appears on the right side of the window. You can use the scroll bar to scroll through results that are no longer in the Stata Results window. While Stata does not allow you to do this with a keystroke, you can use the scroll wheel found on some mice. We find this very convenient.

8.7   Conclusions

Our goal in writing this book was to make it routine to carry out the complex calculations necessary for the full interpretation of regression models for categorical outcomes. While we have gone to great lengths to check the accuracy of our commands and to verify that our instructions are correct, it is possible that there are still some "bugs" in our programs. If you have a problem, here is what we suggest:

1. Make sure that you have the most recent version of the Stata executable and ado-file (select Help→Official Updates from the menus) and the most recent versions of SPost (while online, type `net search spostado`). This is the most common solution to problems people send us.

2. Make sure that you do not have another command from someone else with the same name as one of our commands. If you do, one of them will not work and needs to be removed.

3. Check our FAQ (Frequently Asked Questions) page located at

   http://www.indiana.edu/~jslsoc/spost.htm

   You might find the answer there.

4. Make sure that you do not have anything but letters, numbers, and underscores in your value labels. Numerous programs in Stata get hung up when value labels include other symbols or other special characters.

5. Look at the sample files in the `spostst4` and `spostrm4` packages. These can be obtained when you are online and in Stata. Type `net search spost` and follow the directions you receive. It is sometimes easiest to figure out how to use a command by seeing how others use it.

Next, you can contact us with an e-mail to `spostsup@indiana.edu`. While we cannot guarantee that we can answer every question we get, we will try to help. The best way to have the problem solved is to send us a do-file and sample dataset in which the error occurs. It is very hard to figure out some problems by just seeing the log file. You may not want to send your original data due to size or confidentiality, so you might construct a smaller dataset with a subset of variables and cases.

A Syntax for SPost Commands

This appendix is a quick reference for all commands in SPost. Details on how each statistic is computed and interpreted are provided in the text. The following commands are described:

brant	Perform a Brant test of the parallel regression assumption for the ordered logit model.
fitstat	Fit statistics for regression models.
listcoef	List transformed regression coefficients with guidelines for interpretation.
mlogplot	Create odds ratio and discrete change plots for multinomial logit.
mlogtest	Test for the multinomial logit model.
mlogview	Access a dialog box for using mlogplot interactively.
praccum	Accumulate results from prvalue to construct plots.
prchange	Compute discrete and marginal change for regression models.
prcounts	Compute predicted probabilities and rates for count models.
prgen	Generate variables with predicted values over a range of a continuous independent variable for use in plots.
prtab	Construct tables of predicted values.
prvalue	Compute predicted values for specified values of the independent variables.

A.1 brant

brant Brant test of parallel regression assumption for the ordered logit model fitted by ologit. For a detailed discussion, see Section 5.6 of Chapter 5.

Syntax

brant [, detail]

Description

brant performs a Brant test (Brant 1990) of the parallel regression assumption (also called the proportional odds assumption) for the ordered logit model fitted by ologit. The test compares the slope coefficients from the $J-1$ binary logits implied by the ordered regression model. An omnibus test of all variables and tests for each independent variable are reported. To compute the tests, brant fits the $J-1$ binary logits defined by whether the outcome y is greater than or equal to j. *Note*: If there is perfect prediction in one of these binary logits, the Brant test cannot be computed.

Options

detail specifies that the coefficients for each of the binary logits should be presented.

Examples

brant requires the most recent model to have been fitted with ologit. For example,

```
. use http://www.stata-press.com/data/lfr/ordwarm2,clear
(77 & 89 General Social Survey)
. ologit warm yr89 male white age ed prst, nolog
  (output omitted)
. brant
Brant Test of Parallel Regression Assumption
```

Variable	chi2	p>chi2	df
All	49.18	0.000	12
yr89	13.01	0.001	2
male	22.24	0.000	2
white	1.27	0.531	2
age	7.38	0.025	2
ed	4.31	0.116	2
prst	4.33	0.115	2

```
A significant test statistic provides evidence that the parallel
regression assumption has been violated.
```

Here, we find that the omnibus test rejects the hypothesis of parallel regressions, while the tests for individual coefficients show that the assumption is violated only for the variables yr89, male, and age. To see the results from the binary logits used to compute the Brant test, we add the detail option:

```
. brant, detail
Estimated coefficients from j-1 binary regressions
                y>1         y>2         y>3
   yr89     .9647422    .56540626    .31907316
   male   -.30536425   -.69054232   -1.0837888
  white   -.55265759   -.31427081   -.39299842
    age    -.0164704   -.02533448   -.01859051
     ed    .10479624    .05285265    .05755466
   prst   -.00141118    .00953216    .00553043
  _cons    1.8584045    .73032873   -1.0245168

Brant Test of Parallel Regression Assumption
       Variable |      chi2   p>chi2    df
      ----------+---------------------------
            All |     49.18    0.000    12
      ----------+---------------------------
           yr89 |     13.01    0.001     2
           male |     22.24    0.000     2
          white |      1.27    0.531     2
            age |      7.38    0.025     2
             ed |      4.31    0.116     2
           prst |      4.33    0.115     2
      ----------+---------------------------
```

A significant test statistic provides evidence that the parallel
regression assumption has been violated.

Saved results

Scalars

r(chi2)	chi-squared statistic of omnibus test
r(df)	degrees of freedom of omnibus test
r(p)	*p*-value of omnibus test

Matrices

r(ivtests)	contains the test statistics in the first column and the associated *p*-values in the second column for the tests of each independent variable. The row labels are the names of the independent variables.

A.2   fitstat

fitstat   Compute scalar measures of fit for regression models. For a detailed discussion, see Section 3.4 in Chapter 3.

Syntax

fitstat [, <u>s</u>aving(*name*) <u>us</u>ing(*name*) <u>b</u>ic force save <u>d</u>if]

Description

fitstat is a post-estimation command that computes a variety of measures of fit for the following regression models: clogit, cloglog, cnreg, intreg, logistic, logit, `mlogit`, `nbreg`, `ocratio`, `ologit`, `oprobit`, `poisson`, `probit`, `regress`, `zinb`, and `zip`. With the `saving` and `using` options, it can be used to compare fit from two models. For all models, `fitstat` reports the log likelihoods of the full and intercept-only models, the deviance (D), the likelihood-ratio chi-squared (G^2), Akaike's Information Criterion (AIC), AIC*N, the Bayesian Information Criterion (BIC), and BIC$'$. For OLS regression, `fitstat` reports R^2 and the adjusted R^2. For other models, `fitstat` reports McFadden's R^2, McFadden's adjusted R^2, the maximum likelihood R^2, and Cragg & Uhler's R^2. For categorical outcomes, `fitstat` reports the regular and adjusted count R^2. For ordered, binary, and censored outcomes, McKelvey and Zavoina's R^2 is computed. In addition, Efron's R^2 is computed for binary outcomes. Not all measures are provided for models fitted with `pweights` or `iweights`.

Options

saving(*name*) saves the computed measures in a matrix for subsequent comparisons. *name* cannot be longer than 4 characters.

using(*name*) compares the fit measures for the current model with those of the model saved as *name*.

bic presents only information measures.

force compares two models, even if the number of observations or the type of model differs for the models.

save and dif are equivalents of saving(0) and using(0) and do not require the user to provide model names.

Examples

For a binary logit,

```
. use http://www.stata-press.com/data/lfr/binlfp2, clear
(Data from 1976 PSID-T Mroz)

. logit lfp k5 k618 age wc hc lwg inc, nolog
 (output omitted)

. fitstat
Measures of Fit for logit of lfp
Log-Lik Intercept Only:      -514.873    Log-Lik Full Model:        -452.633
D(745):                       905.266    LR(7):                      124.480
                                         Prob > LR:                    0.000
McFadden's R2:                  0.121    McFadden's Adj R2:            0.105
Maximum Likelihood R2:          0.152    Cragg & Uhler's R2:           0.204
McKelvey and Zavoina's R2:      0.217    Efron's R2:                   0.155
Variance of y*:                 4.203    Variance of error:            3.290
Count R2:                       0.693    Adj Count R2:                 0.289
AIC:                            1.223    AIC*n:                      921.266
BIC:                        -4029.663    BIC':                       -78.112
```

To compare the fit statistics for two models:

```
. logit lfp k5 k618 age wc hc lwg inc, nolog
  (output omitted)

. quietly fitstat, saving(mod1)

. gen age2 = age*age

. logit lfp k5 age age2 wc inc, nolog
  (output omitted)

. fitstat, using(mod1)

Measures of Fit for logit of lfp
                              Current          Saved      Difference
Model:                          logit          logit
N:                                753            753               0
Log-Lik Intercept Only:      -514.873       -514.873           0.000
Log-Lik Full Model:          -461.653       -452.633          -9.020
D:                       923.306(747)   905.266(745)       18.040(2)
LR:                       106.441(5)     124.480(7)         18.040(2)
Prob > LR:                      0.000          0.000           0.000
McFadden's R2:                  0.103          0.121          -0.018
McFadden's Adj R2:              0.092          0.105          -0.014
Maximum Likelihood R2:          0.132          0.152          -0.021
Cragg & Uhler's R2:             0.177          0.204          -0.028
McKelvey and Zavoina's R2:      0.182          0.217          -0.035
Efron's R2:                     0.135          0.155          -0.020
Variance of y*:                 4.023          4.203          -0.180
Variance of error:              3.290          3.290           0.000
Count R2:                       0.677          0.693          -0.016
Adj Count R2:                   0.252          0.289          -0.037
AIC:                            1.242          1.223           0.019
AIC*n:                        935.306        921.266          14.040
BIC:                        -4024.871      -4029.663           4.791
BIC':                         -73.321        -78.112           4.791
Difference of    4.791 in BIC' provides positive support for saved model.

Note: p-value for difference in LR is only valid if models are nested.
```

Note that the **save** and **dif** options could have been used instead of **saving**(mod1) and **using**(mod1).

(Continued on next page)

Saved results

The following results are saved. If a statistic is not appropriate for a given model, a missing value is returned.

r(aic)	AIC	r(n_rhs)	number of right-hand-side
r(aic_n)	AIC×N		variables
r(bic)	BIC	r(r2)	R^2 for linear regression
r(bic_p)	BIC$'$		model
r(dev)	deviance	r(r2_adj)	adjusted R^2 for linear
r(dev_df)	degrees of freedom for deviance		regression model
r(ll)	log likelihood for full model	r(r2_ct)	count R^2
r(ll_0)	log likelihood for model with	r(r2_ctadj)	adjusted count R^2
	only intercept	r(r2_cu)	Cragg & Uhler's R^2
r(lrx2)	likelihood ratio chi-squared	r(r2_ef)	Efron's R^2
r(lrx2_df)	degrees of freedom for likelihood	r(r2_mz)	McKelvey and Zavoina's R^2
	ratio chi-squared	r(r2_mf)	McFadden's R^2
r(lrx2_p)	probability level of chi-squared	r(r2_mfadj)	McFadden's adjusted R^2
	test	r(r2_ml)	maximum likelihood R^2
r(N)	number of observations	r(v_error)	variance of error term
r(n_parm)	number of parameters	r(v_ystar)	variance of y^*

When **saving**(*name*) is specified, the fit statistics are saved in matrix **fs_***name*. The column names are the names of the measures; the row name is the command used to fit the model. Values of -9999 in the matrix indicate that a measure is not appropriate for the given model.

A.3   listcoef

listcoef   List transformed regression coefficients with guidelines for interpretation. For a detailed discussion, see Section 3.1.7 of Chapter 3.

Syntax

listcoef [*varlist*] [, p̲value(#) [f̲actor | p̲ercent | s̲td] r̲everse c̲onstant
    m̲atrix h̲elp]

If *varlist* is provided, only coefficients for these variables are to be listed. Otherwise, coefficients for all variables are listed.

Description

listcoef lists the estimated coefficients for a variety of regression models, with options that allow you to specify different transformations of the coefficients, such as factor change and percent change. Coefficients can be standardized to a unit variance in the independent and dependent variables. For mlogit, coefficients for all comparisons among outcomes are included. The help option provides a guide for interpreting the coefficients. The listcoef command can be used with clogit, cloglog, cnreg, intreg, logistic, logit, mlogit, nbreg, ologit, oprobit, poisson, probit, regress, tobit, zinb, and zip. For models with categorical outcomes, the output is easier to understand if you assign value labels to the dependent variable.

Options

pvalue(#) specifies that only coefficients that are significant at this level or smaller will be printed.

factor requests that factor change coefficients (i.e., odds ratios) be listed.

percent requests that percent change coefficients be listed.

std requests that coefficients standardized to a unit variance for the dependent variable be listed.

reverse reverses the order of comparison for factor or percent change coefficients for logit, ologit, or clogit; that is, it presents results indicating the change in the odds of b versus a instead of a versus b.

constant includes the constant(s) in the output.

matrix returns results in r-class matrices.

help includes details on the meaning of each coefficient.

Examples

listcoef can be used to list only coefficients significant at a given level:

```
. use http://www.stata-press.com/data/lfr/regjob2, clear
(Academic Biochemists / S Long)

. * regress
. regress job fem phd ment fel art cit
  (output omitted)

. listcoef, pv(.05)
regress (N=408): Unstandardized and Standardized Estimates when P>|t| < 0.05

  Observed SD: .97360294
  SD of Error: .8717482
```

job	b	t	P>|t|	bStdX	bStdY	bStdXY	SDofX
phd	0.27268	5.529	0.000	0.2601	0.2801	0.2671	0.9538
fel	0.23414	2.469	0.014	0.1139	0.2405	0.1170	0.4866
cit	0.00448	2.275	0.023	0.1481	0.0046	0.1521	33.0599

For models with a latent variable, coefficients for the effect on the standardized y^* can be listed. Here we request only the coefficients for age:

```
. use http://www.stata-press.com/data/lfr/ordwarm2, clear
(77 & 89 General Social Survey)
```

```
. ologit warm age ed prst male yr89 white, nolog
  (output omitted)
. listcoef age, std
```

ologit (N=2293): Unstandardized and Standardized Estimates

  Observed SD: .9282156
    Latent SD: 1.9410634

warm	b	z	P>|z|	bStdX	bStdY	bStdXY	SDofX
age	-0.02167	-8.778	0.000	-0.3635	-0.0112	-0.1873	16.7790

or coefficients for the odds of outcomes $> m$ versus $\leq m$ can be listed; with the `reverse` option, the coefficients are listed for the odds of $\leq m$ versus $> m$:

```
. listcoef age
```
ologit (N=2293): Factor Change in Odds

  Odds of: >m vs <=m

warm	b	z	P>|z|	e^b	e^bStdX	SDofX
age	-0.02167	-8.778	0.000	0.9786	0.6952	16.7790

```
. listcoef age, rev
```
ologit (N=2293): Factor Change in Odds

  Odds of: <=m vs >m

warm	b	z	P>|z|	e^b	e^bStdX	SDofX
age	-0.02167	-8.778	0.000	1.0219	1.4384	16.7790

For `mlogit`, coefficients for all comparisons of outcomes can be listed. For example, we can request that only the significant percent change coefficients for `age` be listed:

```
. use http://www.stata-press.com/data/lfr/nomocc2, clear
(1982 General Social Survey)
. mlogit occ white ed exp, nolog
  (output omitted)
. listcoef white, percent pv(.05)
```
mlogit (N=337): Percentage Change in the Odds of occ when P>\|z\| < 0.05

Variable: white (sd= .276423)

Odds comparing Group 1 vs Group 2	b	z	P>|z|	%	%StdX
Menial -Prof	-1.77431	-2.350	0.019	-83.0	-38.8
Craft -Prof	-1.30196	-2.011	0.044	-72.8	-30.2
Prof -Menial	1.77431	2.350	0.019	489.6	63.3
Prof -Craft	1.30196	2.011	0.044	267.7	43.3

Saved results

Scalars
  r(pvalue)   1 or value specified with `pvalue(#)`

Macros
  r(cmd)   name of most recent estimation command

Matrices

When the `matrix` option is specified, `listcoef` saves in `r()` all of the statistics that have been computed for the given model. These are only saved when the `matrix` option is specified because saving these matrices can slow execution for `mlogit`. The row and column labels indicate the variable and type of coefficient that has been saved:

r(b)	slope or regression coefficients
r(b_fact)	factor change coefficients (i.e., $\exp\left(\widehat{\beta}\right)$)
r(b_facts)	x-standardized factor change coefficients
r(b_p)	p-values for test of regression coefficients
r(b_pct)	percent change coefficients
r(b_pcts)	x-standardized percent change coefficients
r(b_sdx)	standard deviations for independent variables used to compute x-standardized and fully standardized coefficients
r(b_std)	fully standardized coefficients
r(b_xs)	x-standardized coefficients
r(b_ys)	y or y^*-standardized coefficients
r(b_z)	z-values or t-values for regression coefficients
r(cons)	constant(s)
r(cons_p)	p-value(s) for constant(s)
r(cons_z)	z-value(s) or t-value(s) for constant(s)
r(contrast)	all contrasts from `mlogit`

For `zip` and `zinb`, the matrices `cons2`, `cons2_p`, `cons2_z`, `b2`, `b2_p`, `b2_z`, `b2_fact`, `b2_facts`, `b2_pct`, and `b2_pcts` contain corresponding results for the binary equation.

A.4   mlogplot

mlogplot   Odds ratio and discrete change plots for multinomial logit. For a detailed discussion, see Sections 6.6.7–6.6.10 of Chapter 6.

Syntax

There are two ways that `mlogplot` can be used. If you have just fitted a model using `mlogit`, you can create a discrete change or odds ratio plot for the coefficients in memory using the following syntax:

mlogplot *varlist* [, <u>or</u>atio <u>d</u>change <u>std</u>([s|u|0]...[s|u|0]) min(#) max(#)

    packed labels <u>prob</u>(#) <u>base</u>category(#) <u>nt</u>ics(#) <u>note</u>(*string*)

    dcadd(#)]

Note that `prchange` must have been previously run if option `dchange` is used.

Important: you must use value labels for the dependent variable, and these labels must begin with a different letter or number for each category; otherwise, the plot will be misleading.

You can also create an odds ratio plot from coefficients that you have placed into matrices (e.g., coefficients from a published paper or another program). The syntax is

mlogplot *varlist* $\left[\right.$, matrix <u>v</u>ars(*varlist*) <u>s</u>td($\left[\text{s|u|0}\right]$...$\left[\text{s|u|0}\right]$) min(#) max(#)
    packed <u>b</u>asecategory(#) <u>n</u>tics(#) <u>n</u>ote(*string*) $\left.\right]$

Details on the matrices are given below. Discrete change plots cannot be computed from matrices.

Description

mlogplot facilitates the interpretation of the multinomial logit model. It can plot the odds ratios coefficients (i.e., $\exp\left(\widehat{\beta}_{A|B}\right)$) from mlogit or coefficients that have been saved in matrices. If prchange has been run after mlogit, mlogplot can create either a plot of discrete change coefficients or add information about discrete change to the odds ratio plots. Odds ratios can be plotted for either a unit change or a standard deviation change. Discrete change coefficients can be plotted for either a unit change, a standard deviation change, or a change from 0 to 1 (for dummy variables). In the plots, you can list the same variable more than once if you want to plot the effects of different amounts of change for the same variable. A variety of options control the way the graph looks. The program mlogview provides a convenient dialog box that allows you to use mlogplot interactively.

Options

varlist is the list of variables to be plotted. The same variable can be listed more than once if you want to plot its coefficients for different amounts of change.

oratio indicates that you want to plot the odds ratios (i.e., $\exp(\beta)$). This option cannot be used if coefficients are entered with matrices.

dchange indicates that you want to plot the discrete change. To use this option, you must have first run prchange.

- If only oratio is specified, an odds ratio plot is drawn.

- If only dchange is specified, a discrete change plot is drawn.

- If both are specified, an odds ratio plot is drawn where the size of the letters are proportional to the size of the corresponding discrete change coefficient.

std($\left[\text{s}|\text{u}|0\right]\ldots\left[\text{s}|\text{u}|0\right]$) specifies the type of coefficient to plot for each variable: s indicates standardized coefficients; u indicates unstandardized coefficients; 0 indicates changes from 0 to 1 in discrete change plots. For example, std(u0su) indicates that the first variable is unstandardized, the second is binary, the third is standardized, and the last is unstandardized.

min(#) and max(#) specify the minimum value and the maximum value on the plotting axis. This is useful if you want to control the labeling of the tick marks or if you want to compare coefficients across plots.

packed removes the vertical spacing among the outcome categories in an odds ratio plot. This allows up to 11 variables on a single graph; the maximum otherwise is 6.

labels uses variable labels to label each row of the plot. You might need to revise your variable labels to make them fit the graph (i.e., your current labels might be too long). This option cannot be used with the matrix option.

prob(#) specifies for an odds ratio plot that if a coefficient contrasting two outcomes is *not* significant at this level, a line is to be drawn connecting the letters. This option cannot be used with the matrix option.

basecategory(#) is used for an odds ratio plot to specify which category of the outcome measure is to be used as the reference point.

note(*string*) adds a title at the top of the plot.

ntics(#) sets the number of tick marks on the axes. Used along with min() and max(), this allows you to determine the numbering on the axes and the location of tick marks.

dcadd(#) is rarely used. In odds ratio plots with the dchange option, the size of the letter corresponds to the square root of the size of the discrete change coefficient. dcadd adds an amount to each discrete change to make the size of all letters larger, making it easier to see the letters for small discrete changes. By default, this quantity is 0. If your letters are too small (because the discrete change is small), you might want to increase this by a small amount, say, dcadd(.03).

matrix indicates that the data for the graph will be taken from the following matrices and global macros, which must be previously created by the user:

mnlbeta is a matrix that contains estimated βs, where element (i, j) is the jth variable for comparison i relative to the reference category. That is, columns are for variables; rows for different contrasts. Constants are *not* included.

mnlname is a matrix that contains the names of the variables corresponding to the columns of *mnlbeta*.

mnlsd is a matrix that contains the standard deviations for the variables that correspond to columns of *mnlbeta*.

mnlcatnm is a macro string with labels for the outcome categories. The first category corresponds to the first column of *mnlbeta*, the second to the second, etc. The label for the reference category should be last.

mnlrefn is a macro with the number of the category that is the reference category for the contrasts contained in *mnlbeta*.

vars(*varlist*) is required when **matrix** is specified. *varlist* contains the names of the variables listed in *mnlname* whose coefficients you want to plot, in the order that you want to plot them.

Examples

Plotting coefficients from mlogit: After fitting a multinomial logit model with mlogit, you can create a discrete change plot with

```
. mlogit occ white ed exper
. prchange
. mlogplot white ed exper, dc std(0ss) min(-.5) max(.5)
```

A plot with only the odds ratios is created with

```
. mlogplot white ed exper, or std(0ss) min(-2.75) max(.5)
```

A plot with the size of the letters in the odds ratios being proportional to the amount of discrete change is created with

```
. mlogplot white ed exper, or  dchange std(0ss) base(4) min(-1.75) max(1.75)
```

Plotting coefficients contained in matrices: To use the **matrix** options, you must create matrices and macros that contain the values that you wish to plot. For example,

```
. matrix mnlsd = (2.946427, 13.95936, 2.946427, 13.95936)
. global mnlname = "W_Educ W_Exper NW_Educ NW_Exper"
. global mnlrefn = 5
. global mnlcatnm = "Menial BlueC Craft WhiteC Prof"
. matrix mnlbeta = (-.83075, -.92255, -.68761, -.41964  -.03380, -.03145,
      -.00026,  .00085 -.70126, -.56070, -.88250, -.53115   -.11084,
      -.02611,  -.15979, -.05209 )
. matrix mnlbeta = mnlbeta'
```

Then, the plot can be constructed using mlogplot:

```
. mlogplot, vars(W_Educ NW_Educ W_Exper NW_Exper) matrix
      std(ssss) note("Effects of Education")
```

A.5   mlogtest

mlogtest   Statistical tests for the multinomial logit model. For a detailed discussion, see Section 6.3 of Chapter 6.

Syntax

mlogtest [, <u>a</u>ll <u>l</u>r <u>w</u>ald <u>c</u>ombine <u>lrc</u>omb <u>s</u>et(*varlist* [\ *varlist*]) <u>ii</u>a <u>h</u>ausman
    <u>sm</u>hsiao <u>d</u>etail <u>b</u>ase]

Description

mlogtest computes a variety of tests for the multinomial logit model. Users select the tests they want by specifying the appropriate options. For each independent variable, mlogtest can perform either a likelihood-ratio test or a Wald test of the null hypothesis that the coefficients for the variable equal zero across all equations. mlogtest can also perform Wald or likelihood-ratio tests of whether any pair of outcome categories can be combined. In addition, mlogtest computes both the Hausman and Small–Hsiao tests of the assumption of the independence of irrelevance alternatives (IIA) for each possible omitted category.

Options

all specifies that all available tests should be performed.

lr conducts likelihood-ratio tests for each independent variable.

wald performs Wald tests for each independent variable.

combine computes Wald tests of whether two outcomes in the mlogit model can be combined.

lrcomb conducts likelihood-ratio tests of whether two outcomes can be combined.

set(*varlist* [\ *varlist*]) specifies that a set of variables is to be considered together for the Wald test or the likelihood-ratio test. The slash \ is used to specify multiple sets of variables. This option is useful, for example, when a categorical independent variable has been included as a set of dummy variables.

iia specifies that both tests of the IIA assumption should be performed.

hausman computes Hausman tests of the IIA assumption.

smhsiao performs Small–Hsiao tests of the IIA assumption.

detail reports full hausman output from IIA test (the default is to provide only a summary of results).

base also conducts an IIA test omitting the base category of the original mlogit estimation. This is done by refitting the model using the largest remaining category as the base category, although the original estimates are restored to memory afterward.

Examples

If all tests are requested, the following results are obtained:

```
. mlogit occ white ed exper
  (output omitted)
. mlogtest, all
```

**** Likelihood-ratio tests for independent variables

Ho: All coefficients associated with given variable(s) are 0.

occ	chi2	df	P>chi2
white	8.095	4	0.088
ed	156.937	4	0.000
exper	8.561	4	0.073

**** Wald tests for independent variables

Ho: All coefficients associated with given variable(s) are 0.

occ	chi2	df	P>chi2
white	8.149	4	0.086
ed	84.968	4	0.000
exper	7.995	4	0.092

**** Hausman tests of IIA assumption

Ho: Odds(Outcome-J vs Outcome-K) are independent of other alternatives.

Omitted	chi2	df	P>chi2	evidence
Menial	7.324	12	0.835	for Ho
BlueCol	0.320	12	1.000	for Ho
Craft	-14.436	12	1.000	for Ho
WhiteCol	-5.541	11	1.000	for Ho

**** Small-Hsiao tests of IIA assumption

Ho: Odds(Outcome-J vs Outcome-K) are independent of other alternatives.

Omitted	lnL(full)	lnL(omit)	chi2	df	P>chi2	evidence
Menial	-181.981	-175.597	12.769	4	0.012	against Ho
BlueCol	-141.007	-138.554	4.905	4	0.297	for Ho
Craft	-139.094	-135.464	7.261	4	0.123	for Ho
WhiteCol	-172.969	-167.333	11.271	4	0.024	against Ho

**** Wald tests for combining outcome categories

Ho: All coefficients except intercepts associated with given pair
    of outcomes are 0 (i.e., categories can be collapsed).

Categories tested	chi2	df	P>chi2
Menial- BlueCol	3.994	3	0.262
Menial- Craft	3.203	3	0.361
Menial-WhiteCol	11.951	3	0.008
Menial- Prof	48.190	3	0.000
BlueCol- Craft	8.441	3	0.038
BlueCol-WhiteCol	20.055	3	0.000
BlueCol- Prof	76.393	3	0.000
Craft-WhiteCol	8.892	3	0.031
Craft- Prof	60.583	3	0.000
WhiteCol- Prof	22.203	3	0.000

```
**** LR tests for combining outcome categories
Ho: All coefficients except intercepts associated with given pair
    of outcomes are 0 (i.e., categories can be collapsed).
```

Categories tested	chi2	df	P>chi2
Menial- BlueCol	4.095	3	0.251
Menial- Craft	3.376	3	0.337
Menial-WhiteCol	13.223	3	0.004
Menial- Prof	64.607	3	0.000
BlueCol- Craft	9.176	3	0.027
BlueCol-WhiteCol	22.803	3	0.000
BlueCol- Prof	125.699	3	0.000
Craft-WhiteCol	9.992	3	0.019
Craft- Prof	95.889	3	0.000
WhiteCol- Prof	26.736	3	0.000

Note, however, that the tests of IIA do not include a test based on the base category. These are obtained with the **base** option:

```
. mlogtest, iia base
**** Hausman tests of IIA assumption
Ho: Odds(Outcome-J vs Outcome-K) are independent of other alternatives.
```

Omitted	chi2	df	P>chi2	evidence
Menial	7.324	12	0.835	for Ho
BlueCol	0.320	12	1.000	for Ho
Craft	-14.436	12	1.000	for Ho
WhiteCol	-5.541	11	1.000	for Ho
Prof	-0.119	12	1.000	for Ho

```
**** Small-Hsiao tests of IIA assumption
Ho: Odds(Outcome-J vs Outcome-K) are independent of other alternatives.
```

Omitted	lnL(full)	lnL(omit)	chi2	df	P>chi2	evidence
Menial	-165.418	-160.130	10.577	4	0.032	against Ho
BlueCol	-144.752	-140.906	7.693	4	0.104	for Ho
Craft	-123.924	-120.640	6.570	4	0.160	for Ho
WhiteCol	-152.486	-148.324	8.323	4	0.080	for Ho
Prof	-131.571	-127.534	8.073	4	0.089	for Ho

The **set** option allows you to test coefficients for two or more variables simultaneously:

(Continued on next page)

```
. mlogtest, lr set(ed exper white ed white ed exper)
**** Likelihood-ratio tests for independent variables
Ho: All coefficients associated with given variable(s) are 0.
          occ │     chi2   df   P>chi2
   ───────────┼──────────────────────────
        white │    8.095    4    0.088
           ed │  156.937    4    0.000
        exper │    8.561    4    0.073
   ───────────┼──────────────────────────
       set_1: │  166.087   12    0.000
           ed │
        exper │
        white │
           ed │
        white │
           ed │
        exper │
   ───────────┼──────────────────────────
```

Saved results

`mlogtest` saves whichever of the following results were computed. The row and column labels of the matrix identify the specific elements of the matrices:

Matrices
    `r(wald)`   results of Wald test that all coefficients of an independent variable
                        equal zero
    `r(lrtest)`   results of likelihood-ratio test that all coefficients of an independent
                        variable equal zero
    `r(combine)`   results of Wald tests for combining categories
    `r(lrcomb)`   results of likelihood-ratio tests for combining categories
    `r(hausman)`   results of Hausman tests of IIA assumption
    `r(smhsiao)`   results of Small–Hsiao tests of IIA assumption

Acknowledgment

The code for the Small–Hsiao test is based on a program by Nick Winter.

A.6   mlogview

mlogview   Dialog box for using `mlogplot` interactively. For a detailed discussion, see Sections 6.6.7–6.6.10 of Chapter 6.

Syntax

`mlogview`

Description

mlogview creates a dialog box for creating an odds ratio plot or discrete change plot of the coefficients from a model fitted with mlogit. To make a discrete change plot, you must run prchange before opening the dialog box. The dialog box settings are translated into the appropriate mlogplot command, which is executed when the user clicks one of the plot buttons in mlogview. The resulting mlogplot command is listed in the Results window and can be used later in a do-file.

Dialog box controls

The dialog box has the following controls for selecting variables and the amount of change to plot for each variable:

Select variables: select independent variables to include in plot.

Select amount of change: for a given independent variable, select the amount of change to be plotted:

+1: unit change.

+SD: standard deviation change.

0->1: change from 0 to 1 (for use with discrete change).

Don't plot: exclude variable from plot.

The buttons determine the type of plot to create:

DC Plot: draw discrete change plot.

OR Plot: draw odds ratio plot.

OR+DC Plot: draw odds ratio plot in which the size of the letters indicates the discrete change.

For odds ratio plots, only six variables can be plotted at one time. To automatically create a plot for the next six variables in the model, click

Next 6: list next 6 independent variables for an odds ratio plot.

Characteristics of the graph are controlled by

Note: title for plot to be printed at the top.

Number of ticks: number of tick marks on x-axis of plot.

Plot from: minimum and maximum values of x-axis for plot.

Connect if: for odds ratio plots, connect categories if the coefficient for the odds of those two categories is not significant at the level specified in this box.

Base category: value of category to use as base category for odds ratio plot.

Pack odds ratio plot: eliminate extra vertical space in odds ratio plot.

Use variable labels: use variable labels instead of names to identify variables in plot.

Finally, to print the graph,

Print: send plot to printer.

A.7   Overview of prchange, prgen, prtab, and prvalue

prchange, **prgen**, **prtab**, and **prvalue**   Setting values of the independent variables using x() and rest(). For a detailed discussion, see Section 3.5.3 to 3.5.7 of Chapter 3.

Syntax

prchange, prgen, prtab, and prvalue compute predicted probabilities for specified values of the independent variables. These values are set using x()and rest(). This entry provides a detailed description of how to use these options to set the values for the independent variables for any of these commands.

x(*variable1=value1* [*variable2=value2*]...)   assigns *variable1* to *value1*, *variable2* to *value2*, and so on. You can assign values to as many or as few variables as you want. The assigned value is either a specific number (e.g., female=1) or a mnemonic specifying a descriptive statistic (e.g., phd=mean to set variable phd to the sample mean, and pub3=max to assign pub3 to the maximum value in the sample). The mnemonics for descriptive statistics are discussed below. Only numeric values can be used if a group statistic (i.e., statistics that begin with *gr*) has been specified with rest(). While the equal signs are optional, they make the command easier to read.

rest(*stat*) sets the values of all variables not specified in x() to the sample statistic indicated by *stat*. If x() is not specified, all variables are set to *stat*. For example, rest(mean) sets all variables to their mean. The choices for *stat* can be unconditional or conditional on a group specified by x(). For example, x(female=0) rest(grmean) sets female to 0 and all other variables to their mean for those where female is 0. If rest() is not specified, it is assumed to be rest(mean). The available types of *stat* are mean, median, min, and max refer to the mean, median, minimum, and maximum, conditional on any if or in conditions that are specified. Descriptive statistics are calculated using the estimation sample unless the all option has been specified, in which case all observations in memory are used.

previous sets the variables to their values from a previous command that set x() or rest(). For example, if the previous command was prvalue, x(a=1) rest(mean), then
prvalue, rest(previous) is equivalent to prvalue, x(a=1) rest(mean).

upper and lower can only be used with binary models, and refer to the values of the independent variables that yield either the highest (upper) or the lowest (lower) probability of a positive outcome.

grmean, grmedian, grmin, grmax refer to the group mean, median, minimum, and maximum. These statistics are calculated for whatever group of observations is specified in x(). For example, prvalue, x(male=0 white=0) rest(grmean) calculates predicted probabilities, where male and white are held to 0 and the other independent variables are held to the means of the observations in which male==0 and white==0. This is the same as typing prvalue if male==0 & white==0, x(male=0 white=0) rest(mean).

all specifies that all observations in memory are to be used to calculate descriptive statistics, excepting cases excluded by if or in conditions or the use of rest(gr*). The default is to use only the cases in the estimation sample.

Examples

```
. use http://www.stata-press.com/data/lfr/ordwarm2, clear
(77 & 89 General Social Survey)
. ologit warm male white age ed prst if yr89==1, nolog
  (output omitted )
. * set independent variables to mean in estimation sample
. prvalue
  (output omitted )
. * set independent variables to mean for all cases in memory
. prvalue, all
  (output omitted )
. * set male to 0, ed to 12, age to 30, everything else to mean
. prvalue, x(male=0 ed=12 age=30) rest(mean)
  (output omitted )
. * set male to 0, all others to mean for males
. prvalue, x(male=0) rest(grmean)
  (output omitted )
. * set male to 0, age to median, all others to mean
. prvalue, x(male=0 age=median) rest(mean)
  (output omitted )
. * set to statistics in full sample instead of estimation sample
. prvalue, x(male=0 age=median) rest(mean) all
  (output omitted )
```

A.8   praccum

praccum   Accumulates results from a series of calls to `prvalue`; these results can then be used to plot predictions.

Syntax

praccum , { <u>u</u>sing(*matrixname*) | <u>s</u>aving(*matrixname*) } [xis(*value*)
  <u>gen</u>erate(*prefix*)]

Description

`praccum` accumulates the predictions generated by a series of calls to `prvalue` and optionally saves the accumulated values to new variables for plotting predicted values. The command allows you to plot predicted values in situations (e.g., nonlinearities) that cannot be handled by `prgen`. `praccum` works with `cloglog`, `cnreg`, `intreg`, `logistic`, `logit`, `mlogit`, `nbreg`, `ologit`, `oprobit`, `poisson`, `probit`, `regress`, `tobit`, `zinb`, and `zip`.

Predicted values of interest are produced by a series of paired calls to `prvalue` and `praccum`. The sequence begins with using `prvalue` to compute the first set of predicted values, followed by the initial call of `praccum`, where predicted values are saved in a new matrix defined by `saving()` or `using()`. After each subsequent call to `prvalue` and `praccum`, the probabilities are appended to the matrix specified by `using()`. When all the predicted probabilities have been computed, the `generate()` option is used to create new variables that contain the information from the matrix of saved results.

Options

using(*matrixname*) specifies the name of the matrix to which the predictions from the previous call to `prvalue` should be added. An error is generated if the matrix does not have the correct number of columns. This can happen if you try to append values to a matrix generated from calls to `praccum` based on a different model. Matrix *matrixname* will be created if it does not already exist.

saving(*matrixname*) specifies that a new matrix should be generated and should contain the predicted values from the previous call to `prvalue`. You only use this option when you initially create the matrix. After the matrix is created, you add to it with `using()`. The difference between `saving()` and `using()` is that `saving()` will overwrite *matrixname* if it exists, while `using()` will append results to it.

xis(*value*) indicates the value of the *x*-variable associated with the predicted values that are accumulated. If this is not specified, no new values will be added to the matrix generate(*prefix*) indicates that new variables should be generated from the matrix specified by using(). The names of the new variables begin with *prefix*. Details are given below.

Examples

First, we fit a model that includes age and age^2, and then we call prvalue:

```
. use http://www.stata-press.com/data/lfr/binlfp2,clear
(PSID 1976 / T Mroz)
. gen age2 = age*age
. logit lfp k5 k618 age age2 wc hc lwg inc, nolog
(output omitted)
. prvalue, x(age 20 age2 400) rest(mean) brief
Pr(y=inLF|x):        0.7324   95% ci: (0.3972,0.9191)
Pr(y=NotInLF|x):     0.2676   95% ci: (0.0809,0.6028)
```

The first call of praccum saves the predictions generated by prvalue to the new matrix m_age:

```
. praccum, saving(m_age) xis(20)
```

Additional calls to praccum append to this matrix the new predictions that have been generated by new calls to prvalue:

```
. prvalue, x(age 25 age2 625) rest(mean) brief
(output omitted)
. praccum, using(m_age) xis(25)
. prvalue, x(age 30 age2 900) rest(mean) brief
(output omitted)
. praccum, using(m_age) xis(30)
. prvalue, x(age 35 age2 1225) rest(mean) brief
(output omitted)
. praccum, using(m_age) xis(35)
. prvalue, x(age 40 age2 1600) rest(mean) brief
(output omitted)
. praccum, using(m_age) xis(40)
. prvalue, x(age 45 age2 2025) rest(mean) brief
(output omitted)
. praccum, using(m_age) xis(45)
. prvalue, x(age 50 age2 2500) rest(mean) brief
(output omitted)
. praccum, using(m_age) xis(50)
. prvalue, x(age 55 age2 3025) rest(mean) brief
(output omitted)
. praccum, using(m_age) xis(55)
. prvalue, x(age 60 age2 3600) rest(mean) brief
(output omitted)
```

The last call to praccum generates new variables, which are listed:

```
. praccum, using(m_age) xis(60) gen(agsq)
New variables created by praccum:
```

Variable	Obs	Mean	Std. Dev.	Min	Max
agsqx	9	40	13.69306	20	60
agsqp0	9	.4282142	.1752595	.2676314	.7479599
agsqp1	9	.5717858	.1752595	.2520402	.7323686

```
. * these variables were created
. list agsq* in 1/9
```

	agsqx	agsqp0	agsqp1
1.	20	.2676314	.7323686
2.	25	.2682353	.7317647
3.	30	.2836163	.7163837
4.	35	.3152536	.6847464
5.	40	.3656723	.6343277
6.	45	.4373158	.5626842
7.	50	.5301194	.4698806
8.	55	.6381241	.3618759
9.	60	.7479599	.2520402

`praccum` is less cumbersome when it is used in conjunction with `forvalues`. Including the following in a do-file produces the same results as the example above:

```
capture matrix drop m_age
forvalues count = 20(5)60 {
    local countsq = "`count'"^2
    prvalue, x(age "`count'" age2 "`countsq'") rest(mean) brief
    praccum, using(m_age) xis("`count'")
}
praccum, using(m_age) gen(agsq)
```

New variables generated

The new variables created by `praccum` are the same as those created by `prgen` (see the table under the entry for `prgen`). The only difference is that while *namex* for `prgen` represents the values of x as specified by the `from()` and `to()` options, for `praccum` each value of *namex* must be specified by the user by specifying the `xis()` option.

A.9   prchange

prchange   Discrete and marginal change for regression models for categorical and count variables. For a detailed discussion, see Section 3.5.4 of Chapter 3.

Syntax

prchange [*varlist*] [if *exp*] [in *range*] [, x(*variable1=value1* [...]) r̲est(*stat*)
  o̲utcome(#) f̲romto b̲rief n̲obase n̲olabel h̲elp all u̲ncentered
  d̲elta(#)]

See page A.7 for details on x() and rest().

Description

prchange computes marginal and discrete change for regression models for categorical
and count variables: cloglog, cnreg, gologit, intreg, logistic, logit, mlogit,
nbreg, ologit, oprobit, poisson, probit, regress, tobit, zinb, and zip. Marginal
change is the partial derivative of the predicted probability or rate with respect to the
independent variables. Discrete change is the difference in the predicted value as one
independent variable changes values while all others are held at specified values. By
default, changes are calculated holding all other variables at their mean. Alternatively,
values for the independent variables can be set with the x() and rest() options (see
Section A.7 for details).

By default, discrete change is computed for a change in each variable from its minimum to its maximum (Min->Max), from 0 to 1 (0->1), from its specified value minus .5
units to its specified value plus .5 (-+1/2), and from its specified value minus .5 standard
deviations to its value plus .5 standard deviations (-+sd/2). If the uncentered option is
chosen, the last two quantities are computed from the specified value plus one unit (+1)
and from the specified value plus one standard deviation (+sd). With the delta(#)
option, changes of a standard deviation are replaced with changes of # units.

Options

outcome(#) specifies that changes should be printed only for the outcome indicated; for
    example, outcome(2). For ologit, oprobit and mlogit, the default is to provide
    results for all outcomes. For the count models, the default is to present results with
    respect to the predicted rate; specifying an outcome number will present changes in
    the probability of that outcome.

fromto specifies that the starting and ending probabilities from which the discrete
    change is calculated should be displayed.

brief prints only limited output.

nobase suppresses inclusion of the base values of independent variables.

nolabel uses values rather than value labels in the output.

help provides information explaining the headings in the output.

`all` specifies that calculations of means, medians, etc., should use the entire sample instead of the sample used to fit the model.

`uncentered` requests that changes of one unit and one standard deviation (or the amount specified by `delta()`) should begin at the value specified by `rest()` or `x()` rather than be centered around the specified value.

`delta(#)` indicates that the changes are to be # units rather than a standard deviation change.

Examples

```
. * fit the model
. use http://www.stata-press.com/data/lfr/ordwarm2, clear
(77 & 89 General Social Survey)
. ologit warm male white age ed prst if yr89==1, nolog
  (output omitted)
. * holding all variables to the estimation sample means
. prchange age white
ologit: Changes in Predicted Probabilities for warm
age
            Avg|Chg|         SD          D          A         SA
Min->Max    .18353666   .11080101   .25627232  -.13218805  -.23488527
   -+1/2    .00257031   .00124054   .00390008  -.00145978  -.00368083
  -+sd/2    .04392809   .02133921    .066517   -.02479631  -.06305985
MargEfct    .00257032   .00124052   .00390013  -.00145983  -.00368082
white
            Avg|Chg|         SD          D          A         SA
    0->1    .03860572   .01742518   .05978625  -.01481366  -.06239779
                SD          D          A         SA
Pr(y|x)  .05785507  .28683352  .45240733  .20290408
            male     white      age       ed      prst
   x=    .443107   .866521   45.1269   12.8523   41.5449
sd(x)=   .497025   .340278   17.129    3.0222    14.8981
. * specific set of base values
. prchange, x(male=1 white=1 age=30) rest(median)
ologit: Changes in Predicted Probabilities for warm
male
            Avg|Chg|         SD          D          A         SA
    0->1    .07925584   .03510679   .12340491  -.02130097  -.13721071
white
            Avg|Chg|         SD          D          A         SA
    0->1    .04032505   .01954348   .06110662  -.0226658   -.05798429
age
            Avg|Chg|         SD          D          A         SA
Min->Max    .1914724    .16630837   .21663642  -.21676943  -.16617538
   -+1/2    .00269591   .00144775   .00394407  -.00209403  -.00329781
  -+sd/2    .04605574   .02489513   .06721637  -.03557727  -.0565342
MargEfct    .00269594   .00144773   .00394415  -.00209407  -.00329781
```

```
ed
              Avg|Chg|           SD             D            A            SA
Min->Max     .10027184   -.05862336   -.14192033    .08446175    .11608193
    -+1/2    .00499106   -.00268043   -.00730169    .00387657    .00610557
   -+sd/2    .01508015    -.0081038   -.02205652    .01170659    .01845369
MargEfct     .00499121   -.00268031   -.00730212    .00387692    .00610551

prst
              Avg|Chg|           SD             D            A            SA
Min->Max     .07671434   -.04028213   -.11314654    .05304095    .10038772
    -+1/2    .00112195    -.0006025   -.00164139    .00087148    .00137244
   -+sd/2    .01670909   -.00898061   -.02443758    .01296943    .02044876
MargEfct     .00112195   -.00060249   -.00164141    .00087148    .00137243

                    SD            D            A           SA
Pr(y|x)   .06827433   .31734699    .4385626   .17581607

                male      white       age        ed      prst
     x=            1          1        30        12        40
sd(x)=    .497025    .340278    17.129    3.0222   14.8981
```

A.10   prcounts

prcounts   Predicted probabilities and rates for count models. For a detailed discussion, see Sections 7.1.2, 7.1.3 and 7.5.1 of Chapter 7.

Syntax

prcounts *prefix* [if *exp*] [in *range*] [, max(*max*) plot]

Description

prcounts computes the predicted rate and probabilities of counts from 0 through the specified maximum count based on the last estimates from the count model fitted by poisson, nbreg, zip, or zinb. The predictions for each observation are stored in new variables. Optionally, you can generate variables for the graphical comparison of observed and expected counts.

Options

max(*max*) is the maximum count for which predicted probabilities should be provided. The default is 9.

plot specifies that variables for plotting expected counts should be generated

New variables generated

The command prcounts *prefix* generates the following variables, where *max* is the value specified as the maximum:

*prefix*rate	Predicted count or rate.
*prefix*pr0	Predicted probability of a count of 0.
*prefix*pr1	Predicted probability of a count of 1.
:::	
*prefix*pr*max*	Predicted probability of a count equal to *max* (default is 9).
*prefix*prgt	Predicted probability of a count > *max*.
*prefix*cu0	Predicted probability of a count of 0.
*prefix*cu1	Predicted probability of a count of ≤ 1.
:::	
*prefix*cumax	Predicted probability of a count of ≤ *max*.

When the `plot` option is specified, variables are created that contain the average predicted probability over the sample for each count from 0 to *max*. For these variables, each observation provides information about the probabilities of a given count for the entire sample. For these variables, only the first *max*+1 observations in the dataset are used, with observation $y + 1$ corresponding to count y. The additional variables created when the `plot` option is specified are

*name*val	The specific value of the count y ranging from 0 to *max*.
*name*obeq	Observed probability of counts = y.
*name*oble	Cumulative observed probability of counts ≤ y.
*name*preq	Predicted probability of counts = y.
*name*prle	Predicted cumulative probability of counts ≤ y.

Examples

`prcounts` can be used to compute predictions for each observation in the sample,

```
. zinb art fem mar kid5 phd ment, inf(fem mar kid5 phd ment)
(output omitted)
. prcounts znb
. summarize znb*
```

Variable	Obs	Mean	Std. Dev.	Min	Max
znbrate	915	1.697666	.7165823	.468232	9.678316
znball0	915	.0585168	.1215135	1.01e-30	.6024709
znbpr0	915	.3119487	.1249968	.0169493	.7454953
znbpr1	915	.2556639	.0418208	.0353106	.3098438
(output omitted)					
znbpr9	915	.0032089	.0061532	.0000589	.0610584
znbcu0	915	.3119487	.1249968	.0169493	.7454953
znbcu1	915	.5676126	.1256956	.0522599	.8773996
(output omitted)					
znbcu9	915	.9948109	.0214203	.5720913	.9999722
znbprgt	915	.0051891	.0214203	.0000278	.4279087

or it can be used to compute average predictions that can be graphed:

```
. poisson art fem mar kid5 phd ment
(output omitted)
```

```
. prcounts prm, max(8) plot
. sum prm*
```

Variable	Obs	Mean	Std. Dev.	Min	Max
prmrate	915	1.692896	.6685824	.8883344	9.627207
prmpr0	915	.2092071	.0794247	.0000659	.4113403
prmpr1	915	.3098447	.0634931	.0006345	.3678775
(output omitted)					
prmpr8	915	.001877	.0094055	3.96e-06	.1206255
prmcu0	915	.2092071	.0794247	.0000659	.4113403
prmcu1	915	.5190518	.1395755	.0007004	.7767481
(output omitted)					
prmcu8	915	.9978884	.023188	.3763166	.9999995
prmprgt	915	.0021116	.023188	4.77e-07	.6236834
prmval	9	4	2.738613	0	8
prmobeq	9	.1101396	.1153559	.0010929	.3005464
prmpreq	9	.1108765	.1174511	.001877	.3098447
prmoble	9	.8150577	.2373893	.3005464	.9912568
prmprle	9	.8122127	.2760109	.2092071	.9978884

```
. line prmpreq prmobeq prmval
```
(graph omitted)

A.11   prgen

prgen   Generate variables with predicted values for regression models in a way that is useful for making plots. For a detailed discussion, see Section 3.5.4 of Chapter 3 and Section 4.6.4 of Chapter 4.

Syntax

prgen *varname* [if *exp*] [in *range*] , generate(*prefix*) [from(#) to(#)
    ncases(#) x(*variable1=value1* [...]) rest(*stat*) maxcnt(#) brief all]

See Section A.7 for details on x() and rest().

Description

prgen computes predicted values for regression models in a way that is useful for making plots. Predicted values are computed for the case in which *varname* varies over a specified range while other independent variables are held constant at values set by x() and rest(). New variables are added to the existing dataset that contain these predicted values. These new variables begin with the name *prefix*. The new variables contain data only for the first k observations in the dataset, where k is 11 by default or can be specified with the ncases() option.

Options

generate(*prefix*) is required and sets the prefix for the new variables that are created. Choosing a prefix that is different than the beginning letters of any of the variables in your dataset makes it easier to examine the results. For example, if you choose the prefix `abcd`, then you can use the command `sum abcd*` to examine all newly created variables.

`from(#)` and `to(#)` are the start value and end values for *varname*. The default is for *varname* to range from the observed minimum to the observed maximum of *varname*.

`ncases(#)` specifies the number of predicted values to be computed as *varname* varies from
    `from(#)` to `to(#)`. The default is 11.

`maxcnt(#)` is the maximum count value for which a predicted probability is computed for count models. The default is 9.

`brief` suppresses output; variables are still generated.

`all` specifies that any calculations of means, medians, etc. should use the entire sample instead of the sample used to fit the model.

Examples

To compute predicted probabilities as age varies from 20 to 80, where `list` reproduces the values of some of the new variables created by `prgen`,

```
. use http://www.stata-press.com/data/lfr/ordwarm2, clear
(77 & 89 General Social Survey)
. oprobit warm yr89 male white age ed prst, nolog
  (output omitted )
. * predicted probabilities as age changes from 20 to 80
. prgen age, f(20) t(80) gen(wrm)
oprobit: Predicted values as age varies from 20 to 80.
             yr89        male       white         age          ed        prst
x=     .39860445    .46489315    .8765809   44.935456   12.218055   39.585259
. list wrmx wrmp1 wrmp2 wrmp3 wrmp4 in 1/11
```

	wrmx	wrmp1	wrmp2	wrmp3	wrmp4
1.	20	.0640175	.2609412	.4251544	.2498869
2.	26	.0737291	.2780399	.4210533	.2271777
3.	32	.0845284	.2948091	.4149821	.2056804
4.	38	.0964731	.3110608	.4070269	.1854392
5.	44	.1096137	.326604	.3972993	.166483
6.	50	.1239925	.3412487	.3859335	.1488253
7.	56	.1396419	.3548089	.3730837	.1324654
8.	62	.1565829	.3671075	.3589206	.1173891
9.	68	.1748236	.3779791	.3436272	.1035701
10.	74	.1943586	.3872746	.3273952	.0909716
11.	80	.2151676	.3948639	.3104213	.0795472

or, we can compute predictions at values that are average for male respondents, which can then be plotted:

```
. prgen age, x(male=1) rest(grmean) f(20) t(80) gen(mal)
oprobit: Predicted values as age varies from 20 to 80.
          yr89       male      white        age         ed       prst
x=   .37992495          1   .89212008  44.113508  12.337711  40.366792
. line malp1 malp2 malp3 malp4 malx, c(s ..)
```
(graph omitted)

New variables generated

Models	Name	Contains
All models	*prefix*x	The values of x from from(#) to to(#).
logit, probit	*prefix*p0	Predicted probability $\Pr(y = 0)$.
	*prefix*p1	Predicted probability $\Pr(y = 1)$.
ologit, oprobit	*prefix*pk	Predicted probability $\Pr(y = k)$ for all outcomes.
	*prefix*sk	Cumulative (summed) probability $\Pr(y \leq k)$ for all outcomes.
mlogit	*prefix*pk	Predicted probability $\Pr(y = k)$ for all outcomes.
poisson, nbreg	*prefix*mu	Predicted rate μ.
	*prefix*pk	Predicted probability $\Pr(y = k)$, for $0 \leq k \leq$ maxcnt().
	*prefix*sk	Cumulative probability $\Pr(y \leq k)$, for $0 \leq k \leq$ maxcnt().
zip, zinb	All for poisson, plus:	
	*prefix*inf	Predicted probability $\Pr(\text{Always } 0 = 1) = \Pr(\mathit{inflate})$
regress, tobit,	*prefix*xb	Predicted value of y.
cnreg, & intreg		

A.12   prtab

prtab   Construct tables of predicted values. For a detailed discussion, see Section 3.5.6 of Chapter 3 and Section 4.6.3 of Chapter 4.

Syntax

prtab *rowvar* [*colvar* [*supercolvar*]] [if *exp*] [in *range*] [, by(*superrowvar*)
   x(*variable1=value1* [...]) rest(*stat*) outcome(#) nobase nolabel novarlbl
   brief all]

See Section A.7 for details on x() and rest().

Description

After fitting a regression model, prtab presents up to a four-way table of the predicted values, either probabilities or rates, for different combinations of values of the independent variable(s). The command works with cloglog, cnreg, gologit, intreg, logistic, logit, mlogit, nbreg, ologit, oprobit, poisson, probit, regress, tobit, zinb, and zip.

Options

by(*superrowvar*) specifies a numeric variable to be treated as a superrow. Only one
superrow variable is allowed.

outcome(*value*) presents results for the specified outcome (e.g., outcome(2) requests
values only for outcome 2). For ordered models or mlogit, the default is to provide
results for all outcomes, each in a separate table; for count models, the default is to
present changes in the predicted rate.

nobase suppresses the list of the values of the independent variables.

nolabel causes the numeric codes to be displayed rather than value labels.

novarlbl causes the variable name to be displayed rather than the variable label.

brief prints only limited output.

all specifies that any calculations of means, medians, etc., should use the entire sample
instead of the sample used to fit the model.

Examples

To compute predicted probabilities of labor force participation by whether the husband
or wife attended college,

```
. use http://www.stata-press.com/data/lfr/binlfp2, clear
(Data from 1976 PSID-T Mroz)
. probit lfp k5 k618 age wc hc lwg inc, nolog
  (output omitted)
. prtab wc hc
probit: Predicted probabilities of positive outcome for lfp
```

Wife College: 1=yes 0=no	Husband College: 1=yes 0=no	
	NoCol	College
NoCol	0.5149	0.5376
College	0.7004	0.7200

```
            k5       k618        age         wc         hc        lwg
x=    .2377158  1.3532537  42.537849  .2815405  .39176627  1.0971148

           inc
x=    20.128965
```

In models for ordinal or nominal outcomes, a separate table is produced for each outcome:

```
. use http://www.stata-press.com/data/lfr/ordwarm2, clear
(77 & 89 General Social Survey)
. ologit warm yr89 male white age ed prst, nolog
  (output omitted)
```

```
. prtab male white, x(prst=min)
ologit: Predicted probabilities for warm
Predicted probability of outcome 1 (SD)
```

Gender: 1=male 0=female	Race: 1=white 0=not white White Not Whit
Women	0.0695 0.0995
Men	0.1346 0.1870

```
Predicted probability of outcome 2 (D)
```

Gender: 1=male 0=female	Race: 1=white 0=not white White Not Whit
Women	0.2492 0.3094
Men	0.3588 0.4032

(*and so on*)

For count models, by default the table contains predicted rates,

```
. use http://www.stata-press.com/data/lfr/couart2, clear
(Academic Biochemists / S Long)
. poisson art fem mar kid5 phd ment, nolog
Poisson regression                          Number of obs    =        915
                                            LR chi2(5)       =     183.03
                                            Prob > chi2      =     0.0000
Log likelihood = -1651.0563                 Pseudo R2        =     0.0525
```

art	Coef.	Std. Err.	z	P>|z|	[95% Conf. Interval]
fem	-.2245942	.0546138	-4.11	0.000	-.3316352 -.1175532
mar	.1552434	.0613747	2.53	0.011	.0349512 .2755356
kid5	-.1848827	.0401272	-4.61	0.000	-.2635305 -.1062349
phd	.0128226	.0263972	0.49	0.627	-.038915 .0645601
ment	.0255427	.0020061	12.73	0.000	.0216109 .0294746
_cons	.3046168	.1029822	2.96	0.003	.1027755 .5064581

```
. prtab fem mar
poisson: Predicted rates for art
```

Gender: 1=female 0=male	Married: 1=yes 0=no Single Married
Men	1.6109 1.8815
Women	1.2869 1.5030

```
          fem        mar       kid5        phd       ment
x=   .46010929  .66229508  .49508197  3.1031093  8.7672131
```

or you can request predicted probabilities for a specified count:

```
. prtab fem mar, outcome(0)
poisson: Predicted probabilities of count = 0 for art
```

Gender: 1=female 0=male	Married: 1=yes 0=no Single Married
Men	0.1997 0.1524
Women	0.2761 0.2225

```
          fem          mar          kid5          phd          ment
x=   .46010929   .66229508   .49508197   3.1031093   8.7672131
```

A.13   prvalue

prvalue   Compute predicted values for specified values of the independent variables. For a detailed discussion, see Section 3.5.7 of Chapter 3 and Section 4.6.2 of Chapter 4.

Syntax

prvalue [if *exp*] [in *range*] [, x(*variable1=value1* [...]) rest(*stat*) level(*#*)
   maxcnt(*#*) save dif ystar nobase nolabel brief all]

See Section A.7 for details on x() and rest().

Description

After fitting a regression model, prvalue computes the predicted values at specific values of the independent variables. When appropriate, predicted probabilities are provided. The command works with cloglog, cnreg, gologit, intreg, logistic, logit, mlogit, nbreg, ologit, oprobit, poisson, probit, regress, tobit, zinb, and zip.

Options

level(*#*) sets the level of the confidence interval for predicted values or probabilities for the estimation commands for which these are provided.

maxcnt(*#*) is the maximum count value for which the probability is computed in count models. The default is 9.

save preserves the current values of independent variables and predictions for later comparisons.

dif computes the differences between current predictions and those that were previously saved.

ystar prints the predicted value of y^* for binary and ordinal models.

nobase suppresses printing of the values of the independent variables.

nolabel uses values rather than value labels in output.

brief prints only limited output.

all specifies that any calculations of means, medians, etc., should use the entire sample instead of the sample used to fit the model.

Examples

```
. use http://www.stata-press.com/data/lfr/ordwarm2, clear
(77 & 89 General Social Survey)
. oprobit warm yr89 male white age ed prst, nolog
  (output omitted)
. * by default, hold all independent variables to their means
. prvalue

oprobit: Predictions for warm
  Pr(y=SD|x):        0.1118
  Pr(y=D|x):         0.3290
  Pr(y=A|x):         0.3956
  Pr(y=SA|x):        0.1636

           yr89       male      white       age        ed       prst
x=    .39860445  .46489315  .8765809  44.935456  12.218055  39.585259
. * to compute all variables at their minimum
. prvalue, rest(min)

oprobit: Predictions for warm
  Pr(y=SD|x):        0.1060
  Pr(y=D|x):         0.3226
  Pr(y=A|x):         0.4000
  Pr(y=SA|x):        0.1714

      yr89   male  white   age   ed  prst
x=       0      0      0    18    0    12
. * predictions for white females, with other variables at the median
. prvalue, x(white=1 male=0) rest(median)

oprobit: Predictions for warm
  Pr(y=SD|x):        0.1012
  Pr(y=D|x):         0.3169
  Pr(y=A|x):         0.4036
  Pr(y=SA|x):        0.1783

      yr89   male  white   age   ed  prst
x=       0      0      1    42    12    37
```

(Continued on next page)

```
. * or with other variables at median for white females
. prvalue, x(white=1 male=0) rest(grmedian)

oprobit: Predictions for warm
  Pr(y=SD|x):          0.1056
  Pr(y=D|x):           0.3220
  Pr(y=A|x):           0.4004
  Pr(y=SA|x):          0.1720

       yr89   male  white    age    ed   prst
x=        0      0      1     44    12     37

. * to compare predictions for males and females
. prvalue, x(male=0) save

oprobit: Predictions for warm
  Pr(y=SD|x):          0.0791
  Pr(y=D|x):           0.2867
  Pr(y=A|x):           0.4182
  Pr(y=SA|x):          0.2160

         yr89      male      white        age        ed       prst
x=  .39860445         0  .8765809  44.935456  12.218055  39.585259

. prvalue, x(male=1) dif

oprobit: Change in Predictions for  warm
                      Current     Saved  Difference
  Pr(y=SD|x):          0.1601    0.0791      0.0810
  Pr(y=D|x):           0.3694    0.2867      0.0827
  Pr(y=A|x):           0.3560    0.4182     -0.0622
  Pr(y=SA|x):          0.1145    0.2160     -0.1015

                 yr89      male      white        age        ed       prst
Current=    .39860445         1  .8765809  44.935456  12.218055  39.585259
  Saved=    .39860445         0  .8765809  44.935456  12.218055  39.585259
   Diff=            0         1         0          0          0          0
```

Saved results

Scalars
r(xb)	the linear combination of the bs and the specified base value
r(xb_hi), r(xb_lo)	upper and lower limits of confidence interval for xb
r(p0), r(p1)	predicted probability $y = 0$ and $y = 1$ for binary models
r(p0_hi), r(p1_hi)	upper limits of predicted probabilities for binary models
r(p0_lo), r(p1_lo)	lower limits of predicted probabilities for binary models
r(mu)	predicted rate for count models
r(mu_hi), r(mu_lo)	upper and lower limits of confidence interval for predicted rate for poisson model
r(always0)	predicted probability of being in the always zero (inflate==1) category in zero-inflated count model

Macros
r(level)	level of confidence intervals (when appropriate)

Matrices
r(x)	base values for independent variables
r(x2)	base values for independent variables in the binary equation of zero-inflated count models
r(probs)	predicted probabilities
r(values)	category values corresponding to predicted probabilities in r(probs)

B   Description of Datasets

The following datasets are used as examples in the book:

1. `binlfp2`: Data on labor force participation.

2. `couart2`: Data on scientific productivity.

3. `gsskidvalue2`: Data on parental values from the General Social Survey.

4. `nomocc2`: Data on occupations from the General Social Survey.

5. `ordwarm2`: Data on attitudes toward working mothers from the General Social Survey.

6. `science2`: Data on the careers of biochemists.

7. `travel2`: Data on travel mode choice.

Variable labels and descriptive statistics are provided for each dataset.

B.1   binlfp2: Data on labor force participation

```
. use http://www.stata-press.com/data/lfr/binlfp2, clear
(Data from 1976 PSID-T Mroz)

. describe

Contains data from http://www.stata-press.com/data/lfr/binlfp2.dta
  obs:            753                          Data from 1976 PSID-T Mroz
  vars:             8                          30 Apr 2001 16:17
  size:        13,554 (99.5% of memory free)   (_dta has notes)

              storage   display    value
variable name   type    format     label      variable label

lfp            byte     %9.0g      lfplbl      Paid Labor Force: 1=yes 0=no
k5             byte     %9.0g                  # kids < 6
k618           byte     %9.0g                  # kids 6-18
age            byte     %9.0g                  Wife's age in years
wc             byte     %9.0g      collbl      Wife College: 1=yes 0=no
hc             byte     %9.0g      collbl      Husband College: 1=yes 0=no
lwg            float    %9.0g                  Log of wife's estimated wages
inc            float    %9.0g                  Family income excluding wife's

Sorted by:  lfp
```

```
. summarize
        Variable │      Obs        Mean    Std. Dev.        Min         Max
─────────────────┼──────────────────────────────────────────────────────────
             lfp │      753    .5683931    .4956295          0           1
              k5 │      753    .2377158     .523959          0           3
            k618 │      753    1.353254    1.319874          0           8
             age │      753    42.53785    8.072574         30          60
              wc │      753    .2815405    .4500494          0           1
─────────────────┼──────────────────────────────────────────────────────────
              hc │      753    .3917663    .4884694          0           1
             lwg │      753    1.097115    .5875564   -2.054124    3.218876
             inc │      753    20.12897      11.6348   -.0290001          96
```

B.2   couart2: Data on scientific productivity

```
. use http://www.stata-press.com/data/lfr/couart2, clear
(Academic Biochemists / S Long)

. describe

Contains data from http://www.stata-press.com/data/lfr/couart2.dta
  obs:           915                          Academic Biochemists / S Long
  vars:            6                          30 Jan 2001 10:49
  size:        11,895 (99.5% of memory free)  (_dta has notes)
─────────────────────────────────────────────────────────────────────────────
              storage   display    value
variable name   type    format     label      variable label
─────────────────────────────────────────────────────────────────────────────
art             byte    %9.0g                 Articles in last 3 yrs of PhD
fem             byte    %9.0g      sexlbl      Gender: 1=female 0=male
mar             byte    %9.0g      marlbl      Married: 1=yes 0=no
kid5            byte    %9.0g                 Number of children < 6
phd             float   %9.0g                 PhD prestige
ment            byte    %9.0g                 Article by mentor in last 3 yrs
─────────────────────────────────────────────────────────────────────────────
Sorted by:  art

. summarize, sep(0)
        Variable │      Obs        Mean    Std. Dev.        Min         Max
─────────────────┼──────────────────────────────────────────────────────────
             art │      915    1.692896    1.926069          0          19
             fem │      915    .4601093    .4986788          0           1
             mar │      915    .6622951     .473186          0           1
            kid5 │      915     .495082     .76488          0           3
             phd │      915    3.103109    .9842491       .755        4.62
            ment │      915    8.767213    9.483916          0          77
```

B.3   gsskidvalue2: Data on parental values from the General Social Survey

```
. use http://www.stata-press.com/data/lfr/gsskidvalue2, clear
(1993 and 1994 General Social Survey)
```

```
. describe
Contains data from http://www.stata-press.com/data/lfr/gsskidvalue2.dta
  obs:           4,598                          1993 and 1994 General Social
                                                Survey
  vars:             11                          9 Jan 2001 10:28
  size:         91,960 (99.3% of memory free)
```

variable name	storage type	display format	value label	variable label
year	int	%8.0g		gss year for this respondent
id	int	%8.0g		respondent id number
female	byte	%9.0g	dummy	Female
black	byte	%9.0g	dummy	Black
othrrace	byte	%9.0g	dummy	Nonblack/Nonwhite
degree	byte	%14.0g	EDdeg	rs highest degree
anykids	byte	%9.0g	dummy	R have any children?
kidvalue	byte	%9.0g	kidvalue	Which is most important for a child to learn?
income	long	%8.0g	income91	total family income
age	byte	%8.0g	age	age of respondent
income91	byte	%8.0g	income91	total family income

```
Sorted by:
. summarize
```

Variable	Obs	Mean	Std. Dev.	Min	Max
year	4598	1993.651	.4767952	1993	1994
id	4598	1254.447	818.4198	1	2992
female	4598	.5704654	.4950636	0	1
black	4598	.1233145	.3288336	0	1
othrrace	4598	.0437147	.2044817	0	1
degree	4584	1.430628	1.165915	0	4
anykids	4584	.7236038	.4472639	0	1
kidvalue	2989	2.22449	.8944211	1	4
income	4103	34790.7	22387.45	1000	75000
age	4598	46.12375	17.33162	18	99
income91	4598	19.44411	19.50733	1	99

B.4   nomocc2: Data on occupations from the General Social Survey

```
. use http://www.stata-press.com/data/lfr/nomocc2, clear
(1982 General Social Survey)
```

(Continued on next page)

```
. describe
Contains data from http://www.stata-press.com/data/lfr/nomocc2.dta
  obs:            337                          1982 General Social Survey
  vars:             4                          15 Jan 2001 15:24
  size:         2,696 (99.5% of memory free)   (_dta has notes)
```

variable name	storage type	display format	value label	variable label
occ	byte	%10.0g	occlbl	Occupation
white	byte	%10.0g		Race: 1=white 0=nonwhite
ed	byte	%10.0g		Years of education
exper	byte	%10.0g		Years of work experience

```
Sorted by:  occ
```

```
. summarize
```

Variable	Obs	Mean	Std. Dev.	Min	Max
occ	337	3.397626	1.367913	1	5
white	337	.9169139	.2764227	0	1
ed	337	13.09496	2.946427	3	20
exper	337	20.50148	13.95936	2	66

B.5   ordwarm2: Data on attitudes toward working mothers from the General Social Survey

```
. use http://www.stata-press.com/data/lfr/ordwarm2, clear
(77 & 89 General Social Survey)

. describe
Contains data from http://www.stata-press.com/data/lfr/ordwarm2.dta
  obs:          2,293                          77 & 89 General Social Survey
  vars:            10                          3 May 2001 09:54
  size:        32,102 (99.4% of memory free)   (_dta has notes)
```

variable name	storage type	display format	value label	variable label
warm	byte	%10.0g	SD2SA	Mom can have warm relations with child
yr89	byte	%10.0g	yrlbl	Survey year: 1=1989 0=1977
male	byte	%10.0g	sexlbl	Gender: 1=male 0=female
white	byte	%10.0g	racelbl	Race: 1=white 0=not white
age	byte	%10.0g		Age in years
ed	byte	%10.0g		Years of education
prst	byte	%10.0g		Occupational prestige
warmlt2	byte	%10.0g	SD	1=SD; 0=D,A,SA
warmlt3	byte	%10.0g	SDD	1=SD,D; 0=A,SA
warmlt4	byte	%10.0g	SDDA	1=SD,D,A; 0=SA

```
Sorted by:  warm
```

```
. summarize
    Variable |        Obs        Mean    Std. Dev.        Min        Max
-------------+--------------------------------------------------------
        warm |       2293    2.607501    .9282156          1          4
        yr89 |       2293    .3986044    .4897178          0          1
        male |       2293    .4648932    .4988748          0          1
       white |       2293    .8765809    .3289894          0          1
         age |       2293    44.93546    16.77903         18         89
-------------+--------------------------------------------------------
          ed |       2293    12.21805    3.160827          0         20
        prst |       2293    39.58526    14.49226         12         82
     warmlt2 |       2293    .1295246    .3358529          0          1
     warmlt3 |       2293    .4448321    .4970556          0          1
     warmlt4 |       2293    .8181422    .3858114          0          1
```

B.6   science2: Data on the careers of biochemists

```
. use http://www.stata-press.com/data/lfr/science2, clear
(Note that some of the variables have been artificially constructed.)

. describe

Contains data from http://www.stata-press.com/data/lfr/science2.dta
  obs:           308                          Note that some of the variables
                                                have been artificially
                                                constructed.
  vars:           35                          10 Mar 2001 05:51
  size:        17,556 (99.5% of memory free)  (_dta has notes)
-------------------------------------------------------------------------------
              storage   display    value
variable name   type    format     label      variable label
-------------------------------------------------------------------------------
id            float    %9.0g                  ID Number.
cit1          int      %9.0g                  Citations: PhD yr -1 to 1.
cit3          int      %9.0g                  Citations: PhD yr 1 to 3.
cit6          int      %9.0g                  Citations: PhD yr 4 to 6.
cit9          int      %9.0g                  Citations: PhD yr 7 to  9.
enrol         byte     %9.0g                  Years from BA to PhD.
fel           float    %9.0g                  Fellow or PhD prestige.
felclass      byte     %9.0g      prstlb    * Fellow or PhD prestige class.
fellow        byte     %9.0g      fellbl      Postdoctoral fellow: 1=y,0=n.
female        byte     %9.0g      femlbl      Female: 1=female,0=male.
job           float    %9.0g                  Prestige of 1st univ job.
jobclass      byte     %9.0g      prstlb    * Prestige class of 1st job.
mcit3         int      %9.0g                  Mentor´s 3 yr citation.
mcitt         int      %9.0g                  Mentor´s total citations.
mmale         byte     %9.0g      malelb      Mentor male: 1=male,0=female.
mnas          byte     %9.0g      naslb       Mentor NAS: 1=yes,0=no.
mpub3         byte     %9.0g                  Mentor´s 3 year publications.
nopub1        byte     %9.0g      nopublb     1=No pubs PhD yr -1 to 1.
nopub3        byte     %9.0g      nopublb     1=No pubs PhD yr 1 to 3.
nopub6        byte     %9.0g      nopublb     1=No pubs PhD yr 4 to 6.
nopub9        byte     %9.0g      nopublb     1=No pubs PhD yr 7 to 9.
phd           float    %9.0g                  Prestige of Ph.D. department.
phdclass      byte     %9.0g      prstlb    * Prestige class of Ph.D. dept.
pub1          byte     %9.0g                  Publications: PhD yr -1 to 1.
pub3          byte     %9.0g                  Publications: PhD yr 1 to 3.
```

pub6	byte	%9.0g		Publications: PhD yr 4 to 6.
pub9	byte	%9.0g		Publications: PhD yr 7 to 9.
work	byte	%9.0g	worklbl	Type of first job.
workadmn	byte	%9.0g	wadmnlb	Admin: 1=yes; 0=no.
worktch	byte	%9.0g	wtchlb	* Teaching: 1=yes; 0=no.
workuniv	byte	%9.0g	wunivlb	* Univ Work: 1=yes; 0=no.
wt	byte	%9.0g		
faculty	byte	%9.0g	faclbl	1=Faculty in University
jobrank	byte	%9.0g	joblbl	Rankings of University Job.
totpub	byte	%9.0g		Total Pubs in 9 Yrs post-Ph.D.
				* indicated variables have notes

Sorted by:

. summarize

Variable	Obs	Mean	Std. Dev.	Min	Max
id	308	58654.49	2283.465	57001	62420
cit1	308	11.60714	18.37658	0	137
cit3	308	14.97078	21.37068	0	196
cit6	308	18.37013	23.34766	0	143
cit9	308	21.07143	25.8195	0	214
enrol	278	5.564748	1.467253	3	14
fel	308	3.190877	.9872379	1	4.77
felclass	308	2.694805	1.019855	1	4
fellow	308	.3928571	.4891803	0	1
female	308	.3474026	.4769198	0	1
job	163	2.967117	.880396	1.01	4.69
jobclass	163	2.417178	.8945373	1	4
mcit3	306	20.95098	26.26862	0	133
mcitt	303	44.58086	56.05372	0	223
mmale	303	.9867987	.1143249	0	1
mnas	304	.0789474	.2701012	0	1
mpub3	306	11.02614	9.08571	0	48
nopub1	308	.25	.4337174	0	1
nopub3	308	.2045455	.4040255	0	1
nopub6	308	.1915584	.3941678	0	1
nopub9	308	.1980519	.3991801	0	1
phd	308	3.177987	1.012738	1	4.77
phdclass	308	2.675325	1.057827	1	4
pub1	308	2.545455	3.092685	0	24
pub3	308	3.185065	3.908752	0	31
pub6	308	4.165584	4.780714	0	29
pub9	308	4.512987	5.315134	0	33
work	302	2.062914	1.37829	1	5
workadmn	302	.089404	.2857995	0	1
worktch	302	.615894	.4871904	0	1
workuniv	302	.705298	.4566654	0	1
wt	308	3.402597	1.288989	1	6
faculty	302	.5298013	.4999395	0	1
jobrank	163	2.417178	.8945373	1	4
totpub	308	11.86364	12.77623	0	84

B.7 travel2: Data on travel mode choice

```
. use http://www.stata-press.com/data/lfr/travel2, clear
(Greene & Hensher 1997 data on travel mode choice)
. describe
Contains data from http://www.stata-press.com/data/lfr/travel2.dta
  obs:          456                       Greene & Hensher 1997 data on
                                            travel mode choice
  vars:          13                       10 Mar 2001 06:07
  size:       9,576 (99.5% of memory free)
```

variable name	storage type	display format	value label	variable label
id	int	%9.0g		Identification number
mode	byte	%9.0g	mode	Mode of transportation
train	byte	%9.0g		1=train 0=other mode
bus	byte	%9.0g		1=bus 0=other mode
car	byte	%9.0g		1=car 0=other mode
time	int	%9.0g		Total traveling time
invc	byte	%8.0g		In-vehicle costs
choice	byte	%8.0g		1=option selected 0=not selected
ttme	byte	%8.0g		Terminal time (0 for car)
invt	int	%8.0g		in-vehicle time
gc	int	%8.0g		Generalized costs includeing lost wages
hinc	byte	%8.0g		Household income
psize	byte	%8.0g		Size of traveling party

```
Sorted by:  id
. summarize
```

Variable	Obs	Mean	Std. Dev.	Min	Max
id	456	103.8882	61.03044	1	210
mode	456	2	.8173933	1	3
train	456	.3333333	.4719223	0	1
bus	456	.3333333	.4719223	0	1
car	456	.3333333	.4719223	0	1
time	456	632.1096	270.2547	180	1440
invc	456	33.95175	21.795	2	109
choice	456	.3333333	.4719223	0	1
ttme	456	25.24781	21.15744	0	99
invt	456	606.8618	265.8235	180	1440
gc	456	113.4912	53.7725	30	269
hinc	456	31.80921	19.25813	2	72
psize	456	1.809211	1.069457	1	6

References

Akaike, H. 1973. Information theory and an extension of the maximum likelihood principle. In *Second International Symposium on Information Theory*, ed. B. Petrov and F. Csaki, 267–281. Budapest: Akademiai Kiado.

Allison, P. D. 2001. *Missing Data*. Thousand Oaks, CA: Sage Publications.

Alvarez, R. M. and J. Nagler. 1998. When politics and models collide: Estimating models of multiparty elections. *American Journal of Political Science* 42: 55–96.

Amemiya, T. 1981. Qualitative response models: A survey. *Journal of Economic Literature* 19: 1483–1536.

Anderson, J. 1984. Regression and ordered categorical variables (with discussion). *Journal of the Royal Statistical Society Series B* 46: 1–30.

Arminger, G. 1995. Specification and estimation of mean structures: Regression models. In *Handbook of Statistical Modeling for the Social and Behavioral Sciences*, ed. G. Arminger, C. C. Clogg, and M. E. Sobel, 77–183. New York: Plenum Press.

Brant, R. 1990. Assessing proportionality in the proportional odds model for ordinal logistic regression. *Biometrics* 46: 1171–1178.

Cameron, A. C. and P. K. Trivedi. 1986. Econometric models based on count data: Comparisons and applications of some estimators and tests. *Journal of Applied Econometrics* 1: 29–53.

—. 1998. *Regression Analysis of Count Data*, vol. 30 of *Econometric Society Monograph*. New York: Cambridge University Press.

Clogg, C. C. and E. S. Shihadeh. 1994. *Statistical Models for Ordinal Variables*. Thousand Oaks, CA: Sage Publications.

Cook, R. D. and S. Weisberg. 1999. *Applied Regression Including Computing and Graphics*. New York: John Wiley & Sons.

Cramer, J. 1986. *Econometric Applications of Maximum Likelihood Methods*. Cambridge: Cambridge University Press.

Cytel Software Corporation. 2000. *LogXact Version 4*. Cambridge, MA.

DiPrete, T. A. 1990. Adding covariates to loglinear models for the study of social mobility. *American Sociological Review* 55: 757–773.

Eliason, S. 1993. *Maximum Likelihood Estimation.* Newbury Park, CA: Sage Publications.

Fahrmeir, L. and G. Tutz. 1994. *Multivariate Statistical Modeling Based on Generalized Linear Models.* New York: Springer-Verlag.

Fienberg, S. E. 1980. *The Analysis of Cross-Classified Categorical Data.* 2d ed. Cambridge, MA: MIT Press.

Fox, J. 1991. *Regression Diagnostics: An Introduction.* Newbury Park, CA: Sage Publications.

Freese, J. and J. S. Long. 2000. sg155: Tests for the multinomial logit model. *Stata Technical Bulletin* 58: 19–25.

Fu, V. K. 1998. sg88: Estimating generalized ordered logit models. *Stata Technical Bulletin* 44: 27–30.

Gallup, J. L. 2001. sg97.3: Update to formatting regression output. *Stata Technical Bulletin* 59: 23.

Gould, W., J. Pitblado, and W. Sribney. 2003. *Maximum Likelihood Estimation with Stata.* 2d ed. College Station, TX: Stata Press.

Greene, W. H. 1994. Accounting for excess zeros and sample selection in Poisson and negative binomial regression models. *Stern School of Business, New York University, Department of Economics, Working Paper Number 94-10* .

—. 2003. *Econometric Analysis.* 5th ed. Upper Saddle River, NJ: Prentice–Hall.

Greene, W. H. and D. Hensher. 1995. Multimonial logit and discrete choice models. In *LIMDEP Version 7.0,* ed. W. Greene. Bellport, NY: Econometric Software, Inc. CDA.

Gutierrez, R. G., S. L. Carter, and D. M. Drukker. 2001. sg160: On boundary-value likelihood-ratio tests. *Stata Technical Bulletin* 60: 15–18.

Hagle, T. and G. Mitchell II. 1992. Goodness-of-fit measures for probit and logit. *American Journal of Political Science* 36: 762–784.

Hausman, J. and D. McFadden. 1984. Specification tests for the multinomial logit model. *Econometrica* 52(5): 1219–1240.

Hendrickx, J. 2000. sbe37: Special restrictions in multinomial logistic regression. *Stata Technical Bulletin* 56: 18–26.

Hensher, D. 1986. Simultaneous Estimation of Hierarchical Logit Mode Choice Models. Working Paper 34, Macquarie University, School of Economic and Financial Studies.

Hosmer, D. W. and S. Lemeshow. 2000. *Applied Logistic Regression.* 2d ed. New York: John Wiley & Sons.

Keane, M. P. 1992. A note on identification in the multinomial probit model. *Journal of Business and Economic Statistics* 10: 193–200.

Lambert, D. 1992. Zero-inflated Poisson regression with an application to defects in manufacturing. *Technometrics* 34: 1–14.

Landwehr, J., D. Pregibon, and A. Shoemaker. 1984. Graphical methods for assessing logistic regression models. *Journal of the American Statistical Association* 79: 61–71.

Little, R. and D. Rubin. 1987. *Statistical Analysis with Missing Data.* New York: John Wiley & Sons.

Long, J. S. 1990. The origins of sex differences in science. *Social Forces* 68: 1297–1315.

—. 1997. *Regression Models for Categorical and Limited Dependent Variables*, vol. Volume 7 of *Advanced Quantitative Techniques in the Social Sciences.* Thousand Oaks, CA: Sage Publications.

Long, J. S. and L. H. Ervin. 2000. Using heteroscedasticity consistent standard errors in the linear regression model. *The American Statistician* 54: 217–224.

McCullagh, P. 1980. Regression models for ordinal data (with discussion). *Journal of the Royal Statistical Society* 42: 109–142.

McCullagh, P. and J. Nelder. 1989. *Generalized Linear Models.* 2d ed. New York: Chapman and Hall.

McFadden, D. 1973. Conditional logit analysis of qualitative choice behavior. In *Frontiers of Econometrics*, ed. P. Zarembka, 105–142. New York: Academic Press.

McFadden, D., W. Tye, and K. Train. 1976. An application of diagnostic tests for the irrelevant alternatives property of the multinomial logit model. *Transportation Research Record* 637: 39–46.

McKelvey, R. D. and W. Zavoina. 1975. A statistical model for the analysis of ordinal level dependent variables. *Journal of Mathematical Sociology* 4: 103–120.

Miller, P. W. and P. A. Volker. 1985. On the determination of occupational attainment and mobility. *The Journal of Human Resources* 20: 197–213.

Mroz, T. A. 1987. The sensitivity of an empirical model of married women's hours of work to economic and statistical assumptions. *Econometrica* 55(4): 765–799.

Powers, D. A. and Y. Xie. 2000. *Statistical Methods for Categorical Data Analysis.* San Diego: Academic Press.

Pregibon, D. 1981. Logistic regression diagnostics. *The Annals of Statistics* 9(4): 705–724.

Pudney, S. 1989. *Modelling Individual Choice: The Econometrics of Corners, Kinks and Holes.* Cambridge, MA: Basil Blackwell.

Raftery, A. E. 1996. Bayesian model selection in social research. In *Sociological Methodology*, ed. P. V. Marsden, vol. 26, 111–163. Oxford: Basil Blackwell.

Rothenberg, T. 1984. Hypothesis testing in linear models when the error covariance matrix is nonscalar. *Econometrica* 52: 827–842.

Schafer, J. L. 1997. *Analysis of Incomplete Multivariate Data*. London: Chapman and Hall.

Small, K. and C. Hsiao. 1985. Multinomial logit specification tests. *International Economic Review* 26: 619–627.

Theil, H. 1970. On the estimation of relationships involving qualitative variables. *American Journal of Sociology* 76: 103–154.

Tobias, A. and M. J. Campbell. 1998. sg90: Akaike's information criterion and Schwarz's criterion. *Stata Technical Bulletin* 45: 23–25.

Vuong, Q. H. 1989. Likelihood ratio tests for model selection and non-nested hypotheses. *Econometrica* 57: 307–333.

Weisberg, S. 1980. *Applied Linear Regression*. New York: John Wiley & Sons.

White, H. 1982. Maximum likelihood estimation of misspecified models. *Econometrica* 50(1): 1–24.

Windmeijer, F. A. 1995. Goodness-of-fit measures in binary choice models. *Econometric Reviews* 14(1): 101–116.

Winship, C. and R. D. Mare. 1984. Regression models with ordinal variables. *American Sociological Review* 49: 512–525.

Winship, C. and L. Radbill. 1994. Sampling weights and regression analysis. *Sociological Methods and Research* 23: 230–257.

Wolfe, R. 1998. sg86: Continuation-ratio models for ordinal response data. *Stata Technical Bulletin* 44: 18–21.

Wolfe, R. and W. Gould. 1998. sg76: An approximate likelihood-ratio test for ordinal response models. *Stata Technical Bulletin* 42: 24–27.

Zhang, J. and S. D. Hoffman. 1993. Discrete-choice logit models: Testing the IIA property. *Sociological Methods and Research* 22: 193–213.

Author index

Subject index